Two Sexes. Why?

The Evolutionary Theory of Sex

Wilmington 2012

Two Sexes. Why? The Evolutionary Theory of Sex — Wilmington, 2012.

What do we need two sexes for? What is the evolutionary significance of this phenomenon? What is the nature of the differences between the sexes? What is sexual dimorphism and what does it mean? How is it connected to other life phenomena? The new evolutionary theory of sex differentiation developed by Dr. Biol. Sci. Vigen Geodakian has the answers. It provides a scientific justification for the existence of men and women, and the evolutionary roles that they play in family and society.

Pages: x + 244. Ill. 53, tabl. 33, bibliography 751.

ISBN: 978-1482376135   (Paperback)
        978-0-9856620-4-2   (PDF)

Compiled and edited by S. V. Geodakyan.

The materials presented in the book, are submitted selectively and if they contain any omissions, the responsibility for them bears the originator of this work. In case of any questions and uncertainties readers should refer to the original papers listed in the bibliography.

E-mail: sgeod@yahoo.com

The first publication of Vigen Geodakian devoted to the problem of sex differentiation, appeared in 1965 in cybernetic magazine "Problems of Information Transmission". Then came an article in popular science magazine "Science and Life". Due to multimillion distribution this article received much attention. Many scientists still remember just this one small aspect of the theory—how to influence the sex of offspring. However, over 45 years of its existence, the theory was greatly expanded. Since then, more than 150 articles have been published devoted to longevity, differentiation of a brain and hands, sex chromosomes, mechanisms of sex ratio regulation in plants and animals, congenital heart anomalies, and even culture. Newspapers and magazines repeatedly wrote about the theory. V. Geodakian made presentations at many Russian and international congresses, conferences and symposia, and lectured hundreds of students. Two conferences have been devoted exclusively to the theory (Saint Petersburg, Russia, 1990, 1992).

The theory has already entered the textbooks (V. Vasiltchenko, 1977, 1986, 2005; A. Tkatshenko et al., 2001; Ilin, 2003; D. Zhukov, 2007; A. Palij, 2010), study guides (V. Kagan, 1991; S. Borinskaya; 1995; E. Tyugashev, 2002,2006; E. Tyugashev, T. Popkova, 2003; S. Nartova-Bashaver, 2003; T. Andreeva, 2004) and was also included in study programs of some Russian (Physical-Technical College, Moscow College of Physical Engineers, Russian Humanitarian State University) and foreign (Tel-Aviv University) colleges and universities.

The theory is cited in many monographs (S. Afonkin 2010, A. Belkin, 2000; F. Ballyuzek et al, 2009; N. Bragina, T. Dobrokhotova, 1988; D. Vysotsky 2004; A. Katsenelinboĭgen, 1997) and is used in various fields of research: the evolution of sexual reproduction (A. Kondrashov, 1993 ; A. Katsenelinboĭgen, 1997), biology of plants (A. Pokhilenko 2012) and animals (P. Panteleev, 2003), medicine (P. Rajewski, A. Sherman, 1976; V. Razumov, 2011), social psychology (M. Zhuravleva, T. Soboleva, 2005; T. Kochetova 2010; I. Trofimova, 2011,2012; A. Pankratov, I. Vorkunova 2011; Yu. Zasyad'-Volk, 2011) and education (V. Yeremeeva, T. Hrizman, 1998) and others.

The breadth of the theory, it's explanatory and predictive capacity allow it to substitute many limited and imperfect theories of gender and related fields, in particular—C. Darwin's theory of sexual selection. A look at dioecy as an effective way of evolution, and at males as an evolutionary "vanguard" of a population, allows one to explain from the unified point of view many mysteries related to sex. The basic idea of asynchronous evolution led to the creation of two new theories—brain asymmetry and laterality (in 1993), and the evolutionary role of sex chromosomes (in 1996), as well as new interpretations of hormonal sex, homosexuality, cancer, and culture.

Despite the large number of publications, the most complete presentation of the theory is still the V. Geodakian's doctorate thesis (Institute of Developmental Biology, 1987). This book attempts to describe the current state of the theory including new developments and accumulated experimental data.

General composition of the book is similar to that of the dissertation. The first six chapters cover the current state of the problem of sex with its unclear phenomena and riddles, existing theories and their criticism. Chapters 7 and 8 are devoted to the classification and analysis of reproduction types, chapter 9—involvement of the sexes in the transmission of information to the progeny, chapter 10 to sex ratio. Sexual dimorphism, mechanisms of its occurence, forms and rules are described in chapters 11–14. Sexual dimorphism in pathology is examined in chapter 15.

Chapter 16 covers possible mechanisms of discovered relationships. A separate section contains the predictions arising from the theory. The summary has all statements of the new theory and the general scheme of the trait's evolution in ontogeny and phylogeny. The main goal of the book was to consistently describe the theory while maintaining the style of V. Geodakian's articles. The book does not attempt to make a full review of the area of sex research, which can be found in several monographs. The areas where the theory was already applied received the most coverage.

The book is intended to scientists, teachers, students, and other professionals working in the fields of biology, medicine, psychology, sociology, anthropology, and many other related branches of science. Since sex-related questions are of great interest to almost everybody, a more simplle description of the theory can be found on official theory's web site (*www.geodakian.com*).

Table of Contents

THE EVOLUTIONARY THEORY OF SEX

Introduction

*"One of the primary aims of a theoretical study in any field
of human knowledge is establishing the viewpoint
that reflects the studied object with maximum simplicity."*

Willard Gibbs (Frankfurt W.I., 1964)

Self-reproduction is the main feature distinguishing living systems from the inanimate. It is typical on all levels of life organization: from molecules up to populations and communities of organisms. It can be considered as the "basic essence of a life".

On the level of organism self-reproduction—*duplication*—shows a vast variety of forms; from simple cell *fission* at the level of bacteria and other monocelled organisms, up to extremely different and complex forms of *sexual duplication* at the level of high plants and animals. The dioecious form of reproduction occupies an important place among them, because it is used by all evolutionary progressive forms of animals and plants. It is known, that during the evolution dioecious type of reproduction appeared repeatedly in species that are very distant from each other, independently and convergently. It allows one to think, that underneath dioecious type of reproduction there is a certain fundamental, evolutionary "logic".

Existing theories however aren't capable of explaining either an evolutionary origin of dioecious reproduction, or its selective advantages as compared to other types, for example the *hermaphrodite* one.

Over the last 150 years, the issue of sex has continued to be the central problem of evolutionary biology. It was dealt with by eminent biologists of the XIX–XX century, such as Darwin, Walles, Weismann, Goldshmidt, Fisher, and Müller. Despite this, present-day leading figures continue to write about the "crisis" in evolutionary biology concerning the issue of sex. During the last quarter-century, the problem of sex has enjoyed another period of renaissance. A dozen books have been published, the titles of which contain two words: "Sex" and "Evolution" (Williams, 1975; Maynard Smith, 1978; Bell, 1982; Bull, 1983; Karlin and Lessard, 1986; Hoekstra, 1987; Michod and Levin, 1988; Dawley and Bogart, 1989; Harvey et al., 1991; Mooney, 1992). The first of these begins with the statement that the "...prevalence of sexual reproduction in higher plants and animals is incompatible with modern evolutionary theory..." In the second, we read: "We do not have a satisfactory explanation as to how sex has emerged and how it is maintained." In the third monograph dealing with sex evolution and genetics, the author states that "... sex is the main challenge to the modern evolutionary theory... . Queen of problems... . Intuitions of Darwin and Mendel, which have resolved so many riddles, were unable to cope with the central riddle of sexual reproduction." Another leading figure on the problem of sex writes: "It is rather surprising, but scientists cannot satisfactorily explain, why sex exists." (Crews, 1994). Many articles and reviews dealing with this issue have been published. During the last few years, at least two leading genetic journals have dedicated special issues to this problem *(Heredity,* 1993, vol. 84(5), pp. 321–424; *Developmental Genetics,* 1994, vol. 15(3), pp. 201-312). All this provides evidence that the problem of sex, a central problem of evolutionary biology and genetics, still remains unsolved. The main question why do we need sex and what its adaptive significance, remains so far without an answer (see Crow, 1994).

In his book about evolution published in 2001 E. Mayr writes: "Since the 1880's evolutionists have argued over the selective advantage of sexual reproduction. So far, no clear-cut winner has emerged from this controversy." M. Ridley (2001) "The Cooperative Gene": "Sex is a puzzle that has not yet been solved; no one knows why it exists". In the book of Karl Zimmer (2001) "Sex is not only unnecessary, but it ought to be a recipe for evolutionary disaster. For one thing, it is an inefficient way to reproduce.... And sex carries other costs as well.... By all rights, any group of animals that evolves sexual reproduction should be promptly outcompeted by nonsexual ones. And yet sex reigns... Why is sex a success, despite all its disadvantages?"

Many theories answering the question "Why sex exists?" attempt to explain advantages of sexual types of reproduction (hermaphrodite and dioecious) over asexual one. They consider the effects of two sexes and crossing in the organism level. Thansition from hermaphrodite to dioecious reproduction and the effects related to sex differentiation usually escape their attention. The most obvious, however incomplete answer— sex is needed for reproduction. But asexual and hermaphrodite forms reproduce more effectively. In populations without separation into two sexes all individuals can produce progeny, and asexual organisms do not even have to spend time to find a partner.

Misunderstanding of evolutionary bases and laws creates many riddles. The problem of sex touches many, very important areas of human interests such as demography and medicine, psychology and pedagogy, the law and ethics. Criminal problems and the problems of drug and alcohol addiction are affected too. Lastly there are agricultural problems: genetics and selection in plant and animal industries, hence problems of economy and many others. The correct social concept of gender is necessary to solve the problems of birth rate and mortality, family and education, professional guidance and career counseling.

Such concept should be based on a correct biological basis. It is impossible to correctly define the social roles of males and females without understanding of their biological, evolutionary roles. Males and females differ fundamentally in many ways, including reproductive organs, behaviour, and morphology. Frequently however a completely correct and fair idea of equality between men and women is transformed, treated and preached as "similarity" and "interchangeability" in all areas of human activity. "Asexual" psychology, pedagogy and sociology create too many dead ends. For example, in the US women have many opportunities to fight for their rights to be soldies or firefighters. Men are catching up to the women by being day-care service providers and nursing assistants. In order to increase the competitiveness of the American science, money is spent on the programs that should increase numbers of women in science, technology, engineering and mathematics (STEM). The urge to implement social concepts and decisions that do not have a biological foundation have frequently had bad consequences for the society. Examples are the prohibition of alcohol, the war on drugs, organ transplants, the treatment of congenital defects, and the preservation (or extinction) of existing plant and animal species.

Other sex-related problems (homosexuality, prostitution, abortions) are frequently addressed without considering their biological roots, but rather are based on culture, politics or history. For example, for homosexuality in different periods of time and in different countries one can find the whole spectra of solutions ranging from the death penalty and imprisonment to the right to marry and preach in the church.

The evolutionary theory of sex allows a new view at gender distinctions and provides a scientific basis for replacing the ideas of social equality and the interchangeability of sexes with the idea of their mutual additivity.

Chapter 1

Analysis of Three Main Ways of Reproduction

"The main question is —why sex?"
Bell (1982)

"After more than a century of debate, the major factors
of the evolution of reproduction are still obscure.
Kondrashov, (1993)

T he problem of sex has a lot of aspects, each of them with it's own theories. First of all we will analyze common theories dealing with evolutionary meaning and compare the advantages of the two main types of reproduction—*asexual* and *sexual*. After that we will narrow the analysis to sexual forms only and will compare theories of *hermaphroditism* and *dioecy*—two main types of sexual reproduction.

Comparison of Sexual Ways of Reproduction with Asexual

The earliest life on Earth, dating back some 3.8 billion years, or maybe more, undoubtedly reproduced simply by making copies of itself. Many primitive organisms—viruses, bacteria, single-celled algae, and some animals—still reproduce in this way: amoebae, sponges, and sea anemones divide; hydra bud. Many plants, including higher plants have exclusive or partial asexual reproduction through the vegetative production of stolons, rhizomes, leaflets, and tillers.

Later, much later, perhaps about 1.5 billion years ago, an alternative form of reproduction arose, which allowed some exchange of genetic material between similar individuals. This involved the temporary fusion of two individuals in such a way that genetic material could reciprocally migrate from one to the other before reproduction. This was the beginning of sex.

Initially, sex involved the fusion of whole individuals that were approximately the same size. The individuals in question were small, comprising just a single cell. At some stage in the early evolution of sexual reproduction, differences in the surface chemistry of individuals arose, whereby only individuals with different surface chemistries were able to fuse. As soon as sexuality had been "invented", asexuality becomes relatively rare among eukaryotes. Above the level of the genus there are only three higher taxa of animals that consist exclusively of uniparentally reproducing clones. Strict asexuality is rare in plants but common in some groups of fungi. Asexual forms make less than 1% of 250,000 plant species (Asker, Jerling, 1992; Whitton et al., 2008). Only about 0.1 % of all known animal species are exclusively asexual (Vrijenhoek, 1998). In general, about 1% of all species are asexual, and they tend to represent evolutionary dead ends (Maynard Smith, 1978).

Sexual reproduction is now the almost universal mode of eukaryote reproduction. Every case of uniparental reproduction found in animals and plants is obviously a secondary (derived) condition, usually being restricted to a single species in a genus or to an isolated genus. There are only a few cases of entire families of animals being parthenogenetic. In animals uniparental reproduction has been invented again and again, but the asexual clones always become extinct after a relatively short time.

The existence of sex comes with a price. First there are genetic costs. Evolution favors traits that effectively transfer their underlying genes to future generations. Each sexually reproducing parent passes on only one-half of its genes to its offspring. The other half of its offspring's genes comes from its mate. An asexual parent passes on all of her genes to her offspring.

Sex is costly in terms of energy, time and resources required to find a suitable mate. This applies not only to dioecious populations, but also to hermaphrodites in which self-replication is usually restricted. In many organisms, mating requires intimate contact between the two parents, which provides an opportunity for parasites, viruses and bacteria to move from one body to another.

Bonner (1958) proposed the following common definition of sex: "Sex is adaptation of parents to the possible meeting of their progeny with changed or unusual conditions". Different species, capable to asexual as well as sexual reproduction, are usually sexless, and only under certain conditions switch to sexual reproduction. One can say that sex is adaptation to those conditions. This is quite a reasonable conclusion, even if we do not know why it is adaptive.

There is some regularity in the transition from asexual to sexual reproduction. For example, if there are several asexual and one sexual generation in the life cycle, the last one occurs when two consecutive generations experienced the greatest ecological differences. When asexual and sexual reproduction both occur at the same time, the asexual progeny should grow immediately and close to the parents, while progeny from sexual reproduction can have latency or disseminate far away from parents.

Bonner lists many examples confirming this picture and there is no doubt that it is right. Many high plants produce genetically identical progeny by vegetation. At the same time they sexually produce seeds, which spread far away and have long latency. Many parasites are reproduced asexually in one host but sexually when changing hosts. Free living lower plants, animals and microorganisms, which regularly are reproduced asexually, and only occasionally use sexual reproduction, always switch to sexual reproduction in response to environmental changes.

According to Bell (1982) sex is elicited in populations in which the density of organisms is high. In dense populations it is easier to find a mate and the costs of mating are reduced.

Fisher (1930) was the first to propose that sexual reproduction appeared because of group selection. Moreover, he suggested that sexual reproduction is the only adaptation that appeared as a result of group selection. He also considered, that recombination process can be explained without group selection. The problem of the selection factors responsible for the appearance and preservation of sexual reproduction is still catching the attention of many researchers. Williams (1975) thought that the most important difference between asexual and sexual offspring is that asexual progeny is standardized (produced as a result of mitosis), while sexual progeny is diversified (because of meiosis). He mentioned also some other features of both types (**Table 1.1**).

Table 1.1 Some features of asexual and sexual offspring (Williams, 1975)

Offspring	
Asexual	Sexual
Large initial size	Small initial size
Produced all the time	Have seasonal limitations
Grow close to the parents	Spreads wide
Develops immediately	Have latency
Develops directly into adult stage	Have series of embryonic and larvae stages
Based on parents, the environment and optimal genotype can be predicted	The environment and genotype are unpredictable
Low mortality	High mortality
Natural selection is weak	Natural selection is intense

Sexual reproduction is limited to the part where the conditions are less predictable. Because of their bigger size and low mobility, asexual progeny is located near the parents. Plants have seeds that are relatively small and have special features for wide spreadability. The same plant can produce different kind of seeds with different latency from a month to many years. High seeds mortality is related to the size (big seeds have more resources) and also with latency and wide distribution in space.

Maynard Smith argued, that the main problem is, that it is not known what factors of selection support sexual reproduction. At an explanation of the mechanism of maintenance of sexual duplication with the help of group selection, he accepted, that asexual duplication (parthenogenesis) has a quickly realized (short-term) advantage, and asserted that this momentary advantage overcomes long-term advantages of sexual reproduction. A short-term advantage of parthenogenesis means its double efficiency related to the absence of male progeny. A double increase of speed of duplication for parthenogenesis can give advantages in conditions of a population expansion, when it explores the new ecological niches. It also gives an advantage in conditions of "deviation" of a population when it is compelled "to concede" an ecological niche, and is exposed to strong selection. On this basis Maynard Smith considered, that new restatement of a question is necessary, namely, "Why don't parthenogenetic versions supersede species using sexual reproduction?"

Most authors however, continued to ask the question in the form of: "Why are some parthenogenetic species able to exist for a long time?" Their view was based on the fact that the benefits of sexual forms in the variation of offspring under unpredictable environmental conditions greatly exceed the benefits of parthenogenesis (larger quantity of offspring). So, the problem is to explain why parthenogenesis exists at all, rather than to explain why it is not a widespread phenomenon.

This formulation of the question, why one form of reproduction does not displace the other, seems fruitless and inappropriate. We did not raise the question: why don't evolutionarily advanced forms displace primitive forms? We understand that they occupy different ecological niches. The same considerations give the answer to all questions related to the coexistence of different forms of reproduction.

Williams examined the "balanced" coexistence of sexual and parthenogenetic ways of reproduction in populations with facultative parthenogenesis, at which individuals can multiply sexually, as well as asexually. He came to a conclusion that because sexual reproduction continues to exist alongside with parthenogenesis, it also has some short-term benefits (Williams, 1975).

The advantage of asexual reproduction is its simplicity and high effectiveness. There is no need to find a partner and any asexual organism can reproduce in any place. Another advantage is that gene combinations created as a result of mutations are transmitted from generation to generation without change.

The disadvantage of asexual reproduction is that mutations are the only source of diversity. Therefore if two beneficial mutations a → A and b → B are created in different organisms of asexual population, they will never combine. Organism with AB genotype can appear only if mutation B will appear in some of the descendants of the organism with mutation A, and vice versa (Fisher, 1930; Müller, 1932).

One more disadvantage of asexual reproduction is that it lacks mechanism that prevents accumulation of harmful mutations. The asexual population will continually deteriorate and have no way to return to initial state with no mutations, similar to a ratchet that can only turn one way and once clicked can't turn back (**The "Ratchet hypothesis"**). There can be a limitation that ratchet sets on genome size for asexual organisms. Sexual recombination can eliminate mutations by crossing-over between chromosomes. The "Ratchet hypothesis" may explain why asexual populations go extinct more often than sexual populations; however it does not explain why asexual individuals without recombination don't continually arise and spread in sexual populations.

Hypotheses and Theories

Many bright scientists have attempted to determine the role of sexual reproduction in evolution—Weisman (1889), Goldschmidt (1927), Fisher (1930), Müller (1932), Crow, Kimura (1965, 1970), Bodmer (1970), Eshel, Feldman (1970).

During the past 25-30 years hypotheses have become so numerous that their classification was required. Felsenstein (1988) and later Kondrashov (1993) proposed to divide them into two main groups: *immediate benefit* and *variation and selection* hypotheses. Detailed description of these hypotheses can be found in many reviews and monographs (Kondrashov, 1993; Michod, Levin, 1987; Ridley, 1994). We will discuss some examples only.

"VARIATION AND SELECTION" HYPOTHESES

Variation and selection hypotheses discuss the results of recombination, which leads to increased variation of the progeny.

The Lottery Principle. The Lottery Principle was first suggested by Williams (1975) and uses a lottery analogy to illustrate diversity created as a result of sexual reproduction. One strategy would be to buy a large amount of tickets with the same number. The other strategy would be to buy a small number of tickets, but giving each of them a different number. Asexual forms use the first strategy, sexual—the second. The Lottery

Principle suggests that sex would be favored by a variable environment. However, in stable environments (such as in the tropics) sexual reproduction is most common. Contrary, in the areas where conditions are unstable (such as at high altitudes or in small ponds), asexual reproduction is rife.

R. Fisher—H.T. Müller theory. According to this theory sex accelerates evolution by combining beneficial mutations (Fisher, 1930; Müller, 1932). In asexual populations beneficial mutations will compete with each other and only one would win. The probability that two beneficial mutations will appear in the same asexual individual is very small. In sexual populations, however, different advantageous mutations can increase at the same time if two organisms of the opposite sex with mutations A and B will produce offspring. However, if both mutations are rare, this mating is also unlikely. In addition sex may undo the beneficial combination once it is created by producing offspring that have one or zero mutations. With asexual reproduction, once an advantageous combination is achieved it will be kept and passed to the progeny. So, it is not clear that sex is a more effective way of combining and keeping beneficial mutations together, and that it accelerates evolution. In small populations mutations occur less often, so the time between beneficial mutations is longer. These mutations are unlikely to be present in the same population at the same time. For this reason sex has little effect on the rate of evolution in small populations. It is no better than asexual reproduction. On the contrary, in large populations different beneficial mutations are likely to be present in significant numbers, so some of them could be brought together by mating. Therefore, according to the theory, the advantage of sex should work better in large populations. However, the theory also assumes population level selection between sexual and asexual populations. This process of group selection is known to be most effective in small populations— just the opposite of what is needed for the theory to work. This internal contradiction indicates the theory is deeply flawed.

The Red Queen Hypothesis. Some scientists think that main purpose of sex is to create variety, so animals can defend themselves from parasites. Parasites are usually adapted to attack the most widespread genotype and gene combinations "confuse" them. The Red Queen hypothesis was named after the character in L. Carroll's novel *Through the Looking Glass* who took Alice on a lengthy run that actually went nowhere. Scientists from British Columbia University (Canada), S. Otto and S. Nuismer (2004) verified this hypothesis on the mathematical model. It was shown that the benefits of sexual reproduction are revealed only when the amount of parasites is very high—much higher than can be found in real life.

"IMMEDIATE BENEFIT" HYPOTHESES

According to "Immediate Benefit" hypotheses the progeny may have the following benefits:
1) increased fitness of offspring not caused by changes in their genotypes, because it has two parents and conjugation of chromosomes during meiosis allows reparation of DNA damages;
2) better offspring genotypes because of a lower deleterious mutation rate (by filling the deletion gaps, cutting insertions or performing methylations);
3) better offspring genotypes because the selection amongst "cheap" males and male gametes is more efficient.

"Gene repair" theory. According to a "gene repair" theory the main task of sex is reparation of DNA damages and mutations that threaten life (Michod, 1995). This mechanism allows sexual forms to avoid the ratchet effect that degrades the genome of asexual forms. Sex is viewed as a cooperative process between mates and they both benefit from it. Genetic variation and reproduction are considered as side effects but not the sole purpose of sex.

Many of the abovementioned effects of sex are important indeed. It's also possible that the applicability of different effects depend on population parameters (size, type of organisms, speed of evolution and other factors). Many of the theories explain why sex exists but can not explain its creation. Cited advantages (except selection amongst males) can be equally applied to hermaphrodite as well as dioecious populations. In order to understand the advantages of dioecy it should be compared with hermaphrodite rather than asexual populations.

Comparison of Dioecious Ways with Hermaphrodite

"For evolutionary biologists, the existence of males and females is in many ways paradoxical. ... The question of differentiation at the individual level still remains. Why do most animals have individuals that are either male or female instead of both?"

D. Saraga (Reflex Magazine, April, 2011)

The establishment of sexual process during evolution has passed a number of stages that are easily traced in plants. R. Goldschmidt (1927) and V. Vendrovsky (1933) both came to a conclusion that low organized forms are hermaphrodites, while dioecy has arisen later in phylogeny. They showed that during the evolution from hermaphroditism to dioecy, species undergo a series of stages. Initially hermaphrodite forms during life cycle develop female and male organs of reproduction (sequential hermaphroditism). In the next stage of evolution one kind of sex organs gradually disappears in ontogeny and separate-sex forms appear (rudimentary hermaphroditism). Many scientists recognized such a step in the evolution of plants from hermaphroditism to dioecy (Vavilov, 1920, Rozanova, 1935; Kreshetovich, 1952; Minina, 1952; Takhtajan, 1964; Zhukovsky, 1964,1967; Dzhaparidze, 1963,1965; Shailahian, Hrianin, 1982).

The presence of numerous transitional forms from hermaphrodite to dioecious in many species serves as a confirmation of such a course of evolution of duplication. In particular, the existence of unisex forms of the castor oil plant (*Ricinus communis*) (Sidorov, Sokolov, 1945,1947) and sunflower (Kuptsov, 1935) were described. Formation of hermaphrodite flowers or intersexes on monoecious plants (Kardo-Sisoeva, 1924, Grisko, 1935, Sukashev, 1936, Minina, 1965) testifies an evolutionary path from hermaphroditism to dioecy. Also the facts indicate that hermaphrodite flowers are more primitive than dioecious (Hutchinson, 1959; Heinz, 1927).

Murray (1964), Tomlinson (1966), Ghiselin (1969,1974), Maynard Smith (1978) and other scientists discussed the evolutionary situations in which there can be a selection on hermaphroditism.

If a species often exists in conditions of low density, if individuals are attached or sedentary, if they explore the new, unpopulated habitat, then in all these cases dioecious forms will have difficulties finding a mate. These are the major factors of selection discussed by different authors, favoring the development of hermaphrodite forms over dioecious. They also favor hermaphrodite self-fertilization. In such situations even hermaphrodites unable to self-fertilization have some selective advantages over dioecious as for them each individual is a potential mating partner whereas for dioecious forms only half of individuals (only an opposite sex) are potential partners. It's difficult to estimate the importance of this effect. However, hermaphroditism is really widely distributed in conditions when the meeting of partners is limited. In plants such hermaphrodites are usually unable

I'm a practicing heterosexual... but bisexuality immediately doubles your chances for a date on Saturday night.
W. Allen New-York Times 10 Dec. 1975.

to self-fertilization. The same, probably, is valid for the animals, although there is limited data available.

The case of alternating hermaphroditism was also discussed. When the individual initially acts as a female, and then—as a male (*protogyny*), or on the contrary, first as a male and then—female (*protoandry*), one can imagine that selection will promote such type of hermaphroditism if for performing female and male functions different sizes are more favorable. For example, if males are fighting each other for females, protogyny becomes more preferable. If larger female size is more favorable when laying a large number of eggs, protoandry gets preference.

Ghiselin (1969) discusses two results that hermaphroditism can cause for the genetic structure of a population. First—facultative hermaphroditism can result in decrease of an inbreeding level for animals and self-pollination for plants. The second—hermaphroditism can increase the effective quantity of a population (Murray, 1964), compared to a dioecious one with a sex ratio deviation from 1 : 1 proportion. Both consequences seem to be of less significance, the second besides is not always true, as the deviation from a proportion 1 : 1 towards excess of females will increase the effective quantity of a population.

For many higher plants hermaphroditism is an initial evolutionary condition, and one and two household versions in many cases arise from hermaphrodite predecessors independently. Out of 240,000 species of flowering plants 6% are dioecios, and from 13,000 genera 7% have dioecios species (Renner, Ricklefs, 1995).

Prevention of inbreeding is probably the main selective factor working for the benefit of one and two household plants. Maynard Smith (1978) considers that it is much easier for the species to evolve from self-fertilization hermaphroditism to one or two household than develop the effective mechanism of self-sterility. He supports this point of view with two groups of the facts. First, two household feature is usually found in families in which self-sterility has not arisen. For example, in the family *Cariophyllaceae* in which self-sterility does not occur, whereas two household versions appeared as a result of evolution, at least in three cases (Baker, 1959). Second, one household plants almost never are self-sterile (Godley, 1955).

Thus, as Maynard Smith considers, for the higher plants the following picture is common. Hermaphroditism is a typical way of duplication because it provides "optimum distribution of resources". One and two household features finally are less effective concerning cross-pollination, than self-sterility at hermaphroditism, but they are more easily achieved. The content of island and wood flora illustrates his arguments. For example, on the Hawaiian Islands 27.5% of species are of two households, whereas for continental flora the typical value is less than 5% (Carlquist, 1965,1966). Baker (1967) considers that originally occupied Hawaii plants were really self-fertilizing hermaphrodites, and the two household feature has arisen secondary during evolution on the island. According to this point of view, the high frequency of two household plants on the islands happens due to the fact that any of first plants had no barriers against self-pollination, and in reply to selection against inbreeding these species have chosen the rather easier evolutionary way—acquiring the two household feature.

C. Darwin already knew, that frequency of two household plants among trees is much more common, than among bushes, and among bushes is much greater, than among grasses (**Table 1.2**). In tropical woods of Costa Rica 22% of all trees and 11% of bushes are of two household, whereas among 300 grass species there is only one, and not present among vaines and epifites (Bawa, Opler, 1975). The conclusion was made that the bigger the plant is, the more probability of self-pollination and the stronger selection preventing self-fertilization (Maynard Smith, 1978).

In a tropical forest two household plants are pollinated mainly by insects, and in a moderate zone they are mainly pollinated by wind. Probably, it reflects the relative prevalence of insects and character of winds in these two habitats. At the same time the very high frequency of one household trees in woods of a middle zone of Northern America is completely unexplainable. A one household feature can be an effective enough mechanism of cross-pollination in trees as staminate and pistillate flowers are usually located on different parts of a tree. One household plants are pollinated, as a rule, by a wind or water, therefore they have lost such organs of insect attraction as petals and nectaries. The percentage of two household plants, among tropical is very high (22–24%), but this more likely mostly reflects the fact that they are trees, instead of that they are tropical (Maynard Smith, 1978).

**Table 1.2 Percentage of forest plants with different character
of duplication (Maynard Smith, 1981)**

Type of forest	Level	Hermaph rodites	Of one household plants	Of two household plants	Reference
Deciduous forests of England (4 types)	Trees	50–57	14–30	20–33	Baker, 1959
	Bushes	60–89	0–38	0–23	
	Grasses	86–93	0–9		
Five South American middle zone forests	Trees	0–27	60–83	6–17	Bawa, Opler, 1975
Seven tropical forests	Trees	41–68	10–22	22–40	Bawa, Opler, 1975

The complete review of hermaphroditism in animals is given in Ghiselin articles (1968,1974). With hermaphroditism the following features of ecology and life cycle of animals are connected: the attached way of life, parasitism, deep-water forms, and a planktonic way of life and care of posterity. It is considered, that factors of selection are also the difficulties of finding the mating partner and the optimum distribution of resources.

Thus, in discussion of the comparative advantages of hermaphroditism and dioecy, the strange situation exists when the majority of authors are trying to find the advantages of hermaphroditism which could be supported by the natural selection on a background of dioecy. Whereas it is necessary to show just the opposite: advantages of dioecious forms over hermaphrodite ones, because the advantages of hermaphroditism before dioecy are obvious:

1. Half of dioecial population does not produce progeny ("two-fold cost of sex"). As a result asexual and hermaphrodite populations grow faster with each generation (Maynard Smith, 1978).

2. In the case of synchronous hermaphroditism, as well as for asexual reproduction, when one organism can produce male and female gametes, there is no need to find a partner.

3. The genetic variety provided by hermaphroditism is approximately 2 times more than that of dioecy. If we compare their combinatory potentials for the populations of equal quantity N, then in the first case we shall get:

$$N * (N - 1) / 2 \approx N^2 / 2 \quad \text{(for large value of } N \text{)}, \qquad [1]$$

and for the second case: $(N / 2) * (N / 2) = N^2 / 4$ [2]

possible variants of crossing.

4. Hermaphrodites can easily use a combination, or alternation, of sexual duplication with parthenogenesis. This allows them, depending on environmental conditions, to flexibly utilize advantages of both ways.

Theories of differentiation

Differentiation of sexes prevents inbreeding depression. In some plant species in which the evolution of separate sexes is not complete, hermaphrodites exist among male and female plants. Sex chromosomes in these plants have two loci—or positions of genes on a chromosome—one that controls sterility and fertility in males and the other in females. Offspring that inherit both fertility versions are hemaphrodites capable of self-breeding. Plants that possess one fertility and one sterility version become either male or female. The single-sex plants breed not only with one another but also with hermaphroditic plants and pass on the mutation, resulting in single-sex offspring. When inbreeding depression in hermaphrodites is considered, a gradual decline in the number of hermaphroditic plants is to be expected. Consequently, fewer chromosomes with both fertility versions of the loci will be passed on and the frequency of single-sex individuals will increase.

Disruptive selection theory. In 1972 Parker, Baker and Smith proposed that if the size of the zygote is important enough for its survival, the evolutionarily stable strategy will be *anisogamy*. In such cases a population consisting of males (microgamete producers) and females (megagamete producers) will be stable. The disruptive selection theory forms a basis for the origin and maintenance of the two sexes, although the initial analysis was performed on the organisms with external fertilization (many plants, and certain animals with external fertilization) (Parker, et al., 1972).

*   *   *

Comparison of three basic forms of duplication shows that forms of asexual duplication as many authors fairly specified, give double overweight by quantity of posterity in comparison with sexual forms. In turn, as previously discussed, inside of sexual forms hermaphrodites have double combinatory prevalence compare to dioecious forms. Progressive evolution of animals and plants was going, as a rule, from asexual forms to hermaphrodite and from hermaphrodite to dioecious. The first part of this transition can be explained. The second part is still unclear.

So, the main task consists not in proving advantages of sexual forms of duplication over asexual, but to show, inside of sexual types of duplication, the advantages of dioecious forms before hermaphrodite ones. So, the question, reflected in the title still has no satisfactory answer.

All advantages of sexual duplication cited in the scientific literature: an opportunity to avoid inbreeding, maintenance of a genetic variety and replacement of harmful mutations (Kalmus, Smith, 1960), are related to combinatory only. Therefore they are related entirely to the process of *crossing*, but not *differentiation*. It turns out that existing theories and representations can explain the advantages of *sexual* types of duplication before *asexual*, but are not capable of explaining advantages of *dioecious* types before *hermaphrodite*.

Moreover, if we will stick to the existing view which is taking into account only crossing-related advantages, we will need to recognize the hermaphrodite way of duplication as evolutionary more favorable than dioecious, because it provides twice the genetic variety of the dioecious one. However it is known, that all evolutionarily progressive forms of animals (mammals, birds, insects, etc.) use the dioecious method of reproduction.

Separate-sex forms make about 2% of plants, but in this case also the same evolutionary trend can be traced—more advanced forms have sex differentiation (of two household plants).

Chapter 2

Mysteries of Dioecy: Variation of the Sexes

It has long been a paradox that there are "more male geniuses, more male criminals, more male mental defectives" in spite of only minor differences between the mean performances of the two sexes.
Heim, 1970, p. 137.

Since the time of Darwin, biologists have noted that in many ways in many species the males are more variable. Thus large variability in phenotypic characters was found in male salmon (Zhivotovsky, 1984; Efremov, 1999; Agapova, 2011). Schüler L. et al. (1976) found differences between male and female mice in the weight of endocrine organs. Proportion of genotypic part in phenotypic variability in all studied characters was greater in males than in females.

Variability in humans

Most of the studies on sex differences have focused on differences in mean values rather than differences in variability. In humans particular attention has been devoted to differences in mental abilities and in behavior. It has long been a paradox that there are "more male geniuses, more male criminals and more male mental defectives" (Heim, 1970, p 137) in spite of only minor differences between the mean performances of the two sexes.

> **"biologists since Darwin have noted that for many traits and many species, males are the more variable gender"**
> Edge, 2005

Study of intrasexual variability in humans in many ways showed great variability of the male sex. For example, the distribution of body weight at birth between the sexes differ (Gage, Therriault, 1998). Boys not only weigh more but also show greater variability in weight at birth than girls (Pethybridge et al, 1974; Lehre et al, 2009). Height and weight showed greater variability in adult men, but body mass index (BMI) was more variable in women. Of the 31 blood parameters analyzed, 13 had a greater male variability, seven showed greater female variability, and 11 had no significant differences between the sexes. Greater male variance was showed in times to run a 60 meter distance (Lehre et al., 2009).

Males also show greater phenotypic and genetic variability for traits influenced by male hormones, and hormone levels are in turn influenced by many factors including social stress and infections (Rowe, Houle, 1996).)

VARIATION IN ONTOGENY

The pace of individual development and the duration of ontogeny stages differ between the sexes. It is known that girls outpace boys in development. Therefore in early age groups (from 2 to 4 years old) girls lead both on the average values of intellectual development as well as on the variability. But at about 10 years, the boys start having higher average values of intellectual development, volatility and are more presented on the tail of the distribution with the higher values of intellectual development (Arden, Plomin, 2006).

VARIATION IN TWINS

Vandenberg et al. (1962) investigated intrapair differences on 23 pairs of male monozygous twins and 21 pair of female ones in various parameters: anthropological, cardio-vascular, hematological, biochemical, psychological etc. As a result, the intrapair scatter proved to be significantly greater in female pairs (on 185 variables). Male pairs showed more difference on 41 variables. On 355 variables no significant differences were observed.

Chovanova et al. (1980) studied somatic types in 29 monozygous and 23 dizygous pairs of brothers and 24 monozygous and 15 dizygous pairs of sisters and obtained the similar results. Among monozygous twins, females have wider variation on all the somatic components, while in the dizygous vice versa. Similar results were obtained by other researches (Nikityuk, 1977).

Testing the degree of variation on 9 parameters (social, behavioral and cognitive measures at ages 2, 3 and 4) using data from around 4000 same-sex twin pairs showed that monozygotic males were generally more similar than monozygotic females. Interestingly, dizygotic twins showed the reverse pattern of correlations for similar variables (Loat et al., 2004).

Cognitive test scores
Wider phenotypic variance of males was detected in a variety of tests to determine the intelligence quotient (IQ). Adult men and women differ little in average values of intellectual development. However, the number of men on both sides of the normal distribution, which was built on the results of the IQ measurements, significantly exceeds the number of women (Hurt, 1978). On the basis of these studies it was concluded that the range of men's mental abilities is much wider (Lehrke, 1978; Benbow, Stanley, 1980; Mosiey, Slan, 1984; Rothman, 1988).

Exceptional talent is more prevalent among boys, the boys also dominated among the winners of various competitions (Lehrke, 1978). Survey of adolescents found sexual differences in the mathematical abilities in favor of boys, and these differences increased with the increase of the result threshold. Since the sex ratio among subjects who have scored 500 points was 2 : 1, 600 points— 4 : 1, and 700 or more— 13 : 1 (Benbow, Stanley, 1980).

Psychologists from the Edinburgh University have checked the cognitive ability distribution of more than 80,000 children born in Scotland in 1921. They were tested at age 11 in 1932. There were no significant mean differences in cognitive test scores between boys and girls, but there was a highly significant difference in their standard deviations. Boys were over-represented at the low and high extremes of cognitive ability (ratio of girls to boys was 1 : 1.4 for IQ ranges from 50 to 60 and from 130 to 140) (Deary et al., 2003).

They also have checked up intellectual level of more than 2500 brothers and sisters, based on computer science, mathematics and English tests. They also tested their ability to perform mechanical operations. Men scored maximum and a minimum quantity of points under all tests; they also have better coped with tasks on computer science and arithmetic. Women have received average grades and are better in language tests. Twice more men than women have appeared on the periphery of the distribution—for both two percent clever, as well as among two percent of the least intellectual examinees (Deary et al., 2007).

The American researchers checked mathematical abilities (SAT test) of 7th graders on extreme right border of distribution (the top 0.01 % of population). In the early eighties there were 13 boys per each girl. By 1991 the difference has decreased to 4 : 1 and remained the same after that. For the test on mathematical and scientific reasoning (ACT test) the ratio was 3 : 1, and 18 out of 19 students with the highest score were boys (Wai et al., 2010).

"Men are either very clever, or very dumb": ...among the most intellectual 2% of the population there are twice more men than women.

...among the least intellectual 2% of the population there are twice more men than women.

Deary et al., 2007.

We have to admit that men prevail in many areas of social life. Even the great advocate of women—A. Montagu (1968) writes: "The fact is that by far the largest number of geniuses, painters, poets, philosophers, scientists, and so on, has been men. Women have made, by comparison, a very poor showing. Where are the Leonardos, the Michelangelos, the Shakespeares, the Galileos, the Bachs, and the Kants of the feminine world? ... How many of them (women) have been great composers...? Of composers—none of the first rank."

"There is no female Mozart because there is no female Jack the Ripper".

C. Paglia, 1990, p. 247)

Men surpass women in a game of chess, in musical composition, invention and other creative activity (Harris, 1978; McGlone, 1980). There are few women in the field of humor: among satirists, humorists, comedians and clowns.

Science and inventions
Inventions by women have never been more than 2 % of the total accepted at the US Patent Office (Montagu, 1968). The explanation narrowed to household area [where women are supposed to be the experts] follows: "…women have been so busy using, among other things, the gadgets invented by men that they have had no time left over for inventing anything. … men often have good ideas about housework chiefly because they don't like it much. The laziest person [men] … He is the one who thinks up the short cuts."

Inventions by women have never been more than 2% of the total accepted at the US Patent Office.

Among Nobel Prize winners there are also fewer women. On all 6 categories the percentage of women does not exceed 11 % (see **Table 2.1.** The Nobel Prize Internet Archive).

Among the biographies of 280 scientists taken from "The biographical dictionary of scientists" (Porter, 1994) 273 (97.5%) were men and only 7 (2.5%) were women, indicating very strong sexual dimorphism (Kanazawa, 2000).

Table 2.1 Nobel Prize Winners

Category	Men	Women	% Women
Physics (1901-2004)	170	2	1.2
Chemistry (1901-2004)	143	3	2.0
Physiology & medicine (1901-2004)	175	7	3.8
Economics (1969-2004)	53	0	0
Peace (1901-2004)	93	12	11.4
Literature (1901-2004)	91	10	9.9

Culture

According to Miller (1999) men produced about 20 times as many total jazz albums (1800 albums by 685 men, and 92 albums by 34 women), 8 times more pictures (2979 pictures made by 644 men and 395 paintings by 95 women), and 3 times more books than women.

Aggression and Asocial Behavior

There are more men among antisocial individuals as well. They are more exposed to all social vices— gambling, drinking, smoking, and drug addiction. More men are among gangsters, criminals and terrorists. It was noticed that boys require more efforts to learn social skills, and need more efforts on their parenting (Maccoby, Jacklin, 1974).

Theories of greater male variability

Sex differences have been explained both by culturally mediated gender roles (Eagly, 1987), intermale competition (Archer, Mehdikhani, 2003), higher mortality (Migeon, 2006) and by Darwin's principles of sexual selection (Andersson, 1994; Geary, 1998).

Greater Male Variability hypothesis (Shields, 1982). Given that mean differences between the sexes are small (Lehre et al., 2009), especially for general and many broad abilities, the observation of large numbers of males in the right and left tail of the ability distribution has led to the variability hypothesis. This hypothesis states that males exhibit greater variation than females in many cognitive abilities.

Feminists call it a "*pernicious hypothesis*" and try to explain the shortage of outstanding females by social factors, discrimination, lack of motivation or methodological and statistical errors (Noddings, 1992). It's

The idea that men are more often deflected towards a good or bad side (more geniuses and more idiots), was introduced in the early 20th century to explain why men are superior to women in many areas despite the small difference in the average values of psychological tests between the sexes.

Shields, 1982

interesting to note that "over-representation of males in the extreme low tail of the distributions is less disputed among scholars" (Arden, Plomin, 2006). One proposed explanation is that women are not imposed to such stringent requirements as to the men and they can carry out their duties more effectively.

Darwin's theory of sexual selection. Darwin's theory of sexual selection explains sex differences in variability so that males can afford more differences depending on environmental conditions and still maintain reproductive success. Males who deviate from average on a trait desired by females will experience increased fitness some of the time, even though they will suffer a loss another time (Archer, 2006; Moore, 1991). This may explain the empirically observed increased variation, in other species, in several traits under direct selection. In several non human species traits under direct selection have been shown to be more variable than morphological traits not under direct selection (Brandt, Greenfield, 2004; Rowe, Houle, 1996; Tomkins et al., 2004).

Miller's courtship model. Miller (1999), developing views of C. Darwin and F. Nietzsche, has suggested that greter achievements of men in science, inventions and culture can help in the competition for mating partners, and their retention. Specific human behavior, such as clothing, dancing, creating new music and ideas are regarded as a demonstration of such innate characteristics, as talent, creativity and taste. In accordance with the predictions of the model there must exist a significant sexual dimorphism in the production of cultural objects, since men are more involved in this kind of demonstration. Secondly, the cultural products should grow rapidly after puberty, reaching a peak in young people when sexual competition is maximal, and gradually decline as the care for children comes to replace the search for a partner. Indeed, there was significant sexual dimorphism in production of music alboms, pictures and books. Age distribution was also consistent with the predictions. Miller believes that this pattern can follow any conduct which is available for viewing by potential partners and can not be easily reproduced by other competitors. Examples include loud music in the car, creating Web pages, risky sports (surfing, parachuting, car racing) and scientific discoveries. So S. Kanazawa revealed the existence of a very strong sexual dimorphism among scientists (39 : 1). The age distribution, corresponding to Miller's model, was noted only in those scientists who were married; unmarried scientists have not showed such a strong reduction in the scientific contribution (Kanazawa, 2000).

*   *   *

The data presented show that greater intrasex variability among males is a fundamental aspect of the differences between sexes. This is true not only for humans but also for many animals. In humans, the variability is not limited to intelligence parameters, and includes birth weight, blood parameters, physical performance and university grades. The differences in the variability of body weight at birth strongly suggest that social factors can not account for all the observed differences in variability.

Chapter 3

Mysteries of Dioecy: Sex Ratio

M ain parameters of the dioecious population are absent in hermaphrodites,—sex ratio, variation of the sexes, and sexual dimorphism. In this chapter we will discuss unclear facts of the higher birth and death rate of males, influence of a genotype and other parents' characteristics on a sex ratio of their progeny, influence of environmental factors on a secondary sex ratio and other phenomena. We will also discuss theories of sex ratio evolution, and more specific theories dealing with the separate phenomena, such as differential mortality of sexes and "phenomenon of war-time".

High primary and secondary sex ratio

Now it is well established, that for the majority of animal and plant species the chromosomal mechanism of sex determination is the basic one.

In a course of gametogenesis the gametes containing X and Y-chromosomes, are made in equal proportion. Therefore, it was considered, that this mechanism provides approximately equal quantities of sexes at conception. However, for a long time it has been noticed, that for many animal species the secondary sex ratio differs from a 1 : 1 proportion towards a surplus of males: approximately 105–106 of males per 100 females. Of course, the most reliable of all data is collected on humans. The average value of a secondary sex ratio on all human populations is 106 (Ellis, 2008).

In the beginning, the observable surplus of boys among newborns was explained by higher death rate of female embryos during uterine development. However, research of embryonic mortality of the sexes has shown that it was not true. Taking the differential death rate of sexes during the embryonic stages of development into account not only has not approached extrapolated value of a primary sex ratio to a proportion 1 : 1, but has even been more removed. All accessible data on gender distribution for human abortions and children born dead show that more boys die in utero, than girls (Stump, 1985). The long-term data on the registered abortions, available for many Europe cities, and also collected by the US Bureau of qualifications, confirm that at earlier stages of uterine development, the death rate of boys exceeds the death rate of girls by 2–4 times (Stern, 1960). Thus, in the first two months of pregnancy there are 7–8 times more male abortions compare to females. At the end of the third month this ratio drops to 1.7 : 1.

However, these estimates did not match the sexual structure of abortions from a known collection of the Embryology Department of the Carnegie Institute in Washington. This collection consists of 6000 embryos delivered for many years by various doctors. The sex ratio of embryos of this collection is constant enough on months of a uterine life and equals on the average 107.9 boys on 100 girls.

Other authors received the extrapolated primary sex ratio values from 117 to 180 (Kukharenko, 1970,1971; Loginov, 1989; Moore, Persaud, 1998; Fukuda et. al., 2002)!

Not discussing specific values of a primary sex ratio reported in numerous works, let's note the general picture. The primary sex ratio in humans, most likely, deviates from a 1 : 1 proportion towards a surplus of male zygotes, and it is probably higher than a secondary sex ratio. The same picture is observed for some animals, for which the data is available. A high mortality of males was discovered at the beginning of pregnancy in deer (Robinette et al., 1957) and cows (Chapman et al., 1938). This is the first group of unclear facts related to a sex ratio—high values of a primary and secondary sex ratio.

Influence of Heredity on Primary and Secondary Sex Ratio

In connection with observable deviations of a human primary and secondary sex ratio from the theoretically expected proportion 1 : 1, the other important question has been raised. Is sex determination and a birth of the child of this or that sex a pure stochastic event?

If the probability of occurrence of the boy is p, and the one for the girl—q it is theoretically possible to calculate the expected probabilities of occurrence of families with a different sex ratio and to compare them to the observable frequencies. Theoretically expected distribution corresponds to a binomial: $(p + q)^n$, where n—number of children. It has been established that families in which one gender prevails appear much more often than theoretically predicted, while families with an equal or close to equal sex ratio—much less often. From this the conclusion has been reached that the probability of a birth of the boy or the girl varies with different individuals of a population (Edwards, 1958). Similar studies that have been conducted later in England, France, the USA and Finland have led to the same results.

Probabilities of a birth of the second boy in a family after the first-born-boy and the first-born-girl have been investigated. For the US population they were equal accordingly 0.5387 and 0.4988, and for Finland—0.5132 and 0.4955. These distinctions are statistically valid. Hence, it has been shown, that there is a weak, but real tendency to make the second child of the same gender, as the first, third children of an identical sex with the second, etc. The analysis of these and other similar data lead Gini to a conclusion that the probability of a birth of the boy varies not only between different families, but also inside the family. Elderly parents tend to produce children of the opposite sex, compared to the sex of children they produced during their young years (Gini, 1951). These works are analyzed in detail in K. Stern's monograph (1960).

It was discovered that sex ratio depends upon different heritable traits. C. Kanazawa (2005) found that large and tall parents produce more boys, while short and small parents have boy's birth-rate lower than expected. He also found that sex ratio is related to physical attractiveness (attractive parents produce more girls) (Kanazawa, 2007), brain type (people with "male (S-type) of brain", such as engineers and mathematicians have more sons, and on the contrary, people with "female (E-type) of brain" —nurses and teachers—have more daughters) (Kanazawa, Vandermassen, 2005), and aggressiveness (more aggressive fathers have more sons) (Kanazawa, 2006).

C. Kanazawa explains his results within the framework of a generalized Trivers-Willard Hypothesis, noting that big height and weight, aggressiveness, and the "male type of brain" increase male's reproductive success, leading to high boy's birth-rate. Female reproductive success gets increased by attractiveness, low height, and "female brain type", which increases girl's birth-rate. The author itself points to the "seemingly logical absurdity" of the assumption that most men (with normal for them S-type of brain) should have more boys, and most women (with also normal for them E-type of brain)—more girls.

Influence of Environment on Primary and Secondary Sex Ratio

The secondary sex ratio depends not only on a proportion of heterogametes, but also from many other factors. For example from the speed of the ageing and elimination of X and Y-bearing sperm in a male body, from their ability to reach the egg and to impregnate it, from the affinity of eggs to X or Y- sperm, at last from stability (viability) of male and female embryos at different stages of embryonic development.

It was considered, that a secondary sex ratio is a specific constant, characteristic for each species. Attention was paid to deviations from a 1 : 1 proportion and, mainly, on shifts of a secondary sex ratio depending on different factors (Svetlov, Svetlova, 1950b; Stern, 1960) such as age, constitution, wealth and even occupation of parents (Bernstein, 1951).

A large amount of data on dependence of a sex ratio of posterity on age of mother is analyzed in V. Bolshakov and B. Kubantsev's monograph (1984). For all species: fox, mink, polar fox, dog, pig, sheep, cattle, horse and human "... newborns of a male sex, as a rule, prevail in posterity of young mothers. For middle age mothers that are in the blossoming of the reproductive function, the relative number of female descendants grows. For mothers of the senior age group the percent of males in posterity increases again."

The data, testifying that the sex ratio of posterity depends also on the age of a father: the older the father is, the less is his chance to have a son. Besides it appears that this chance depends upon the father's body type: men with an athletic constitution have lesser chance to have a son, than thin man (Heath, 1954). The birth rate in rich and poor families of America was specially surveyed. The percent of boys was higher in rich families. On the other hand, for the whole countries, the better living conditions are (wealth, climate and food) the fewer boys are born. The same was also observed for agricultural animals—the better the conditions of their maintenance are, the more females are born, even with artificial insemination from the same donor (Milovanov, 1962). And contrarily, bad conditions lead to increase in secondary sex ratio. When the population decreases due to different reasons, the male sex starts prevailing and boys' birth-rate is increasing, which also was noticed by C. Darwin (cit. by Bednij, 1987).

Some articles were devoted to a relation of a secondary sex ratio and the profession of parents. Russian physicists and geologists working on expeditions located at a height about 3000 m above the sea level (Kamalian, 1958), and also American pilots of high-altitude aircrafts (Snyder, 1961) have an excess of girls in posterity. Sex ratio of posterity for men working in female surrounding gets shifted by 15 % in one direction, and if they are surrounded with other man—the same shift changes direction (Stern, 1960). More boys are born in families of long-livers, and longevity of both father and mother is important (Lawrence, 1940,1941).

Secondary sex ratio is different in different countries. It has the maximal value for Korea—115 and Greece—113, and minimal—for England and Japan—104. In the European part of Russia there are 108 boys on 100 girls, and in Siberia—115. In the US sex ratio for the white population is 105.7, and for others—100 (Lek, 1957; Schienfield, 1944; Kang, 1959). It differs as well for various racial ethnic groups inside the country.

During big natural or social shifts (sharp changes of a climate, a drought, war, famine, resettlements) the secondary sex ratio has a tendency to increase (Novoselsky, 1958; Stern, 1960; Schienfield, 1944). The same picture is observed in animals during stress and the adaptive meaning of this phenomenon is unclear (Cameron, 2004).

"Phenomenon of war years"

Perhaps most discussed in the scientific literature was the phenomenon of boys birth rate increase during the wars. In demography it's called *"a phenomenon of war years"*. The statistics of birth rate during the First and Second World wars confirms it (Lowe, McKeown, 1951; MacMahon, Pugh, 1954; James, 1987, 2003; Van der Broek, 1997; Graffelman, Hoekstra's 2000; Ellis and Bonin, 2004). Cartwright (2000) calls it the "returning soldier effect". For example, in European countries involved in First World War, from 1 to 2.5 % more boys were born, compare to pre-war "peace" years (**Figure 3.1**). The maximal gain was observed in Germany where the sex ratio of newborns increased to 108.5. The same phenomenon was observed also during the Second World War in Austria, Belgium, Denmark, France, Germany, the Netherlands, USA and UK, with the exception of Italy and Spain. In England and France, for example, by 1943 the secondary sex ratio has increased by 1.5-2 % in comparison to a peace time (Novoselsky, 1958; Stern, 1960; Schienfield, 1944; Heath, 1954). Sex ratio among whites in the USA rose during World War II from 51.406 to 51.481 (MacMahon, Pugh, 1954). In the UK during wartime (1914–1920, 1939–1948) sex ratio increased to 51.365 compare to 51.2 during peacetime (Graffelman, Hoekstra's 2000).

It's interesting to note that sex ratio was not changed during more recent wars, such as the Iran–Iraq wars in 1980–1988 (Ansari-Lari, Saadat, 2002) and 1991 war in Slovenia (Zorn et al., 2002), which can be related to lower rate of mobilization. Lower rate of mobilization can also explain the absence of the *"phenomenon of the war years"* during World War I in the USA. (MacMahon, Pugh, 1954; Graffelman, Hoekstra, 2000).

"PHENOMENON OF WAR YEARS"— THEORIES

Several theories were offered for an explanation of this phenomenon. One of them pointed to the fact that during or after wars younger contingents of people get married, and with young mothers and fathers as it was already noted, the secondary sex ratio is higher, than at elderly ages. The second theory considers that during wars the number of first-time birth mothers grows. They, as was already mentioned, more frequently have boys. For example, among the first-borns who have been born within the first 18 months of marriage, a secondary sex ratio was 124, while among children who have been born after this period—99 (Stern, 1960). According to the third theory when the husbands serve in the army, wives are "having a rest" from pregnancy and this rest raises their chance to give birth to the boy. The fourth theory sees the reason of boys' birth rate increase in the changes of diet, in particular in a reduction of meat and other proteins. None of these hypotheses have collected enough proofs.

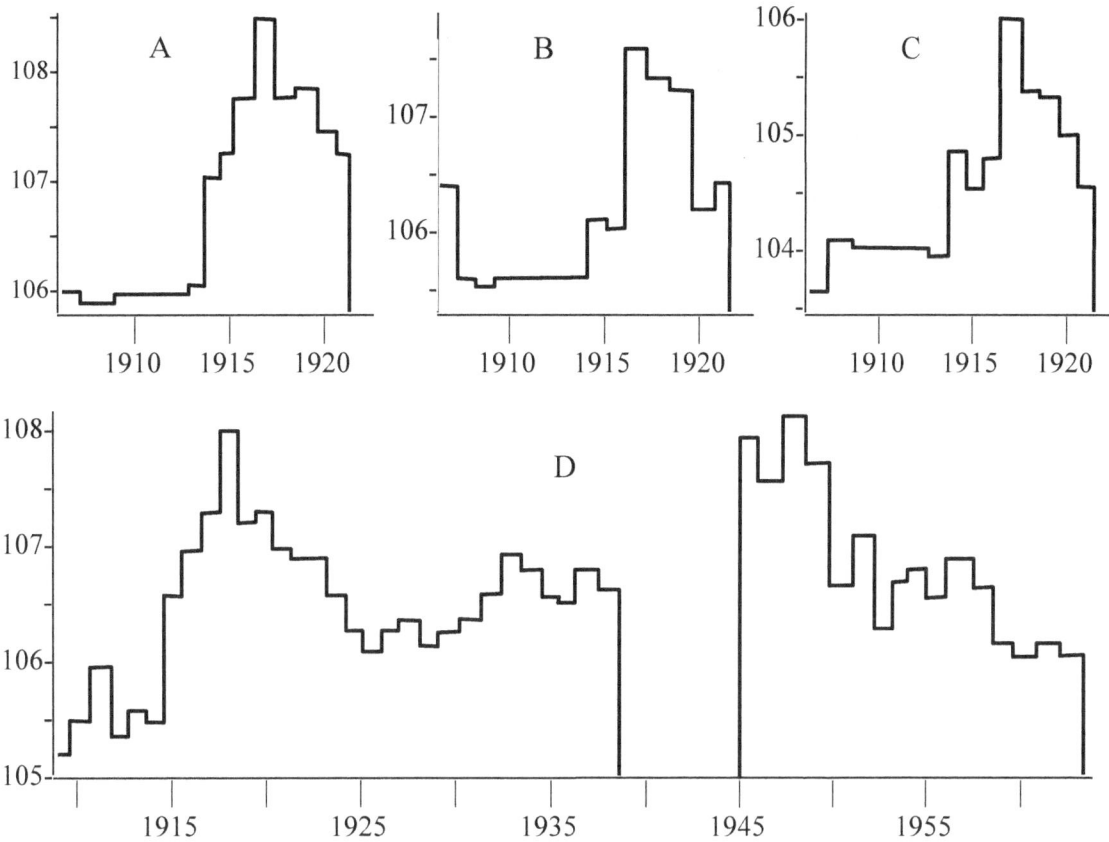

Figure 3.1 Increase of boys' birth rate during the first and second world wars.
A—Germany, B—France, C—England, D—The German Federal Republic.
Abscissa—years; Ordinate—number of boys per 100 girls.

Grant's "maternal dominance hypothesis". According to the "maternal dominance hypothesis" (Grant, 1998, 2003), the mother determines the sex of the offspring, and dominant, 'tough' women high in testosterone are more likely to have sons. Grant argues that women become 'tougher' during wars because they have to take over some of the traditionally male roles in society in the absence of men, and they as a result have more boys, presumably because their testosterone levels increase during wars. However, there is very little human evidence for the maternal dominance hypothesis in general, and as an explanation for the returning soldier effect in particular. It also cannot explain why the secondary sex ratio remains high for a few years after the end of the war.

James hypothesis (James, 2003). James explains *phenomenon of war years* with the assumption that couples are reunited only during short leaves from the armed services. They are expected to have frequent intercourse and, as a result, are more likely to conceive during the early phase of the cycle when the estrogen/gonadotropin ratio in women is high. Mammalian (including human) sex ratio is higher when the maternal estrogen level is higher at the time of conception (James, 1996). James (1981, 1983) provides some indirect evidence in favor of his main assumption; there is no direct quantitative evidence that coital frequency is higher during wartime and remains high for a few years after that.

Trivers–Willard hypothesis (TWH) (Trivers, Willard, 1973). In their classic paper, Trivers and Willard suggested that parents might under some circumstances be able to vary the sex ratio of their offspring in order to maximize their reproductive success. The Trivers–Willard hypothesis proposes that, for all species for which male fitness variance exceeds female fitness variance, male offspring of parents in better material condition are expected to have greater reproductive success than their female siblings. Since during wars many men die, it is possible to expect that sons will have more reproductive success during and right after wars than daughters. In this situation the hypothesis predicts that parents will have more sons. However parents cannot know how long the war will last and it is not so clear how their expectations can be related with the sex of their children. Initially a Trivers–Willard hypothesis dealt with material (or for the humans, with economic) conditions of parents, and said nothing about which parent influences sex of offspring.

Kanazawa hypothesis ("survival of the big and tall soldiers") (Kanazawa, 2007). In his work of 2007 S. Kanazawa writes that though there are no doubts in the existence of "a phenomenon of war years", it still does not have a satisfactory explanation. Generalizing the Trivers–Willard hypothesis, he tried to find an inherited trait which would increase the reproductive success of men. Then the parents that have such a trait will have more sons. And on the contrary, the parents that have traits, increasing the reproductive success of women will have more daughters. Height influences the reproductive success of men and women in a different way: tallness raises success in men, but lowers it in women. Kanazawa was able to show that big and tall soldiers have more chances to survive and come back home and he thinks that a prevalence of sons in their posterity can explain the effect.

Many noted factors, including stress, are responsible for the existence of "a phenomenon of war years" however the problem is still far from resolution.

Riddles of Longevity

Decreasing of Tertiary Sex Ratio in Ontogeny

In a course of ontogeny the sex ratio for many species of plants, animals and humans goes down. It is related to the raised death rate and damageability of male's systems in comparison with female ones at almost all ontogeny stages and at all levels of organization. Whether we study various species (humans, animals or plants), different levels of the organization (an individual, organ, tissue or a cell) or stability in the face of different harmful environmental factors (low and high temperature, starvation, poisons, parasites, diseases, etc.)—anywhere the same picture is observed: the raised death rate or damageability of male's systems in comparison with corresponding female's (Levin, 1949,1951a,b; Lek, 1957; Svetlov, 1943a,b,1945,1949; Svetlov, Svetlova, 1950; Svetlov, Shekanovskaja, 1945).

Hamilton (1948) reviewed the differential death rate of the sexes for 70 species, including such various forms of life, as nematodes, mollusks, crustaceans, insects, arachnoidea, birds, reptiles, fishes and mammals. According to the data, for 62 species (89 %) the average life of males is shorter, than females; for the majority of the others there is no difference, and only on occasion males live longer, than females.

The most extensive statistical material on death rate and diseases is present on humans. An integrated parameter of organism sensitivity to the environmental conditions is general average life expectancy.

Figure 3.2

The ratio between average life expectancy of men and women (Gavrilov, Gavrilova, 1991). Each point correspond a data on one country (Brook, 1981).

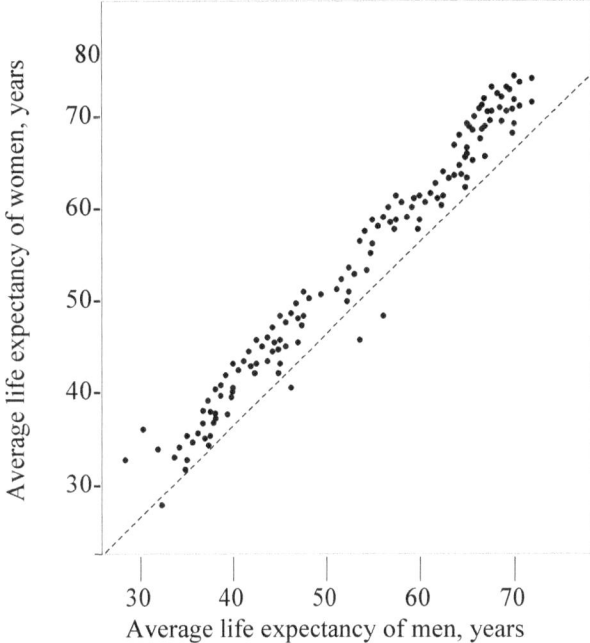

From the 163 countries of the world for which data are available women live longer than men in 152 (Brook, 1981). In five countries (Bangladesh, India, Indonesia, Turkey and Tanzania) there is no difference. The life span of men is greater than that of women only in six countries: Pakistan (by 4.9 years), Liberia (1.8), Upper Volta (1) and Jordan (0.6). In the world as a whole, the average life span of women (59.3 years) is 3.6 years greater than that of men (55.7). This difference is especially great for developed countries. In the USSR it is 10 years; in Finland, 9.1; in France, 8.0; in the USA 7.8, in Austria, Great Britain, and Canada, more than 7 years. The majority of points, each of which correspond to a country during certain time, lay above bisectors of a right angle, which shows equal longevity.

Let's analyze now a picture of death rate depending on age. Male embryos are more susceptible to changes of external conditions, their growth slows down more because of stress, and their mortality at birth is higher (Stinson, 1985; Wells, 2000). The data on intensity of infants' death rate (0–1 year) in many countries are submitted on **Figure 3.3** (Gavrilov, Gavrilova, 1991). The death rate of boys is higher and they are considerably more sensitive to change of external conditions than girls.

Changes in living conditions lead to approximately identical death rate changes of men and women, however man's death rate always exceeds female one (**Figure 3.4**, Gavrilov, Gavrilova, 1991).

Figure 3.3

Intensity of infants' death rate (Gavrilov, Gavrilova, 1991). Coordinates of each point correspond to intensity of death rate of boys (ordinate) and girls (abscissa) in the age of up to 1 yr.
The data from tables of life expectancy of the following countries (Gavrilova, et al., 1983):
● —France 1931–1937, 1946–1965, 1967, 1969 yy.
■ —Sweden 1956–1960, 1961–1965, 1966–1970, 1971–1975, 1975–1979 yy. ▲ —Dutch 1921–1925, 1926–1930, 1936–1940 yy. o —USA 1939–1941, 1949–1951, 1964, 1975, 1977 yy. □ —Norway 1911–1920, 1946–1950, 1971–1975, 1976–1980 yy. Δ —Switzerland 1933–1937, 1941–1950, 1948–1953 yy.

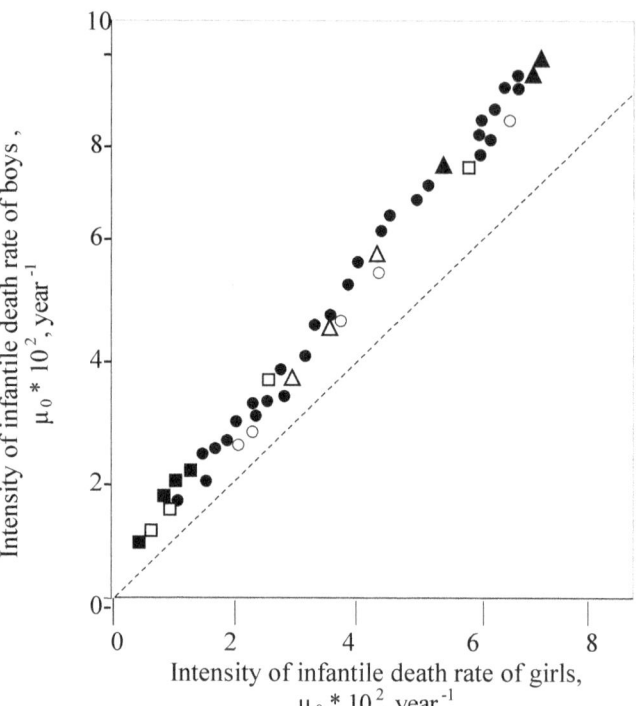

Males also have higher death rates from many diseases (**Table 3.1 (Appendix C);** Waldron, 1976). A higher death rate of men is observed almost from all illnesses, with just a few exceptions (whooping cough, gonococcal infections) (Stern, 1960).

Figure 3.4

Ratio between intensity of death rate of men and women (Gavrilov, Gavrilova, 1991).
Coordinates of points correspond to intensity of death rate of men (ordinate) and women (abscissa) at the age of 40 years. The given life expectancies of the white population of the USA for 1920-1929, 1929-1931, 1939-1941, 1949-1951, 1959-1961, 1964, 1970, 1975, 1977 (Gavrilova, et al., 1983).

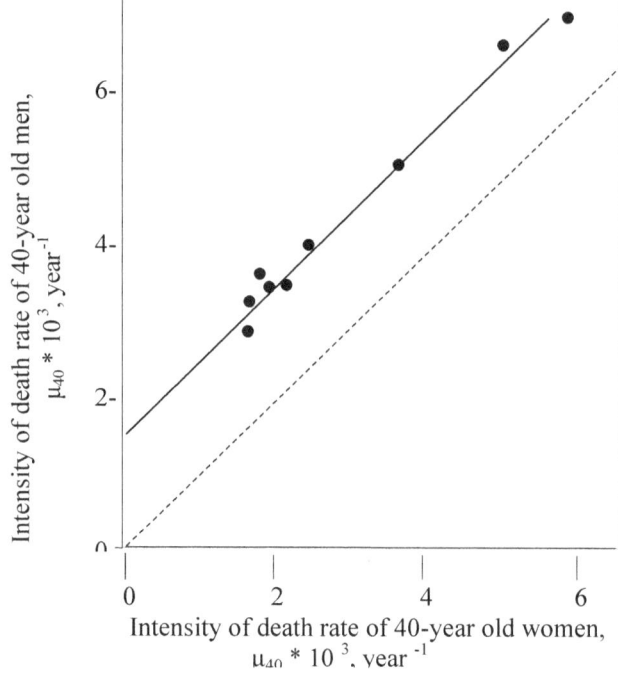

Sharp sexual dimorphism exists on sensitivity to various poisons and chemicals, including drugs. There are significant distinctions in metabolism, efficiency and toxicity of various drugs for males and females (Goble, 1975). For example, to reach the same degree of anesthesia male rats required 4 times higher hexobarbital doze compared to females (Quinn et al., 1954; Streicher, Garbus, 1955).

Males have also higher rates of unintentional, as well as intentional injuries, as a consequence of more risky behavior, aggression and job preference (police, fireman, soldiers) **Table 3.2.** Since the late 1980s, women have started using seat belts more frequently than men.

Table 3.2 Mortality from injuries (WHO Annex Table 10 p. 88, 2000)

Cause	Both sexes *(000)*	Males *(000)*	Females *(000)*	Sex ratio, % ♂♂
All Injuries	**5012**	**3406**	**1693**	**68**
Unintentional Injuries	**3529**	**2270**	**1258**	**64**
Road traffic accidents	1203	856	347	71
Poisonings	344	218	126	63
Falls	383	228	155	60
Fires	311	130	181	42
Drownings	409	282	128	69
Other unintentional Injuries	877	556	322	63
Intentional Injuries	**1601**	**1135**	**466**	**71**
Self-inflicted injuries	855	526	329	61
Violence	485	386	113	80
War	235	212	23	90

Summarizing, it is possible to tell, that the higher death rate of males is a general biological phenomenon, it is observed in plants, animals and humans for all levels of organization from all extreme factors of the environment.

Men are "Champions" of Longevity
Despite the fact that average life-span is longer in women, the "champions" of longevity are men. For example, in the 40ies of the 18-th century in the central part of South Transcaucasus there were 14 men out of 15 long-livers who attained 110-140 years. In the 20ies of the 19-th century in Abkhazia men also occupied all the steps of the age ladder (Arkhipov, 1978). It is also strange that the phenomenon of high longevity is encountered in the populations living under conditions that are far from the optimum (Comfort, 1964, Davidovsky, 1966).

Early termination of fertility in women
In humans, the maximum lifespan is nearly 100 years, but fertility in women usually ends in approximately half that time, well in advance of other aspects of physiological frailty (Hawkes et al., 1998). Long postmenopausal lifespan distinguish humans from all other primates. Human reproduction does not end early

in comparison with other apes. The main difference lies in the low adult mortalities which lead to long average lifespan after menopause. This characteristic is not directly related to medical advances, because low adult mortality is observed in populations that have no access to pharmaceuticals.

Apes live no longer than 50 years. They become frail with age so that all physiological systems, including fertility, fail in tandem. Still aging females of chimpanzees and other primates continue to produce offspring despite the low survival probabilities for late-borns (Hawkes, et. al., 1998). This observation contradicts Williams's assumption (Williams, 1957) that menopause and early termination of fertility evolved in humans when extended maternal care become crucial to offspring survival (the "stopping early hypothesis").

"Grandmothering" hypothesis. The grandmothering hypothesis is based on the assumption that elderly women played, and still play an essential role in the survival of children. The studies conducted in modern societies of hunter-gaverers and early farmers, confirmed this assumption. Really, grandmothers on the mother's side often provide a considerable share of the food supply for their grandsons, bringing vegetables and small invertebrates (Ache of Paraguay, Hadza of Tanzania, and Bushmen of Namibia). Grandmothers also look after the older children (Hawkes, 2004).

Theories of High Male Mortality

The existing theories are not able to explain the phenomenon of the higher males' death rate. Usually on a question "Why females live longer than males?" biologists answer: "because mothers are more necessary to the progeny, than fathers". Of course, group natural selection could lengthen the life of a more caring parent. But in some species mothers take care of posterity, in another species fathers do, and in other species there is no care at all. And of course plants do not take care of their offspring. However, the lowered viability of males is observed anywhere and everywhere.

For animals some theories explain the higher death rate of males, as a result of their bigger or smaller sizes, brighter coloring or the "risky" behavior connected to getting the food, fighting for females, fights with predators, and for humans—with dangerous professions (sailors, pilots, military personnel). However, different animal species have different sexual dimorphism on these attributes (for example, some species have larger males, while others—larger females) or none at all while the higher death rate of males is observed almost for all species. It is not clear as well why these distinctions (sexual dimorphism), as a rule, grow with age, and the difference in death rate, on the contrary, decreases; it is maximal for the young and levels out with age. These reasons allow rejecting the majority of the listed reasons as main or important. Hence, the given group of theories cannot explain the problem as a whole.

Sex-linked lethal theory. The theory of *gene imbalance* (Gunter, 1923; Huxley, 1924; Lenz, 1923; Schirmer, 1929) explains higher male death rate due to its heterogametic constitution, absence of the second X-chromosome in a man's set. Males can receive the defective recessive gene only from one parent and the gene will manifest itself. Contrarily, female should receive the defective gene from both parents. If the theory of chromosomal or gene imbalance is right, then first the differential mortality of sexes should depend upon a share of hemizigous (nonpaired) genes in a particular genome. *Drosophila* has a few autosomes compare to human, so this share is relatively more for *Drosophila*. The whole genome of the haploid hymenoptera males is in a hemizigous state. Therefore according to the theory, maximal difference in death rate can be expected for hymenoptera, much smaller—for *Drosophila* and even less for humans. However, nobody noted such distinctions. Further, as the son receives the X-chromosome only from the mother, a higher correlation in longevity between the son and mother should be observed than between the son and father. But as it has been shown (McArthur, Baillie, 1932), such a difference in correlation is also absent. And most important—for species with the chromosomal structure of *Abraxas* type with homogametic males, the females

> **More and more it appears that biologically men are designed for short, brutal lives and women for long miserable ones.**
>
> E. Ramey: Observer 7 April 1985 "Sayings of the Week."

should have a higher death rate.

Observations and experiments on *Abraxas* species (birds, butterflies, moth, some kinds of fishes) do not leave doubts, that, for many of them despite of heterogametic chromosomal constitution of females, the males also have higher death rate (Latham, 1947). It has been shown for hens (Landauer, Landauer, 1931), a pheasant (Haig-Thomas, Huxley, 1927), twelve species of *Lepidoptera*, and codling moth (*Cydia pomonella*) (McArthur, Baillie, 1932). For example, Howe's data (1977) on the secondary sex ratio in a population of Common Grackle *(Quiscalus quiscula)* in Michigan at the moment of an exit from the nest was 52♂♂ : 83♀♀. Such difference results from the greater death rate of males, because embryos had a sex ratio of 1 : 1.

McArthur and Baillie (1932) specially devoted their work to a question of relative sex death rate for species with *Abraxas* gamety type. After analyzing data on many species they came to the conclusion, that heterogametic constitution lowers viability, but it cannot be considered as a sole or main cause of different sex mortality. And as heterogametic and homogametic males, as a rule, possess a higher level of the basic metabolism than females, authors thought that *metabolic* theory is more comprehensible. This theory explains high male death rate by a high level of their basic metabolism. Certainly, there is a close relationship between level of metabolism and death rate, in particular between heart beat frequency and longevity (McArthur, Baillie, 1929). It is natural and clear. But the replacement of unclear "high male death rate" to also unclear "a high level of metabolism" doesn't resolve the question. These theories simply establish correlation between different attributes, characteristic for one sex or another: the *chromosomal imbalance* theory—between longevity and *gamety* type, and the metabolic theory—between longevity and level of metabolism. But they do not explain the evolutionary sense, logic and expediency of this phenomenon. Why, despite a huge variety of species, do males have a higher death rate (or metabolism)? It is possible to tell that the raised metabolism is a way to provide males with higher death rate (sensitivity). But what is its "purpose" or evolutionary sense?

Results of experiments in which the X-chromosome of *Drosophila* males was damaged by X-rays are indicative. Such damage caused males to live longer up to a level of females. Similar results were observed at castration of cats (Hamilton, et al., 1969). The impression is created, that the raised death rate of males is specially produced during evolution, the adaptation useful for a species, which sense remains unclear.

Evolutionary Mechanisms of Sex Ratio Regulation

Since Darwin times in the scientific literature, the tables of birth rate of sexes (a secondary sex ratio) for the different species were frequently published.

First of all these tables show that a secondary sex ratio was considered a constant, characteristic for the given species. The second circumstance which draws attention in these tables, is that if humans does not interfere with the species' reproduction, the data on a secondary sex ratio at different authors in different years are pretty consistent. If the humans do interfere with the species duplication, having entered the separate maintenance of sexes, different restrictions, castration, artificial crossing or fertilization—the big variations in the given different places, times and authors are observed.

Fisher's *"theory of equal expenses"*. Fisher's theory (Fisher, 1930) provided a genetic explanation for the evolution of a stable sex ratio of 1 : 1. Fisher proposed that because every individual has a mother and a father, females and males must contribute equally, on average, to subsequent generations, and therefore, must have the same average fitnesses. In a population that has more males than females, it is advantageous for an individual female to produce a biased sex ratio favoring daughters. If a mutation causing a biased female sex ratio enters the population, it will thus increase in frequency. Once it reaches a high frequency, and there is now an excess of females, it will be no longer selectively favored. On the contrary, if a new mutation producing a male-biased sex ratio occurs, it will be favored by selection. Finally, the system will reach a stable equilibrium point of 1 : 1 sex ratio.

To explain deviations of a secondary sex ratio from a proportion 1 : 1 observed for many species, Fisher asserted, that parents' expenses to produce offspring of male or a female sex should be equal, rather than the number of born males and females. That is the "more cost" the descendants of the given sex place on their parents, the less of them are made. Modest enough abilities of the Fisher's theory to explain the existing facts and to predict new, forced its followers to make it more complex in order to explain different sex related phenomena.

For example, Charlesworth et al. (1977) with reference to humans, in view of the elevated death rate of men, considers, that the progeny which were lost in childhood, can be replaced with smaller expenses, than the progeny which has lived up to the end of the parental guardianship. He comes to a conclusion, that older women should have fewer boys in posterity.

From Fisher's *"theory of equal expenses"* follows, in particular, that when descendants of a different sex have different sizes, the deviation of a secondary sex ratio should be observed. However Howe's data (1977) on Common Grackle at which males are 20 % heavier than females, and a big study of Newton and Marquiss (1978) on the hawk *(Accipiter nisus)* at which on the contrary, females are twice heavier than males, have not confirmed a prediction of the Fisher's theory. In the first case in progeny there was 52 males and 83 females, a sex ratio of 0.39. Such distinction was observed as a result of the greater male death rate, because for the embryos the sex ratio was 1 : 1. In the second case, the sex ratio was 0.51 (1102 $\male\male$ and 1061 $\female\female$). To somehow coordinate Fisher's theory to the facts, the supporters of the theory said that male hawk nestlings "ate the same amount of food", as their sisters, despite the fact that they are twice smaller and leave a nest 3–4 days earlier thus reducing parental expenses even more. It sounds unpersuasive. And it is difficult to agree and recognize logical Maynard Smith's (1981) conclusion, which considers that "these results are in the consent with Fisher's theory, but do not give heavy arguments in its advantage". Though he acknowledges that these results "despite of all their clearness, cause some disappointment".

Hamilton (1967) was first to pay attention to an inaccuracy of the Fisher's theory, in all cases when the local competition for crossings takes place. In a limiting case of close inbreeding when all crossings occur only between brothers and sisters and when female lay a certain number of eggs, and every male can impregnate all the sisters, the parent who has only one son and many daughters will on the average have more grandsons than the parent with an equal number of sons and daughters. Hence, selection will support not a sex ratio of 1 : 1 according to Fisher's theory, but a proportion shifted towards female surplus.

Hamilton has counted about 25 species of ticks and insects from 16 different families at which constant significant inbreeding is combined with the big female surplus and with *arrhenotoky* system of duplication. The meaning of arrhenotoky thus will be, that in a case when all eggs laid by the female are impregnated, except for one, there will be only one male in the progeny. A tick *Acarophenax*, which females are viviparous, represents the extreme case. The single male in a breed hatches and impregnates its 15 or so sisters and perishes. Here it's not clear, whether the same close relationship between inbreeding and the excess of females can evolutionary arise without arrhenotoky. In plants close inbreeding is usually connected with hermaphroditism.

Hamilton, having applied ideas of games theory created by Von Neumann and Morgenstern, has come to a conclusion that even partial inbreeding strongly shifts the value of an evolutionary stable sex ratio towards a surplus of females. The same shift occurs, if inbreeding is carried out only in some generations, alternating with outbreeding (Maynard Smith, 1981). Hamilton's approach allows interpreting some features of the duplication system of *hymenoptera* social insects. Another interesting conclusion of Hamilton concerns genetic inertness of a Y-chromosome at animals. He thinks that the absence of genes in a Y-chromosome can be explained as a result of the past selection on suppression of the genes linked to a Y-chromosome causing meiotic drive (Hamilton, 1967).

Genes causing meiotic drive in sexual chromosomes can lead also to evolution of the mechanism of sex determination. For example, if the Y-chromosome causing meiotic drive gets widely distributed in a population, the selection for the benefit of the conversion of XY individuals into females begin. If differentiation between X and Y-chromosomes is small, it's possible that the role of "sex-determined" will move to another pair of chromosomes, that is to autosomes. In this aspect the work of Wagoner, McDonald and Shilders (Wagoner et al., 1974) is very interesting, in which the example of existence of two different mechanisms of sex determination for house fly *Musca Domestica* is described.

Fisher considered the distinction in expenses for making male or female progeny, but has not taken into account the fact, that in some cases, it is more favorable to produce offspring of any one sex.

In opinion of Kalmus and Smith (1960), value of a tertiary sex ratio 1 : 1 is optimal, because it facilitates a meeting of individuals of an opposite sex and reduces a degree of inbreeding. Their theory is not capable of explaining observed deviations of a secondary sex ratio for many species from 1 : 1, and also its changes depending on various factors.

Trivers-Willard Hypothesis (TWH) (sex-allocation). Trivers and Willard (1973) have noted that if one gender has greater adaptation variation and if the parental care is higher for the more adapted part of progeny, then parents capable of the greater expenses for progeny will be more favorable to make offspring of a sex with greater adaptation variation. Parents with limited ability to invest in posterity will produce offspring with less adaptation variation. They have applied this idea to polygamous species of mammals in which the males' variation on reproductive success can be higher than that of females'. They have predicted, that females in good conditions for duplication should make more sons, and in less favorable conditions—more daughters. In their work the data on such species as a deer, a pig, a sheep, a dog, a seal and the human show that the sex ratio changes in the direction predicted by the theory.

During the age of the hypothesis, it was extensively researched and a lot of criticisms exist. Most authors only agree that the results are difficult to interpret because of their inconsistency (Frank, 1990; Brown, 2001). Meta-analysis conducted by E. Cameron (2004) showed that out of 422 tests, only 34% support the hypothesis, in 5% support was very small, and 8.5% of the tests showed significant results opposite to its predictions. In most of the studies the results were not significant. The support in favor of the hypothesis was significantly higher (74% and more) if only maternal body condition close to conception was taken into consideration. Similar results were obtained by B. Sheldon and S. West (2004).

Freeman et al. (1976) studied the sex ratio in natural populations of saltgrass *(Distichlis spicata)*, meadow rue *(Thalictrum fendleri)*, ash-leaved maple *(Acer negundo)*, "Mormon Tea" *(Ephedra viridis)* and a shrub Shadscale *(Atriplex confertifolia)*. All plants are from two households. In all five species they have discovered a surplus of male's plants in dry and sunny places and a surplus of female plants in humid and shaded habitats. Distinctions in sex ratio are great enough: for meadow rue, for example it varies from 0.80 in dry places to 0.19 in places damp and shaded. The mechanism of the phenomenon remains unclear.

Maynard Smith (1981) discussing the results of these authors, thinks that it is difficult to imagine how such a system could arise evolutionary due to a different mortality of sexes because death rate of progeny is defined more likely by their genes rather than their parents. If this is the case, then how can the genes that facilitate the destruction of male plants in humid places and female plants in dry places be fixed in a population? And he comes to a more plausible conclusion, that sex is determined by environmental conditions. Then there is a question: why sex determination by external factors is not distributed more widely than is actually found.

And Maynard Smith (1978) comes, at last to a guess, that "to parents it can be favorable to produce a gender, more rare in the given region". To confirm such assumption he cites a remarkable observation of Snyder (1976) on populations of North American marmot *Marmota monax*. In its natural populations a sex ratio of young animals is close to 1 : 1. When Snyder has withdrawn about half of reproductive females from one population, on the next year the sex ratio among young individuals appeared to be 40 males and 89 females, so the compensation of disturbed sex ratio was observed.

*   *   *

Thus, existing theories of evolution of a sex ratio are neither capable of explaining the known facts in this area, nor especially to predict new facts. They lack generality and explanation of all the complexities of the problems connected to a sex ratio, observable deviations of its values, and also dependence of its change on various factors, such as age of parents, life conditions, nutrition, starvation, presence of wars, or climatic conditions.

Weakness of existing theories is caused by a lack of uniform evolutionary logic of the phenomena. As a rule, they proceed from erroneous representations that for a tertiary sex ratio an optimum is always the proportion 1 : 1, and that primary and secondary sex ratio are constants pertaining to a given species and independent of environmental conditions.

Chapter 4

Mysteries of Dioecy: Sexual Dimorphism

"Men are from Mars, Women are from Venus"
J. Grey (1982)

"The "reversed" sexual size dimorphism of predatory birds,
with females in some cases almost twice as large as males,
has long puzzled ecologists. ... A review by Mueller and Meyer
(1985) listed **twenty hypotheses***!"*

Clearly pronounced sexual dimorphism exists in many animals and plants. Some species, such as snakes, have almost no sexual dimorphism; others are very unusual in that respect.

In relation to sex the characters of the organisms can be divided into three groups. The first group includes the characters, which show no difference between males and females. There is no sexual dimorphism for these characters in the norm. Among these are the majority of specific characters (number of organs, extremities plan and general structure of the body, and many others).

The characters in the second group are those inherent only to one sex. Sexual dimorphism of this type can be named the *"organismic"* one. It has an absolute pattern; it distinguishes any male from any female. In this case all the individuals of the given sex normally either have them or not. These include all primary and secondary sexual characters (internal and external sex organs, mammary glands, beard in man, mane in lion, and many economically valuable characters). Quantitative estimation of such character and of its distribution pattern in the population is sensible only for one sex.

The third group of characters are those, which present in both sexes but are differently pronounced or/and are met in the population with different frequency depending on sex. This group of characters according to the extent of sexual dimorphism lies in between of the first two groups. The pattern of sexual dimorphism for these characters is not an absolute, organismic, but the populational one (different distribution of the character in males and females in a population). Populational sexual dimorphism can exist for such characters as stature, weight, size, proportions, and many morpho-physiological and ethologo-psychological characters.

Sexual dimorphism and genetic differences. In the case where the genes controlling the formation of a particular trait are located in the autosomes, the trait is inherited regardless of which parent (mother or father) is a carrier. The presence of sex chromosomes alters the nature of trait inheritance. Genes (and their associated characters) located in the Y-chromosome, are passed only through the male line. Genes located on the homologous region of the X-chromosome are transmitted to both sexes, will show up in males even if they are recessive. Females inherit two X-chromosomes and have the ability to compensate for defective genes. One of the X-chromosomes undergoes inactivation (Barr body) to compensate for gene dosage. In various cells of the body either the same or different X-chromosome can get inactivated, in the latter case, a mosaic manifestation of traits is observed.

There are several general patterns that are associated with sexual dimorphism:

(1) Mating system: the polygamous species have a more pronounced sexual dimorphism than monogamous. Males have larger sizes, because large body helps males to fight for a female. Under monogamous breeding system with a weak competition for females, males and females have similar dimensions.

(2) Sexual dimorphism in size increases with body size when males are bigger than females, and decreases with the increase in average body size when females are bigger (**Rensch's Rule**) (Rensch, 1959). Rule was derived from the analysis of phylogeny of 40 independent lines of terrestrial animals, mostly vertebrates. Examples of the lines that follow the rule include primates, pinnipeds, and artiodactyls (Fairbairn, 1997). Exceptions from Rensch's rule are found mainly in taxa in which females are larger than males (Bunce et al., 2003; Tubaro, Bertelli, 2003; Webb, Freckleton, 2007). The reasons for this pattern are still unclear. Why males are much bigger than females in large animals, and why, in the same phylogenetic line of smaller size species are the males of the same size or even smaller than females (Rensch, 1959; Andersson, 1994; Fairbairn, 1997)?

(3) Terrestrial species are generally more sexually dimorphic than arboreal species. Possible competition between males in terrestrial species is more dependent on body size.

(4) Both within one species, and among related species of animals (vertebrates and invertebrates) and plants, males often more strongly differ from each other whereas females can be practically indistinguishable (Willson, 1991). It was noticed still by C. Darwin. He considered that the occurrence of sexual dimorphism and formation of species can be the result of the same processes.

(5) Social structure. Sexual dimorphism is more expressed in communities where separate males compete among themselves in comparison with societies in which males can unite in groups. Intensity of a competition is probably, the same, but its type changes. In a single competition the sizes of body and canines can be important to acquire and protect the harem. At a group competition, the coalition of co-operating males can be more important in defeating dominant individuals and acquiring mating partners. The intelligence becomes more valuable than the size.

"The "reversed" sexual size dimorphism. In most insects, amphibians, fish, and snakes, and also in some birds, females are bigger than males. In birds they are mainly predators of vertebrates: raptors (*Accipitriformes* and *Falconiformes*), owls (*Strigiformes*), skuas (*Stercorariidae*), frigatebirds (*Frigatidae*), and boobies (*Sulidae*).

For the explanation of reverse size dimorphism it is necessary to search for factors of sexual selection for which the small body sizes are preferable. Such factors can include weight of spiders who should get on high trees (Moya-Laraño et al., 2002) and mobility necessary for demonstration of the art of flight of birds of prey (Székely et al., 2004).

Some authors also list factors promoting selection for bigger female sizes: the necessity to lay the big eggs, to protect the nest, and to care for posterity (Andersson, 1994). There exist more than 20 hypotheses but the question is still far from been resolved. It should be noted that selection operates on both sexes at the same time; therefore it is necessary to explain not only why one of the sexes is relatively bigger, but also why the other sex is smaller (Székely et al., 2004).

Anderson and Vallander (2004) come very close to the formulation of the *"Phylogenetic rule of sexual dimorphism"* (**Ch. 12**) having assumed a correlation of changes between sexes as a result of selection. In their opinion the females follow the changes of males due to genetic correlations. Male's size increase and the subsequent female's size increase leads to the species' growth as a whole and to increase in relative difference between sexes, and vice versa.

Plants. Many dioecious plants have sexual dimorphism not only on sex-related traits (structure of the flowers and their arrangement), but also on morphology, architecture, physiology, and life cycle. Though it is well enough established that male plants are more attractive (have higher amount of larger flowers), it is not absolutely clear how such sexual dimorphism has arisen. The traditional explanation includes selection depending upon pollinators.

Sex Differences in Humans

Anatomy

In humans, apparently, sexual dimorphism decreased in the course of evolution. So the men of some of human's ancestors (*Australopithecus afarensis*), were almost twice bigger than women. Such significant differences in body size between the sexes indicate the polygamous system of reproduction (Clutton-Brock, Harvey 1976). Physical male domination and female subordination may be an ancient phenomenon (Margulis, Sagan, 1991). On average, women have a lower body weight, smaller organ size, reduced blood flow and a higher proportion of fat compared with men. Modern men have a slightly larger size of the cranial bones and proportionately greater length of the arms and legs, which shows sexual dimorphism, corresponding to their larger body size. Muscles account for 42 % of the total body weight in the male and only 36 % in the female.

Brain. The human brain also demonstrates general anatomical features of sexual dimorphism. The female scull looks more like that of an infant than does the male scull. The average brain weight of men is 1,385 g (2 % of the body weight) while that of women is 1,265 g (2.5 % of the body weight). The intermediate region on the side of the brain known as the parietal area (sensory-motor representation) is larger in males. The occipital lobe (the back part of the brain) is of equal size in both sexes. A recent neuropathological study showed that women have larger size of the front and middle commissure (Allen, Gorski, 1991). The cerebellum in women is much larger then it is in men. The cerebellum is the most ancient part of the brain.

Studies of brain hemisphere asymmetry have led to the assumption that sexual differences in verbal and spatial abilities can be related with the different distribution of these functions between hemispheres of men and women. If the left hemisphere is damaged as a result of a hemorrhage or tumor, or surgically removed as an epilepsy treatment, deficiency of verbal functions in men happens more often, than in women. Similar damages to the right hemisphere also lead to a higher deficiency of nonverbal functions in men than women (McGlone, 1978). The conclusion has been drawn that in women language and spatial abilities are presented more bilaterally, than in men (Springer, Deitch, 1989; McGlone, 1980).

Men much more frequently show the prevalence of the right ear and the left hand (in right-handers) on tactile recognition of objects (digaptic stimulation) (Springer, Deitch, 1989). Sexual dimorphism was also found on a ratio of lengths of the left and right temporal planes (Wada, 1975). Sexual distinctions were noticed in anatomic, clinical, dichotic listening, tahistoscopic, electrophysiological and psychological studies of the hemispheres. The majority of authors agree that lateralization of hemispheres is more clearly expressed in men.

There exist a close relationship between brain asymmetry and sex. In the XIX century, Crichton-Brown (1880) hypothesized that the tendency toward asymmetry in the two hemispheres is stronger in men. An opposite hypothesis was expressed by Baffery and Gray (1972). Currently, most authors support the first hypothesis.

With the discovery of sex differences in lateralization of the brain there is hope to understand and explain psychological sexual dimorphism: the different abilities and preferences of men and women, different professional competence, professional preferendum, different learning ability, and ingenuity.

Physiology

Males have a metabolic rate that is between 5 and 6 % higher than that of females, and from the earliest ages males are more active than females. Vital lung capacity in men is on average 25 % more than in women. The mass of the heart, blood volume (5–7 % more), the concentration of red blood cells, hemoglobin and fibrinogen content in the blood are also higher. Men's lungs are working 20 % and heart—15 % more intensively. Men's kidneys are working faster. These features are associated with lower oxygen consumption by adipose tissue with its less intense metabolism than that of other tissues (Tkachenko, 2001).

Women endure all sorts of devitalizing conditions better than men: starvation, exposure, fatigue, shock, illness and the like. American anthropologist Donald Grayson in his work writes that more women survive in extreme situations. For example, after the Spitak earthquake in Armenia, 1.5 times more women were rescued from the rubble alive. As was already noted in **Chapter 3**, female embryos have a greater chance of survival than male ones, the number of male miscarriages is 20 % more than female.

Stress. Women respond to less intensive stress stimuli, but more easily tolerate stress: their blood pressure raises less, and less adrenaline gets released in the blood as compared to men. There is evidence of a greater antistress mechanisms capacity for women, in particular the antistress activity of estrogens (Anishchenko, 1991).

Sex differences in immune response. Increased production of antibodies was found in girls (Michaels, Rogers, 1971), but among the adults in response to the introduction of vaccines, such differences were not found (Vranckx et. al., 1986). The immunoglobulin M (IgM), but not IgG levels were significantly higher among women (Lichtman, Vaughan, 1967).

SENSE OF TOUCH

The sense of touch is developed in women than men much more strongly: even the least sensitive to touch woman is superior in this respect, to the most sensitive men. Women's skin has 15 % more receptors. Therefore they have higher tactile sensitivity: they like to be touched, and love to touch themselves. This habit is mediated by the hormone oxytocin, involved in the stimulation of the brain.

OLFACTION

Olfaction is the oldest and most important of senses by which animals orient themselves in their environment. This type of sense is one of the most important in many animals. "It preceded all the other senses by which the animal could at a distance feel the presence of food, individuals of the opposite sex or approaching danger" (Milne, Milne, 1966; Korytin, 2007). Unlike the other senses, the sense of smell can not be directly measured. The study of olfaction is based on indirect measurements of smell sensations, such as the quantity ("How strong is the smell?"), the determination of the perception threshold (at which force the smell is detected) and comparison to other scents ("what that smell resembles?").

Olfaction depends on sex. On standard tests of smelling ability—including odor detection, discrimination and identification—women consistently score significantly higher than men (Brand, Millot, 2001). With few exceptions (Bailey, Powell, 1885; Amoore, Venstrom, 1966; Venstrom, Amoore, 1968), most olfactory studies have noted that women, on average, out-perform men on tests of odor detection and identification. Women better feel a higher than threshold concentrations of substances (see the review by Doty, 1986), which was demonstrated for vaginal odors (Doty et al., 1975), armpits (Doty et al., 1978), and mouth (Doty et al., 1982). Women are also significantly more likely than men to suffer from 'cacosmia'—feeling ill from the smell of common environmental chemicals such as paint and perfume. The most acute sense of smell becomes in the period shortly before and after ovulation, such as sensitivity to male pheromones increases a thousand times (Navarrete-Palacios, et al., 2003). For example, it is known that female sensitivity to male pheromones (scented sex hormones), for example, is 10,000 times stronger during ovulation than during menstruation. The sense of smell is also significantly increased during pregnancy.

Babies have strongly developed olfaction, but during the first year of life it is lost by 40–50%. The survey of 10.7 million people showed the decrease in olfactory sensitivity with age for all six investigated odors (Gilbert, Wysocki, 1987). The ability to discriminate odors was also reduced. The influence of age was more significant than the influence of gender, and women kept the sense of smell to a later age than men (Doty et al., 1978). The olfactory fibers undergo atrophy with age and their number in the olfactory nerve is steadily declining (Blinkov, Glezer, 1964; Smith, 1942). In the phylogeny the human sense of smell is deteriorating.

VISION

Men have better developed depth perception. It is more difficult for women to determine the distance to objects, especially at dusk. However, peripheral vision is better developed in females, which allows them to notice all the details, without even turning the head. They clearly see the sector for at least 45 degrees on each side of the head, that is, left-right and up and down. Effective peripheral vision for many women approaches full 180 degrees. Men's eyes are bigger than women's. Men's brain gives him "tunnel" vision, which means the ability to see clearly and directly forward at a long distance, like binoculars. Men's vision has evolved to nearly a limited vision.

Evolution of vision. Common ancestors of tetrapods and amniotes (~360 million years ago) had tetrachromatic vision, or the ability to discern four different types of color: near ultraviolet, blue, green, and red parts of spectra. Four photopigment opsins can be found now in birds, reptiles and teleost fish (Yokoyama, Radlwimmer, 2001).

It is assumed that during the *Mesozoic* period, early mammals had smaller sizes and crepuscular lifestyle. In such conditions, vision for orientation in space becomes inferior to the sense of smell and hearing, so the majority of mammals have only two types of color receptors (cones), and does not have color vision. Expansion of the ecological niche of mammals and the transition of some of the species to the daily lifestyle, led to the emergence of the third type of cone cells responsible for the perception of middle range of the spectrum.

Some people may be carriers of the mutant genes on the X chromosome, responsible for the formation of the fourth photopigment. Only women can have the ability to see more colors if one type of red cone will be activated on one X chromosome and the other type of red cone on the other one.

HEARING AND VOICE

The cochlea of the inner ear is a very ancient form, and only serves for the perception of hearing. The emergence of the perception of acoustic signals apparently preceded the emergence of voice emission. A voice came from the structures whose function initially was to provide vital activities (respiration and nutrition), and only secondarily adapted during evolution for voice emission. Therefore, it is possible that for many millions of years hearing was preceded by phonation. Voice has evolved in such a way as to have the physical characteristics corresponding to the auditory window. In humans, this window is between 15 and 20,000 Hz, with an optimal perception of sound in the area around 400–6000 Hz and from 0 to 130 dB.

Evolution of hearing. Masterton et al. (1969) compared the evolution of hearing in humans and phylogenetic sequence of 4 primitive mammals: opossum (*Didelphis virginiana*), hedgehog (*Hemiechinus auritus*), tree shrew (*Tupaia glis*), and bushbaby (*Galago senegalensis*). They also used audiograms from many other species. For most parameters, other mammalian species showed superiority compare to humans. All of the examined species had higher sensitivity at high frequencies (on average by 1.5 octave higher), 35 % of them had greater sensitivity (lower threshold of perception), and 25 % of species had a large area of auditory perception. For example, dog's ears perceive sound in the ultrasonic field with a frequency of 50–100 kHz.

Sensitivity at high frequencies (above 32 kHz) was typical of the ancestors of mammals and primitive mammals alive today. This was due to their small body size and, accordingly, with a small distance between the ears (common to all mammals of the *Cretaceous* period). To determine the direction of the sound they rely on the difference in its frequency, since even with a small head it is possible to distinguish a difference in the frequency at high frequencies. In human evolution increasing the distance between the ears has led to the emergence of another mechanism for determining the direction of the sound—the difference in time of sound registration by the right and left ears. This mechanism has eliminated the selection pressure for the maintenance of hearing at higher frequencies, resulting in a reduction of upper limit to the level of nonmammalian vertebrates.

Sensitivity at low frequencies has been reached in the early stages of human development, remained virtually unchanged at the level achieved in the *Eocene*. At the same time the sensitivity at high frequencies deteriorated. The area of auditory perception, reaching its maximum in the *Eocene* due to the expansion into the low-frequency region, is reduced and continues to decrease until the present time due to loss of sensitivity at high frequencies.

Corso's work (1959) established without any question that (on average) "women have more sensitive hearing than men". They also distinguish high frequency sounds quite well. Women have better aural skills: 6 out of 10 women and only 1 in 10 men can correctly repeat a melody.

During fetal life the baby is better suited to hear the high frequency sounds of the mother's voice, rather than the low-frequency sounds of the father's voice. For both men and women, there is a decrease in average hearing sensitivity with increasing age, and a progressive spreading of the loss from the higher to the lower frequencies. So the child can hear sounds with a frequency of 30 kHz, a teenager (up to twenty years)—up to

20 kHz, and in sixty years—only up to 12 kHz. Men are more affected than women, with the hearing loss occurring at an earlier age and producing a greater degree of auditory impairment.

Voice. Voice development with age is characterized changes in height (the basic frequency), loudness and timbre. The basic frequency can be used as an indicator of a voice maturity. At many species, voice serves as an indicator of a reproductive condition of the individual.

Voice height in humans influences attractiveness of both men and women (Collins, 2000; Collins, Missing, 2003). Men are attracted to women with higher pitched voices, which they probably associate with youth and fertility. Women on the contrary, find men with deep voices more attractive and sexual. During the maximum fertility times within the ovulation cycle the preference for men with deep voices increased (Putz, 2005). Such men had better reproductive success (more children) (Apicella, 2007).

As one of the last evolutionary acquisitions, human voice has maximally expressed sexual, age-related and lateral dimorphism. The range of voice frequencies in women (~180 Hz) is wider, than in men (~120 Hz) and is shifted towards high frequencies: from soprano (340 Hz) to a contralto (160 Hz) in women and from the tenor (200 Hz) to a bass (80 Hz) in men.

Psychological Sexual Dimorphism

Psychological sexual dimorphism in humans is manifested in many aspects: different abilities and inclinations of men and women, different professional fitness and professional preferendum, different learning ability and ingenuity. Examinations of children of different ages showed that in the early stages of ontogeny (up to 7 years), girls are ahead of boys in their intellectual development. Later, these differences are smoothed out, and grown men and women differ a little on the average parameters of intellectual development. However, the number of men on both ends of a normal distribution constructed by measuring intelligent quotient (IQ) greatly exceeds number of women. Based on these studies it was concluded that range of mental abilities for men is much wider (Lehrke, 1978; Benbow, Stanley, 1980; Mosiey, Slan, 1984; Rothman, 1988).

The world record on speed of speech belongs to the woman from New-York— 603 words per minute.

Men on the average can say 225 words in a minute.

Men and women differ not only by the range of mental abilities, but also qualitatively. For example, on verbal (linguistic) abilities in all age groups, higher levels were observed in women. According to MacCoby and Jakline (1972), in early childhood girls develop language skills faster than boys. Girls begin to speak 2–6 weeks earlier than boys, and during all pre-school years, retain the primacy of articulation, clarity, intelligibility of speech closer to the speech of adults (Wellman, 1931; Goodenough, 1957; Darbey, 1961). Girls have more pronounced fluency and better verbal and written language abilities compare to boys (Hall, 1985; Kelly, Britton, 1996). Girls have more developed aesthetic sense, they have better developed speech and more subtle coordination (McNemar, 1942; MacCoby, Jakline, 1974; Watson, 1991).

Advantages of women are established on different parameters: speech as a whole, fluency of speech, spelling, reading, etc. (Blinkov, Glezer, 1964; Harris, 1978; McGlone, 1978,1980; Levy, 1978; Maccoby, Jaklin, 1974). Women have higher results on short-term memory. A higher degree of conformity of thinking in women (McGlone, 1980) is marked also.

On the other hand, boys are better in math and calculus and have better-developed mechanical and spatial-visual abilities. In childhood both boys and girls have approximately the same abilities (MacCoby, Jaklin, 1972), but at school boys begin to show a more pronounced tendency to solve visual and spatial tasks, and continue to engage more successfully in this activity as adults (Levine et al., 1999). Different types of spatial tasks included the mental rotation of objects, the mental preparation of the figures from the elements, the perception of horizontality, the recognition of forms included in other forms, and assessment of the speed of moving objects. Boys understand geometrical problems much better than girls. Men are better guided in visual and tactile labyrinths, read geographical maps, define directions of cities, rivers and roads, define left—right more easily. The greatest differences were noted for mental rotation of objects, as well as three-dimensional problems, compared with two-dimensional (Halpern, 1992; Phillips, Silvermam, 1993). On many tests only 20–25 % of women exceed the average value for men.

Men have higher performance in spatial orientation tests (Harnqvist, 1997). Men are also more likely to use the concept of the parts of the world and estimate the distance to objects, whereas females use relative directions (such as the right and left) and the key points of the area (Ward et al., 1986; Bever, 1992; Galea, Kimura, 1993).

Roles in the society

Work

The degree to which the jobs that women and men hold differ, reflecting a number of factors including: the amount and types of education that workers have completed; personal preferences; societal attitudes about gender roles; and in some cases discrimination. The data that are available (see for ex. Wootton, 1997) for the past two decades clearly indicate two major points. First, the gender distribution of many occupations has shifted substantially. Second, despite these shifts, women and men still tend to be concentrated in different occupations: women are highly overrepresented in clerical and service occupations, for example, while men are disproportionately employed in craft, operator, and laborer jobs.

Currently women and men are most equally represented among managers and professionals. Employment of technicians and sales occupations also was about evenly split between women and men. Gender differences were still pronounced, however, among workers in other major occupational groups. Men were much more likely than women to work in the precision production, craft, and repair (9 out of 10 such jobs). Men also had large majority of employment in protective service (84 %), farming, forestry and fishing (80 %) and operating, fabricating, and laboring occupations. More than 97% of auto mechanics, carpenters, electricians, plumbers, and firefighters were men. Only 1 % of mechanics and carpenters, and also 3 % of firemen were women. Computer systems analysts (70 %) and engineers (90 %) were also dominated by men.

Women held 4 out of 5 administrative support jobs. Women were also represented heavily in services occupations, particularly private household occupations. Women dominate the health care occupations with 93 % of nurses, 94 % of speech therapists and 90 % of occupational therapists, while only 30 % of physicians were female. In the field of medicine women tend to favor such specialties as pediatrics, psychiatry, obstetrics and anesthesia. Men on the other hand prefer surgery, orthopedics, urology, and otolaryngology. "Women physicians are happier when they can give the support, sympathy and understanding, as well as their wisdom and skill, to their patients. The technical accomplishment of a cure or a surgical procedure does not constitute their greatest reward, as seems to be the case with the male. The male is interested in the performance of the task, in the solving of a problem. The female doctor knows what the male doctor seldom remembers: the care of the patients begins with caring for the patient." (Montagu, 1968).

Education is an occupational group dominated by women with 98 % of kindergarten teachers, 83% of elementary school teachers, and 60% of middle and high school teachers. Librarians, social workers and psychologists were 80, 75 and 65 % women.

The occupational differences were least pronounced among teens, those aged 16 to 19, and among those over age 65. The largest difference among teens in the gender distribution in 1995 occurred among occupations that do not require a high level of education. Women aged 16 to 19 tend to hold a disproportionate employment share of sales workers, apparel, cashiers, secretaries, receptionists, clerks, waitresses, food counter, childcare workers, and early childhood teacher's assistants. Men aged 16 to 19, by contrast, tend to be disproportionately employed as cooks, janitors and cleaners, truck drivers, farm workers, ground keepers, and various laborers and handlers.

The distributions of men and women among specific occupations, while still very different from one another, were much less so than 20 years earlier. Some changes occurred during World War I, when for the first time women were called upon to replace men in occupations that were formerly the exclusive preserve of men. They became bus drivers, conductors, factory workers, laborers, supervisors, and executive officers. Many of these professions were traditionally considered masculine.

In some fields, particularly where delicate precision work is involved, women had proved themselves superior to men. Usually, they are more interested in applying their skills and abilities in the field of human relations.

Sport

Sports is a good indicator of the extreme capabilities of a human body. Historically sports was mainly a men's endeavor and was probably initially related to war training. There were initially no women's events in the Olympic Games, and adult women were barred from attending the games on pain of death. There are still kinds of sport that belong to males only—boxing, cycle-racing, martial arts, golf, ice hockey, water polo, weight-lifting, wrestling, and yachting. In some of them women just recently started to participate. There are a few kinds of sports just for women—aerobics and synchronous swimming.

The problem of sexual dimorphism in sports is not new. Almost all authors agree that there exists a big enough significant difference in favor of males on body features, functional capacity, power and speed performance and endurance. Moreover, these differences are observed already from the early years of playing sports (8–10 years), somewhat attenuated in the phase of puberty (12–14 years) and increased again to a time of sportsmanship. The advantage of men in these characteristics sometimes reaches 10–20 % or more (Sadowski, 1999).

In many kinds of sports men outperform women. The fact that there are separate competitions for men and women, however, speaks for itself. This is also the case for some kinds of sports that do not require physical strength, for example chess.

Religion

For decades researchers have pondered a mysterious gender disparity in religious commitment. Rates of religiosity data collected in 57 countries were taken from the World Values Survey. The world's major faiths were included and the data came from such countries as the United States, most European states, Mexico, Brazil, Argentina, Japan, China, India, South Africa and Turkey. For all 57 countries, a higher percentage of women than men said they were religious.

The gender differences show up in many cultures, even in religions that are very male centered, such as Orthodox Judaism. The rates of religiousness are not just a phenomenon of our time. It is true of all ages—ancient Greece, the Roman Empire when the early Christians were mostly women, and medieval Europe.

These results hold across time periods, cohorts, religious traditions and cultures, therefore their explanation also should be general and independent from most social and cultural factors. The popular explanation is that women are raised to be nurturing and submissive and this socialization makes religious acceptance and commitment more likely. Another explanation that women are more religious because they don't work outside the home and have more free time to engage in religious activities. However, a number of studies have shown that career women are just as religious as those who stay at home and both are far more religious than their male peers or spouses.

Recent studies indicate that men's atheism, as well as criminal behavior related to the fact that men have underdeveloped ability to inhibit their impulsive desires, especially those that entail immediate gratification and thrills. Some men do not think far ahead, and thus end up in prison or going to hell has no particular importance to them. "Not being religious is similar to any other shortsighted, risky and impulsive behavior that some men—primarily young males—engage in, such as assault, robbery, burglary, murder and rape." (Stark, Miller, 2003).

Aggression, Wars, Risky and Criminal Behavior

Women's' behavior is characterized by greater caution in their dealings with others; they differ in greater altruism, empathy and collectivism. Men are much more aggressive (Coie, Dodge 1997, Maccoby, Jacklin 1974), 5 times more likely to commit crimes, and are responsible for most murders (Buss, 2005). Nine out of ten inmates in prisons are men. This is one of the most significant gender differences observed in different age groups and in all cultures. Men often resort to physical aggression (Bjorkqvist et al., 1994). Even in the case of suicide, women resort to milder methods such as drugs and sleeping pills, while men shot, hanged themselves, and jumped from tall buildings.

Men are twice as likely as women to drive a car drunk, 9 % less likely to use seat belts, 12 % more likely to smoke, 2 times more likely to become alcoholics, 2.5 times more drug abusers and 70 % more likely to have guns (Kaiser Permanente, 1995).

Aggressive behavior is correlated with the level of testosterone in both men and women, although a direct relationship was not found. Also there were no systematic differences in the levels of testosterone in aggressive and nonaggressive people. Aggression does not increase at puberty, when testosterone levels are rising (Albert et al., 1993).

Theories of sexual dimorphism

For a long time it has been noticed that:
    (1) Polygamous species in which males usually invest very little resources in offspring, have more pronounced sexual dimorphism (Darwin, 1871; Fisher, 1958).
    (2) In such systems, a small number of males mate with most females (Bateman, 1948, LeBoeuf, 1972, 1974; Mackenzie et al., 1995). Many males do not participate in pairing, and therefore do not affect their own survival or the survival of other animals.
    (3) Dimorphic species usually have one of two systems: either the males are fighting each other and the winner mates with the female or males develop characters attractive to females.
    (4) In both situations such male traits limit their survival and, apparently, are developed despite natural selection.
These general patterns are recognized by most experts in the field of sexual dimorphism.

The main theory explaining with the evolutionary position the origination and development of sexual dimorphism is still the theory of Darwinian sexual selection. In his book "Origin of man and sexual selection," C. Darwin (1871) considered known at that time, examples of secondary sexual characteristics of males, and introduced the theory of sexual selection for their explanation.

Darwin's theory of sexual selection

Darwin's theory of sexual selection was a matter of controversy even then. Many authors thought it to be the weakest point of his evolution theory. A. Wallace in his book "Darwinism" (1989) considered theory of sexual selection as incorrect. Recently the problem of sexual selection and dimorphism has again attracted the attention of researchers. Some authors think that there are no specific theoretical problems related to sexual selection (Grant, 1980). Others on the contrary present much evidence, which cannot be interpreted in terms of the existing theories (Maynard Smith, 1981).

Darwin viewed sexual selection as a process supplementing the far more widespread process of natural selection. He thought that natural selection has created all adaptive attributes of a species common for both sexes, and also primary sexual attributes directly connected to reproduction. Outside of this concept was left an extensive class of the special attributes peculiar to one sex only—secondary sexual attributes. These characters often do not represent adaptations favorable to the species as a whole and are not necessary for reproduction. It is believed that a number of such characters reduce fitness and survival of their owners (mostly males), and therefore such features should be eliminated by natural selection.

Darwin considered that secondary sexual attributes which are adaptations neither favorable for a species as a whole nor necessary for duplication, raise the probability of successful mating for their owners. So, the theory of sexual selection has been introduced as an explanation for sexual dimorphism. Darwin distinguished two phenomena, which lead to sexual selection: competition between males in a struggle for a female, and choice of males by a female. [Of course, depending on the conditions and mating system, the opposite situation is possible—the competition between females and choice of females by males.] Therefore the theory of sexual selection can explain the existence of sexual dimorphism only for two categories of secondary sexual attributes of animals: a) male organs of an attack, as instruments of struggle for a female, and b) the attributes attracting and arousing individuals of an opposite sex. In the first case presence of a male struggle for a female is assumed. The winner, stronger and better armed, gets the female and is able to transfer these properties to its progeny. In the second case males try to attract females by bright color, smell, sounds and dancing and the female chooses the most attractive male.

Several theories stress useful features of traits that reduce fitness. For example, large antlers, reduce mobility, but at the same time help to defend from predators, to raise the hierarchy in the group and to improve reproductive success. It is also possible that reproduction, in some cases, is more important than survival; that is, to leave progeny and to die is better than not to reproduce, but to live longer.

Mate choice may be based on two types of criteria: genetic indicators (selection for "good genes" and "healthy children") and aesthetic displays (selection based on "good taste" or for the goal of having "beautiful children"). The A. Zahavi's handicap concept is an example of the first type. The theories based on the "runaway" process belong to the second type.

"...from [the handicap concept] must logically follow the evolution of one-legged and one-eyed males."
R. Dawkins (1993).

The concept of handicap. Was proposed in 1975 by A. Zahavi to explain the sexual dimorphism on the characters that reduce males' fitness. For example, bright coloration of plumage and loud songs of male birds make them more visible to predators. A male with brightly colored plumage shows a female his high fitness that he has survived, despite the "handicap", compared to other, less visible males. Other traits, such as bare skin, may show that the individual has no pests or resistant to them.

"Runaway" theory. Was proposed by Morgan in 1903 and developed by Fisher. In the evolution of traits that are conducive to sexual selection, there may be a positive feedback when the selection pressure from the females moves the development of a trait to the extreme values, which could reduce the fitness of males. The strength of females' preference usually grows exponentially until it is balanced by the pressure of selection. Any mutations of a character in other directions, even if they will improve fitness, have no chance of success, because such males are much less likely to be selected as partners. It was shown that such behavior of females is the evolutionarily stable strategy (Fisher, 1930). Later, verification of the Fisher's theory on quantitative models has shown that positive feedback is too weak to cause the observed in the nature extreme values of such attributes (Ridley, 1994).

"Sexy son" hypothesis. Has been proposed to improve the model of polygyny-threshold, in which females choose males based on the quality of the occupied territory (Orians, 1969). The hypothesis is based on the assumption that females can also distinguish the individual quality of males and choose the "beautiful" males in order to produce a "beautiful sons" who would have greater reproductive success (Weatherhead, Robertson, 1979).

In subsequent years, characters and mechanisms of sexual selection have been greatly expanded, particularly through the inclusion of features that does not require intraspecies competition (see review and classification of Murphy (1998)).

Sure mate choice may be based on many other important qualities: caring for the offspring and resources, fertility, optimal genetic distance and similarity of appearance and behavior (Crawford, Krebs, 1998, p. 92). In addition, most animals have mechanisms for pairing with partners of its species, suitable sex, age, and in a certain place and time (Andersson, 1994). Males typically invest more resources in the development of traits of attraction and winning of females, whereas females—in the development of reproductive traits. The result is the emergence of sexual dimorphism.

Critics of the idea of genetic indicators believe that in some cases, these genes must be spread very quickly in the group (the *lek paradox*). At leks very few males perform most of the pairings. In this case, after a few generations, all males should have the right character, and it should become indifferent to females (Crawford, Krebs, 1998, p. 95). However, recent studies have shown that traits that are subject to sexual selection, have a much greater variability and heritability than traits that are subject to natural selection (Pomiankowski,1995). According to Miller, to be selected, the males are forced, "to have windows that allow females to see the quality of their genes." In this case, males become sort of a genetic "sieve" of a species, which removes bad and retains good genes.

There are a number of weak points in the theory of sexual selection.

1. Sexual dimorphism is often observed for such characters, which are with great difficulty related to sexual selection (e.g., leaf number and shape, branching pattern in plants). According to the theory of sexual selection sexual dimorphism should promote preference in either struggle for the female, or choice by the female. Consequently, at best the theory can be applied only to the animals and characters, which provide such advantages.

2. Sexual selection can help strong, better-equipped or more attractive male in the struggle for female, but it cannot maintain sexual dimorphism for these characters. Thus it is unclear why these characters are inherited by male offspring only.

3. Interpretation of the same phenomenon needs different logics. For example, in birds larger size of males is explained by preference in the struggle for female and larger size of females—by advantage of laying large eggs. But it is unclear why no large eggs is needed in the first case and no struggle for female is needed in the second case. It is still more difficult for the theory to explain large size of females in some mammals, such as bats, rabbits, flying squirrels, spotted hyenas, dwarf mongooses, some whales and seals.

4. Other obstacle for the theory of sexual selection is the dependence of sexual dimorphism on the reproductive structure of the population (monogamy, polygamy, panmixia). In this respect two regularities of sexual dimorphism for size are mentioned: (a) sexual dimorphism is more often found and is more pronounced in polygamous species than in monogamous ones, and (b) sexual dimorphism increases with body weight. There is no satisfactory explanation of these phenomena. (Maynard Smith, 1981).

5. It is also difficult to explain the existence of marked sexual dimorphism in monogamous species with sex ratio 1 : 1. Darwin believed that a male preferred by females starts reproducing earlier compare to others, which provides some advantages. Also females, which were first ready to reproduce, seem to be better mating partners. Such arguments are unconvincing. Each species has optimal reproduction time established in the course of evolution. Deviations both towards earlier or later onset of reproduction are disadvantageous and are eliminated by stabilizing natural selection.

The theory should always be wider than the treated phenomenon ("as a blanket-is better to be longer than the legs").
A. Lubischev

Weakness of the Darwin's theory treating sexual dimorphism as consequence of sexual selection is a result of a methodological mistake: the wide phenomenon cannot be treated as a consequence of the narrow mechanism.

Female Sabotage hypothesis. The theory of "female sabotage" was proposed by Joe Abraham, who noticed the crisis situation in this area (the appearance of a large number of theories). In polygamous species a significant part of males may not engage in breeding, competing at the same time with other individuals for food and other resources. According to Abraham, females choose males that have characters that increase their mortality, since the death of "excess" (suspended from breeding) males leave more resources. The theory allows combining features that help in the fight for females and traits that attract females because they both increase the survival of females and their offspring, and absolve them of intraspecies competition. This solution reduces sexual selection into natural, can explain the whole picture in the terms of original theory of evolution and eliminates theory of sexual selection (Abraham, 1998). Abraham's theory only applies to polygamous animal species. It's not entirely clear why females have to choose so complex and bloody way to get rid of males, when it can be done much cheaper by reducing the primary sex ratio.

Ecological Differentiation of Sexes

Darwin also proposed another hypothesis that evolution of sexual dimorphism in addition to sexual selection, can be a consequence of natural selection as well (Darwin, 1871). If males and females occupy slightly different ecological niches (perhaps due to differences in body size or activity that were selected by sexual selection), natural selection may favor an independent adaptation of each sex in the direction of greater differences or greater similarity. Such ecological differentiation of the sexes, mainly in relation to nutrition, can lead for example, to specialization by the sizes of food consumed. In any situation the food with dimensions typical for the given species is exhausted earlier, therefore the individuals consuming smaller or larger objects get certain advantages. And if sex difference in size initially exists, ecological food differentiation will promote its increase. Sometimes such sexual dimorphism is related only to food organs (e.g., beak size of woodpeckers). The possibility that sexual dimorphism can be created as a result of sex-related adaptation by nutrition without competition between males and females has been shown in some species of snakes by Slatkin (1984) and Shine (1989).

Such interpretation is also unsatisfactory. First, such process can at best explain an increase or maintaining of sexual dimorphism rather than its initiation, and second, it is absolutely unclear why such differentiation, which is basically usual disruptive selection, should be sex-linked. If sexual dimorphism exists for body or beak sizes, it seems most likely that distributions of males and females will be overlapped to a great extent.

Then small and large animals ought to be specialized regardless of their sex and according only to their body or beak sizes.

Other theories

Many theories have been proposed to explain sex differences in humans (Vinogradova, Semenov, 1992). Biological group of theories relates women's traits with childbirth and breastfeeding, and men's traits—with obtaining food and protection. Thus, according to Levy, the fact that men were hunters and directed migrations, led to a better development of their visual-spatial abilities, and verbal superiority of women is due to the fact that they brought up their children, and this requires intensive communication (Levy, 1978). Some scientists are trying to explain men's penchant for inventiveness and dedication to work by their inability to conceive and bear children. Male superiority in many areas, state and political power, is attributed to their physical strength and more aggressive behavior (Montagu, 1968). Weber (1976) suggested that these differences are not associated with sex per se, but with differences in rates of development for men and women. Such interpretation may explain, at best, sexual dimorphism in children and adolescents but not in adults. It is clear that such theories have limited scope and prediction power.

*   *   *

One can conclude that sexual dimorphism is a general biological phenomenon widely spread among dioecious animals and plants. It is observed for a vast number of characters. Therefore the theory, which claims to interpret it, should also be a general one and cover all the characters, which show sexual dimorphism. The existing theories, especially the ones that apply to humans only, do not satisfy this requirement; therefore they have weak explaining and predicting abilities. Usually they are trying to explain the mechanisms: how could sexual dimorphism arise and be maintained? They did not ask the questions revealing its regularities. What is sexual dimorphism? What is its evolutionary significance?

Chapter 5

Mysteries of Dioecy:
Sexual Dimorphism in Pathology

In April 2001 the U.S. Institute of Medicine report "Study of biological contributions into human health" has confirmed that gender differences exist in preferences and intensity on a wide range of diseases, disorders and conditions.

Questions of differential ageing and incidence of the sexes are closely related to their differential mortality. There are a number of diseases which mainly strike people of this or that gender. For example, women suffer from Graves' disease approximately 7 times more often, than men, rheumatic mitral heart defect—3 times, appendicitis—2 times more. On the contrary gout strikes men more than 10 times more frequently, a stomach and a duodenal ulcer—approximately 3 times more often, kidney and stomach cancer, rickets—2 times more. Ankylosing spondylitis, tumors of an oral cavity, tong, throat, and esophagus are more frequently in men, and cholecystites, stones of gall bladder, obesity with diabetes, rheumatoid illnesses—in women. These distinctions are not limited to the reproductive period only. For example, diseases of respiratory system are more fatal to boys of up to one year, than for girls. In humans the sex ratio for the majority of the reasons of death is shifted towards male prevalence (**Table 3.1 Appendix C**; Waldron, 1976).

Among children who stutter, strabismus, left-handers, dyslexics, incontinence of urine and feces, the sex ratio is 1 girl to 8.4 boys. These phenomena are closely related to the asymmetry of the brain and among themselves. For example, when left-handed children were forcibly reeducated to write with right hand, they often developed neurotic disorders, mental retardation, speech defects, stuttering, strabismus, and incontinence (Bianchi, 1985; Harris, 1978; McGlone, 1980).

Infectious and Parasitic Diseases

Sickness and death rate from infectious diseases till 5 years different at boys and girls: we shall tell for the whooping cough caused by *B. pertussis* is higher at girls, and from meningitis—at boys. Higher death rate of men is observed almost from all infectious illnesses, with a few exceptions (whooping cough, some gonococcal infections) (Stern, 1960).

Sex differences in immunocompetence and susceptibility to parasites have been found in mammals, including humans, with males being generally less immunocompetent and more susceptible to parasites than females (Olsen, Kovacs, 1996; Poulin, 1996; Schalk, Forbes, 1997).

Autoimmune Diseases

Women have stronger immune system which raises their resistance to many infections. They have thymus gland 3 times thicker than that of males. They produce more immune antibodies. Rejection of organ transplants in women is stronger. At the same time strong immune system makes women more sensitive to autoimmune diseases, such as Hashimoto's thyroiditis, Graves' disease, multiple sclerosis, scleroderma, systemic lupus erythematosus, rheumatoid arthritis, and type I diabetes mellitus. Almost 75 % of patients for autoimmune diseases are women (*Autoimmune diseases in women*, 2002; **Table 5.1**). It's interesting to note that maternal parent-of-origin effect was observed in multiple sclerosis susceptibility (Ebers et al., 2004). For maternal half-siblings, the risk was 2.35% (34 affected siblings of 1859), and 1.31% for paternal half-siblings (15 of 1577).

The small number of autoimmune illnesses strike men and women equally—type 1 diabetes mellitus, Wegener's granulomatosis, Crohn's disease and psoriasis, and very few diseases, such as ankylosing spondylitis, occur more frequently in men (Porter et al., 2006).

Table 5.1 Sexual differences in susceptibility to autoimmune diseases.

Disease	Sex Ratio, ♂♂ : ♀♀
Hashimoto's thyroiditis	0.1
Graves' disease	0.14
Systemic lupus erythematosus	0.17
Scleroderma	0.33
Rheumatoid arthritis	0.4
Multiple sclerosis	0.5
Type I diabetes mellitus	1.0
Ankylosing spondylitis	3.0

The reasons for the sex role in autoimmunity are unclear and can include such factors as sexual distinctions in the immune response, reaction to an infection, effects of sex hormones and sex-related genetic factors.

Estrogens have different influence on autoimmune diseases. For example, during pregnancy when estrogen levels are high, the condition of women, suffering from systemic lupus erythematosus worsens (Petri et al., 1994). Patients with rheumatoid arthritis and multiple sclerosis have reported improvement (Lockshin, 1989).

Psychological disorders

Men are three to four times more likely to be diagnosed with antisocial personality disorder (80%), psychopathy, and sociopathy than women. They have poor compliance with routine, prescribed behaviour and social regulations (Black 1999; Davidson, Neale, 1994; Eysenck, Gudjonsson, 1989; Zuckerman, 1994). Males also have higher rates of autism (M:F 4:1) and Asperger's syndrome (M:F 7:1) (Chakrabarti & Fombonne, 2001; Fombonne, 2005; Risch et al., 1999), and learning disabilities (Baron-Cohen, 1999; Rommelse et al., 2008), which are disorders related to the perception of messages from other people.

Schizophrenia is 1.4 times more common in men than women. In women, it tends to begin later in life (Picchioni, Murray, 2007; Cullen et al., 2008) and in a milder form compared to men.

Alcoholism

Alcohol consumption and dependence on it is defined both genetic predisposition, and set of external and internal factors of environment.

Ethnic and Cultural Differences in Alcohol Consumption

There are ethnic and cultural differences in prevalence of alcohol dependence. Among the Eastern peoples are more likely than in European populations, found alcohol intolerance, aversive reaction, the so-called "flushing" phenomenon. It depends on the genetic structure of populations, namely, the frequency of alcohol dehydrogenase and aldehyde dehydrogenase isoenzymes. Genetic characteristics are responsible for lower incidence of alcoholism in Oriental populations compare to Western and European ones.

With respect to alcohol consumption one can distinguish four forms of cultures (Bales, 1946):

- Abstinent cultures. All forms of consumption are forbidden.
- Ambivalent cultures. The conflict exists between the existing values and alcohol use.
- Permissive cultures. Alcohol use is permitted, but drunkenness and pathological phenomena associated with the consumption of alcohol, are dismissed.
- Permissive cultures, admitting dysfunction (some Scandinavian and Eastern European countries). Not only "normal" alcohol consumption, but also the alcoholic excesses are acceptable. Alcoholism in this type of cultures is most common.

Probably that frequency of alcoholism in population is inversely proportional the duration of historical time during which alcoholic drinks were accessible to the given society. Thus predisposition to alcoholism depends on frequency of autbreeding (interethnic marriages) in a society with the proceeding use of alcohol. In parallel with this process as a result of inbreeding the factors lowering susceptibility to alcoholism can accumulate. This concept is confirmed, in particular, that in the Mediterranean countries, where alcoholic beverages are consumed since the most ancient times, frequency of alcoholism is lower, than in the populations which were only recently exposed to alcohol, for example, North American Indians and Eskimos. High frequency of alcoholism is observed also among the Northern people, and rather low-among the people of Transcaucasia.

Also described is the phenomenon called secular trends, resulting in a decreasing age of onset of alcoholism and in increasing the risk of its occurrence during the lifetime. For example, for men born in 1938 the risk was determined as 8.9%, while for men born in 1953 it increased to 20.3%. Secular trends in the population are carried out fairly quickly which is hard to explain by any biological changes. This phenomenon is rather a consequence of changes in alcohol consumption and tolerance of heavy drinking (Reich, et al., 1988).

Dynamics of Alcohol Consumption

Statistical data about world dynamics of alcohol consumption in different countries from 1928 to 1974 is presented on **Figure 5.1** in logarithmic scale (Balljuzek et al., 2009).

Straight lines with a slope tgα = -2 indicate the presence of a power law of distribution (Pareto distribution):

$$f(d) = const / d^2 \qquad\qquad [3]$$

where d –dose of alcohol consumption [g / day].

G. Skorobogatov showed that this distribution is a consequence of positive feedback on the parameter d (Skorobogatov, 2005). This link leads to the fact that over time all states on the chart **Fig. 5.1** get shifted to the right.

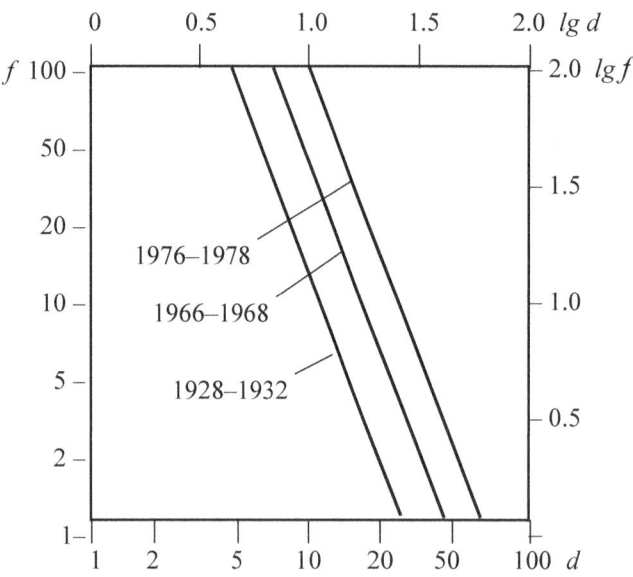

Figure 5.1. Distribution function f (d) numbers of states with average dose of alcohol consumption d (C_2H_5OH /days) (Balljuzek et al., 2009)

Sexual Dimorphism in Alcohol Consumption

Historical, ethnographic studies and special surveys show that since the Roman Empire to the present day and in different cultures, men are much more involved in alcohol consumption and related problems (Gefou-Madianou, 1992; Park et al., 1998; Wilsnack et al., 2000; Wolcott, 1974). Particularly strong difference in alcohol consumption is observed in many developing countries with the regular use of alcohol (11.2 in Korea, 8.4 in India, 20.0 in Costa Rica and 4.4 in Malaysia) (What Drives Underage Drinking? 2004). Women do not drink as much as men, and compared with them are rarely become drunk. Alcoholism and deaths from alcoholism are extremely more common among men than among women. Women are able to control their own problems with drinking more successful than men. More men drink voraciously and, as a result are becoming ill.

In a stomach of young women less alcohol dehydrogenase is produced, therefore the same quantity of alcohol (even normalized to body weight) results in higher alcohol blood level compare to men.

Some researchers believe that it is possible to explain these tendencies, by man's and woman's roles. For example, the alcohol use is less compatible with woman's roles, including responsibility for sexual restraint and care for small children. Men, as a rule, work during a week, on which getting drunk on weekends has less influence.

Age and Alcohol Consumption

At school alcohol consumption is growing steadily. For example a survey of pupils in 1995-96, in the U.S. showed an increase from 30% in the second grade to 75% in the 12-th (Mitchell, 1998). Most of the studies noted a permanent increase in consumption with age. Between 11 and 19 years an alcohol use increases substantially in all countries (Currie et al., 2000). According to Lintonen et al. (2000) increase in consumption with age was greater for the generation born in 1980-s, compared with those born between 1962 and 1966 years.

There also exists a relationship between sex and age. The increase in the alcohol consumption with age was higher at boys than at girls (Choquet et al., 2001).

Population studies show that younger men drink more (Treno et al., 1993). However, it is necessary to notice that chronic alcoholics often do not live till an old age.

Genetical Predisposition

Genetic studies conducted on twins and adopted children show that alcoholism is heritable. So 40 % of children of fathers-alcoholics had very low reaction to alcohol in comparison with 10 % in control group which had no alcoholics in a family. Such reaction was directly related with increase in liver metabolic activity. Another effect is the acquired stability of nervous system which is expressed as better functioning under alcohol intoxication.

Sexual Dimorphism in Oncology

Epidemiology of neoplasms contains many unclear facts connected to gender. The sex ratio of death rate for all kinds of malignant tumors was 1.37 (averaged data on 52 countries for 1973 (Glücksman, 1981)). It tends to grow because the death rate of men from a cancer grows faster, than that of women (**Figure 5.2**). But such generalized data can give a little without discussion and analysis of the contribution which is brought in them by different age groups and tumors of different location.

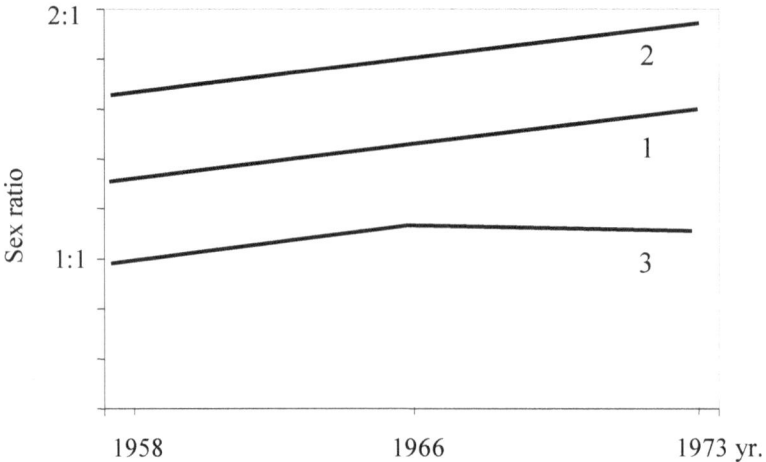

Figure 5.2 Changes in the sex ratio of death rate from a cancer in 22
countries from 1958 to 1973 (Glücksmann, 1981).
1—average sex ratio, 2—country with the highest sex ratio (Finland),
3—country with the lowest sex ratio (Israel).

Data Clemmesen for Denmark for such analysis are very typical and indicative, as they reflect the general picture of the age contribution observable in many countries (**Figure 5.3**). In the same figure age dynamics of a sex ratio of all population and age structure of a population (also typical for many populations) (Glücksman, 1981) are presented. Apparently from figure, the sex ratio of death rate from neoplasms strongly deviates from a sex ratio for the whole population in first and last three decades towards higher values, and falls noticeably from the fourth till sixth decade. So, men more frequently die from a cancer at children's, young and also old age, while women more often die in the middle ages from a cancer of a breast, uterus and ovaries. Middle age men die more from cardiovascular illnesses, accidents and occupational diseases.

Figure 5.3

Changes in the sex ratio of death rate from a
cancer depending on age in the Danish population.
- - - - 1943-1947, ▬ - 1963-1967.
On the histogram distribution on the age, used
for normalization of the data (Glücksmann, 1981)
is shown.

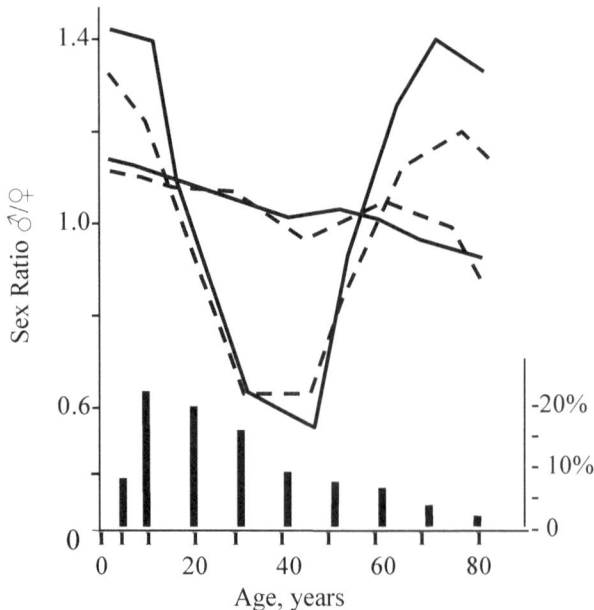

Average for 24 countries percentage distributions of the general death rate from cancer depending on a place of localization of a tumor are presented in **Table 5.2** (Glücksman, 1981). As can be seen from the table, tumors of respiratory system and digestive tract make 2/3 of all tumors for men and less than 1/2 tumors for women, while the cancer of reproductive system and mammary glands forms almost 1/3 for women and only 1/8—for men.

Table 5.2 Men and women death rate as a percentage of the general death rate from cancer (Segi, Kurixara, 1972)

Location	Men	Women	Sex ratio, ♂♂ : ♀♀
Respiratory system	26.0	5.4	4.8
Leukemia	3.8	3.5	1.1
Gastro-intestinal tract	41.2	39.8	1.04
Skin	1.3	1.3	1.0
Sex organs	11.8	16.0	0.74
Thyroid	0.3	0.8	0.37
Chest	0.2	16.9	0.01

Detailed analysis of separate systems reveals big differences between men and women. P. M. Rajewski and A. L. Sherman (1976) have shown that tumors of phylogenetically younger organs or systems are more frequent in men. Tumors of such phylogenetically young formations as lungs, larynx, tong and esophagus more frequently strike men. On the contrary, women have more tumors of the reproductive system and thyroid glands **Table 5.3 Appendix C.**

High sex ratio values are observed also for the organs and tissues contacting with the environment. For example, sex ratio of death rate from a cancer of a digestive tract as a whole equals 1.03. In **Table 5.4** the data on separate sites of the digestive tract, averaged on 24 countries, and in **Table 5.5 Appendix C**—similar more detailed data for Japan 1950-1971 (Glücksman, 1981) have shown. From these data it is visible, that the sex ratio of death rate has the maximal value for the beginning and the end of a digestive tract, passes through a minimum for thin and thick intestines and has the greater value for a pancreas, than for a liver.

For the same localization the sex ratio of death rate varies in the different countries within the limits of 1.36–12.6 for tissues of an oral cavity, 1.4–12.3 for an esophagus, 1.51–2.33 for a stomach, 0.88–1.24 for intestines, 1.01–1.94 for rectum, 0.61–1.58 for a liver and 1.37–1.84 for a pancreas. These numbers reflect very high death rate of men from a cancer of an oral cavity and esophagus in some countries (for example in France) while the corresponding death rate for women varies less between the countries.

Table 5.4 Sex ratio of death rate from a cancer of various gastro-intestinal tract locations (Segi, Kurihara, 1972).

Location	Sex ratio, ♂♂ : ♀♀
Mouth and pharynx	3.6
Esophagus	3.4
Stomach	1.9
Pancreas	1.6
Liver	1.1
Thin and thick intestines	0.9
Rectum	1.5

In the 1970-s death rate from lung cancer in 52 countries was on average 5.8 times higher in man compare to women. The maximal difference of 18 to 20 times was observed in Netherlands, Finland and Luxembourg where however the death rate for women did not differ noticeably from a level in other countries (Glücksman, 1981).

Congenital Anomalies

There is a category of illnesses at which significant distinctions in frequency of defeat of sexes is difficult to explain by specificity of reproductive function or due to social factors. Congenital anomalies can belong to this category, because it is reasonable to consider, that all social factors act equally on embryos of both sexes.

Many studies have found that the frequency of occurrence of certain congenital malformations depends on the sex of the child (**Table 5.6 Appendix B**) (Gittelsohn, Milham, 1964; Fernando, ea, 1978; Lubinsky, 1997; Lary, Paulozzi, 2001; Cui, ea, 2005). For example pyloric stenosis more often occurs at boys, and congenital hip dislocation is 4–5 times more often at girls. Among children with one kidney, there are approximately twice as many boys, whereas among children with three kidneys there are approximately 2.5 times more girls. The same pattern is observed among infants with excessive number of ribs, vertebrae, teeth and other organs which in a process of evolution have undergone reduction—among them there are more girls. Contrary, among the infants with their scarcity, there are more boys. Let's notice that data, received on twins of a different sex, essentially exclude a difference in environmental risk factors (Cui et al., 2005).

Anencephaly is approximately twice frequently occurs at girls (World Health Organization (reports), 1966). Excess muscle was 1.5 times more likely to be found in the corpses of men than women. The number of boys born with 6 fingers is two times higher than the number of girls (Darwin, 1871).

P. M. Rajewski and A. L. Sherman (1976) have analyzed the frequency of congenital anomalies in relation to the system of the organism. Prevalence of men was recorded for the anomalies of phylogenetically younger organs and systems (**Table 5.3 Appendix C**).

In respect of an etiology, sexual distinctions can be divided on appearing before and after differentiation of male's gonads in during embryonic development, which begins from eighteenth week. The testosterone level in male embryos thus raises considerably (Reyes et al., 1974). The subsequent hormonal and physiological distinctions of male and female embryos can explain some sexual differences in frequency of congenital defects.

It is necessary to notice that with the introduction of new diagnostic methods of an embryo and their constant improvement it is possible to interrupt pregnancy at early stages and to prevent birth of children with developmental anomalies. Therefore in last decades, for example, it is not possible to see high birth-rate of girls with anencephaly because of interruption of pregnancy. It's not possible to determine sex of embryos on early stages of pregnancy (0.64 ♂ : ♀ in 1983–1994 in comparison with 1.19 in 1995–1996, 0.9 in 1997–1998 and 1.5 in 1999–2000 (Riley, Halliday, 2002)).

Congenital Malformations of the Heart and Major Blood Vessels

The incidence of many congenital heart anomalies depends upon the sex of the newborn (**Table 15.1 Appendix C**). Since verification of one of positions of the theory was carried out on congenital heart diseases and vessels data, we shall briefly consider existing theories explaining their occurrence.

Rokitansky (1875) explained congenital heart diseases as breaks in heart development at various stages of ontogeny. Spitzer (1923) treats them as returns to one of the phylogeny stages. Krimsky (1963), synthesizing two previous points of view, considers congenital heart diseases as a stop of development at the certain stage of ontogeny, corresponding to this or that stage of the phylogeny.

Hence these theories can explain the defects with high prevalence in females ("feminine" types of defects) and defects with no difference in sex ratio ("neutral" types of defects) only (Zhedenov, 1954; Dzhagaryan, 1961). The defects with high prevalence in males have no analogs compare to the normal embryo or phylogeny stages of human ancestors and therefore cannot be explained by these theories.

*   *   *

In some cases the differences in disease and death rates from different illnesses can be explained by specificity of men and women reproductive function. For example, the fact, that a breast cancer occur much less often in men, than in women, seems natural and clear. But for other organs, such as heart, kidneys, stomach or liver such explanation becomes already flimsy.

In other cases scientists try to explain sexual dimorphism on disease and death rates from different illnesses as a consequence of social factors: different conditions of a life and work, the use of alcohol and nicotine, military service, and participation in warfare. So primary defeat of men by a lung cancer can be naturally connected with smoking, and a stomach ulcer—with alcoholism.

The role of resulted factors in etiology of diseases and a picture sexual dimorphism is certainly important; however they can not explain all observable differences. The comparative data on death rate of non-smokers and all population from cardiovascular, lung and other diseases are shown in **Table 5.7**. It is visible, that sex ratio among non-smokers decreases a little, but significant distinctions between men and women still remain.

Table 5.7 Comparison of sexual dimorphism on death rate for the non-smokers and the whole population (Waldron, 1976)

Cause of Death	Age (years)	Sex ratio, ♂♂ : ♀♀	
		Non-smokers	Whole population
Cardiovascular illnesses	45–54	4.5	7.5
	55–64	3.3	4.4
	65–74	2.1	2.4
Lung tumors	45–64	1.6	7.3
	65–70	1.4	9.4
Emphysema	45–64	4.0	11.7
	65–79	2.2	7.3
All cases of death	45–54	1.3	2.2
	55–64	1.7	2.5
	65–74	1.6	2.0

*   *   *

There is no general biological explanation of these facts, though besides the big theoretical value it would have also a practical value, in particular, helping with diagnostics. Analyzing the above-stated material, it is possible to draw a conclusion, that high death rate and damageability of a male gender is a common biological, universal phenomenon.

Chapter 6

Mysteries of Dioecy: Sex Hormons

fter chromosomal sex determination, gonads in embryo secrete the hormones that continue the process of fetal differentiation into the anatomy of a male or female. After birth, environmental variables play a determining role in shaping the individual's sex identity—usually, but not always, in accordance with his genetic sex.

Forms of Dioecy

"... there are many gradations running from female to male; and ... along that spectrum lie at least five sexes and perhaps even more."

A. Fausto-Sterling "The Five Sexes"

Determination of sex in humans is a complex multistep process, which depends on biological as well as psychosocial factors. A hundred years ago, German psychiatrist O. Weininger (1906) in his book "Sex and Character" noted that in nature there is no such thing as perfect man or perfect woman. Everyone carries a mix of both feminine and masculine traits. The combination of these components create a myriad of transitional forms. If we imagine a straight line, the ends of which will be "ideal" man and woman, then in the middle of it will be a true hermaphrodite, endowed with characteristics of both sexes. Between the marked points is a whole spectrum of sexual orientations: masculine women, feminine men, bisexuals, homosexuals and transsexuals.

Well-known sexologists (Kon, 1988; Vasilchenko, 1990) and psychiatrists (Isaev, Kagan, 1998, Bern, 1974) also confirm the empirical conclusions of O. Weininger about the presence of masculine women and feminine men in the human population, and Fausto-Sterling (2000) even proposed to distinguish 5 types of sexes.

In utero under the influence of sex hormones the brain of the fetus undergoes sexual differentiation (SDB). Deviation of this process in female fetuses causes masculinization of the sex center, which in severe cases leads to a change in the future feminine behavior. The maternal instinct, sex-role behavior, sexual orientation and degree of motor activity become altered. It was found that male animals from an early age are more mobile, which is associated with prenatal effects of sex hormones on the brain. Masculinization of the brain of female fetuses may occur under the influence of sex hormones, certain medications and stress.

Masculinization of the brain can manifest itself in childhood in girls as alteration of sex-role behavior (playing boys' games with cars and guns, or leadership in the boy's group), at a later age—in the choice of "male" kinds of sports (boxing, wrestling, weightlifting, and soccer), and at maturity—in the choice of "male" occupations (pilot, sailor, fireman, geologist, surgeon). In the sexual sphere such women may have decreased libido or change sexual orientation.

Deviations in the formation of sex can lead to the emergence of heterosexual, bisexual, or homosexual patterns of life-styles (Money, Ehrhardt, 1972).

Forms of Dioecy and Homosexuality

Homosexuality—Norm or Pathology?

The phenomenon of homosexuality has a long history. The choice of homo-heterosexual orientation is a matter of taste, inclination and traditions. The ancient Greeks, Romans, Persians and Muslims accepted and understood homosexuality as an integral part of human biology and sexual behavior.

Up until the early 20th century in Europe and the US prevailed view of homosexuality as a sin or heresy. During the Inquisition people accused of heresy, were often also accused of homosexuality and burned at the stake. In those days, almost all mental illnesses were considered as a sin, so the appearance in the 20th century the medical view of mental disorders and homosexuality in particular as an illnesses, can be considered as some improvement.

In the 60's and 70's medical interpretation of homosexuality has been criticized by many scientists (Szasz, 1965), as well as by the communities of homosexuals, who considered it a form of discrimination. In 1973 homosexuality was excluded by American Psychiatric Association from the list of mental illnesses, which instantly "cured" millions of American gays and lesbians. World Health Organization removed homosexuality from the list of psychological and behavioral disorders in 1993, Japan—in 1995, China—in 2001.

The legal side of the issue has also undergone significant changes. The centuries-old laws against homosexuality in England were repealed by Parliament in 1967 stating that homosexual acts between consenting adults in private have nothing to do with the law. In the United States, Illinois in 1961 was the first state to repeal existing statutes against homosexuals and many other states followed.

In some modern societies, homosexuals, particularly men, sometimes get arrested and put in jail on charges of "crimes against nature"(!) and "sodomy", along with various other forms of social disapproval and discrimination. In other aspects of life, homosexuals may be well adapted, well educated and achieve significant success in different occupations. Many have made an outstanding contribution to music, drama and other areas. A lot of famous historical figures—Plato, Alexander the Great, Leonardo da Vinci, Michelangelo, O. Wilde, and P. Tchaikovsky, were considered homosexuals, and among women—Sappho, G. Stein and V. Woolf— lesbians. Of course, high talent and achievements in such cases cannot be considered as an argument in favor of "normality" of homosexuality as such. Rather, it suggests that homosexual orientation can be compatible with high level of functionality (Carson et al., 1988).

C. Ford and F. Beech (1951) found that 2/3 of the societies they studied (total 191) accepted homosexuality, at least to some degree. They also noted three general rules: 1) no matter how the society would refer to homosexuality, this type of behavior is always present at least in some members, and 2) homosexuality is about two times more common among men (3–4 % vs. 1–2 % for women) 3) homosexuality is not the dominant form of behavior. Also, in the interaction of two lesbians or two gay men, one person always performs the male role, and the other—the female role. It is unclear whether the percentage of homosexual individuals is increasing, or is it a reflection of society's more open policy with regard to sex.

According to Kinsey about 2.6 million men and 1.4 million women in the U.S. are exclusively homosexual.

Kinsey et al., (1953) found that homo- and heterosexuality are not separate, disconnected categories. There are different variations of behavior, which occupy different positions on the scale of "exclusive heterosexual – exclusive homosexual." Some people have occasional homosexual (or heterosexual, respectively) contacts, while others have them at a certain age. Anatomical differences between homosexuals and heterosexuals are not detected (Perloff, 1965; Wolff, 1971). Somewhere in the middle are bisexual (people who have contacts with both his own and with the opposite sex). Moreover, homosexuals themselves are not a homogeneous category. There are different kinds (or types) of homosexual behavior (Bell, 1974).

Homosexuality is found not only in humans. Homosexual relationships were described at least in 450 species (Bagemihl, 1998). The author believes that homosexuality in various forms can occur in 10-15% of the total number of species.

Theories of Homosexuality

Genetic theories. Several studies have found 100 % concordance between monozygotic twins and only 15 % among dizygotic (Kallman). However, other researchers were unable to confirm these results. Pillard and Veynrih have found 4-fold increase in the frequency of homosexuality among male twins, at the same time; this effect was not observed in their sisters (see the review by Taylor, 1992). Genetic predisposition to homosexuality, if it exists, should be eliminated by natural selection, although Kirkpatrick noticed that homosexuality could evolve as a means of establishing social contacts (Kirkpatrick, 2000).

Neurohormonal theories. It is known that various prenatal hormonal influences can cause homosexual behavior in some animal species. Excess or deficiency of male sex hormones in the early stages of fetal development may cause changes in sexual behavior (Money et al., 1984; Ehrhardt et al., 1985). The results for the differences in hormone levels in adults are extremely controversial (see Ruse, 1988).

Psychological theories. Were initiated by the works of S. Freud (1920). Based on the innate propensity of people to bisexuality, these theories consider homosexuality as an aberration or an abnormality of psychosexual development of the child and as a consequence of different relationship with the child's mother or father (the powerful mothers and weak-willed, passive fathers). Data of different authors are contradictory and do not allow unambiguous judgments (Bell et al., 1981).

Learning and sociological theories. These theories explain the emergence of homosexuality as a consequence of training or establishing social "roles". They assume that human beings are born "sexual" rather than homo- or hetero-sexual, and are based on observations of bisexual behavior of other species and human children (Masters, Johnson, 1979).

We still do not understand why homosexuality exists and how it appeared. Since homosexual couples do not leave offspring, "genes" of homosexuality (if they exist) must be removed by natural selection.

By analyzing numerous observations and studies Hyde (1979) concludes that it is not possible to identify any single factor that would lead to a permanent homosexual or lesbian behavior. And frankly admits that "...We do not know what causes homosexuality."

Sex and culture

Riddle of hypersexuality. In all animals, other than humans, females are able to mate only during a limited period of time, or only at the time of ovulation, or some time thereafter. Enlarged female breasts during ovulation in primates is one of the signals of readiness for breeding. After the ovulation, breasts decrease in size. In humans, sex is possible even during pregnancy and lactation, when a woman's body has quite different hormonal situation. Unique is also women's lack of visible signs of ovulation. Mammary glands are constantly in an enlarged state, thus signaling a constant readiness. One possible explanation for this hypersexuality is due to the fact that the continuation of mating after conception binds a man to a woman and makes him take care of her and the offspring (Dolnik, 1995).

*   *   *

Difficulties of an explanation of the above-stated groups of the facts are connected by that the classical genetics, being actually genetics of dioecious forms, considers, however, only the results of *crossing*, without considering the results of their *differentiation*. The latter appear mainly as a consequence of specialization of sexes on a population level (for panmictic or polygamous populations). Hence, it can more fully describe *hermaphrodite* way of reproduction, than *dioecious* one. Therefore the phenomena associated with differentiation, with the type of reproduction (hermaphroditism, dioecism), with the scheme of crossing or the structure of the population (mono-or polygamy, panmixia, etc.) can not be explained within the framework of classical genetics despite the fact that special branches of population and evolutionary genetics are already present for a long time. Among these phenomena are first of all differential death rate of sexes, their different susceptibility to many illnesses, various kinds of sexual dimorphism on morphological, physiological, ethological and psychological attributes not directly related to reproductive specificity of a given gender, reciprocal hybridization differences and many others.

It looks like some positions of classical genetics require an essential addition and development in the aspect which takes into account first of all different participation of genders in reception and transfer of the genetic and ecological information to posterity.

The Evolutionary Theory of Sex

"Nothing in biology makes sense except
in the light of evolution. ... Without that light
it becomes a pile of sundry facts—some
of them interesting or curious but making
no meaningful picture as a whole."

T. Dobzhansky (1972)

"The simplest answer [that sex is needed]
for the reproduction cannot
be considered as satisfactory."

S. K. Nartova-Boshaver (2003)

Dioecy is not the best way of reproduction,
—it's the inexpensive way of evolution

V. Geodakian (2000)

Chapter 7

Ways of Reproduction

F irst of all, rational and simple classification of the ways of reproduction should be established. The most basic features of reproduction should be taken into account.

The first program of life is the *reproductive* program (REP) (**Figure 7.1**). It creates discreteness of genetic information in time (generations and other phases of life) and in the morphofunctional space (genes, chromosomes, cells, organisms, populations, and other organizational forms). It is the main criterion of life. It determines the size of populations and provides a foundation for such biological phenomena as replication, reduplication, and asexual (AS) reproduction. For all practical purposes, prebiological systems had no discreteness; there were no generations, nor an abundance of various forms. In fact, there were no intermediary levels of organization between the level of simple molecules and cenoses. This implies that the richness of forms and phases of the living systems is the result of divergent processes (differentiations). Replication errors gave rise to a second program, mutagenesis (MUT), which provides a source of diversity. This was enough for selection and evolution (EV) to act.

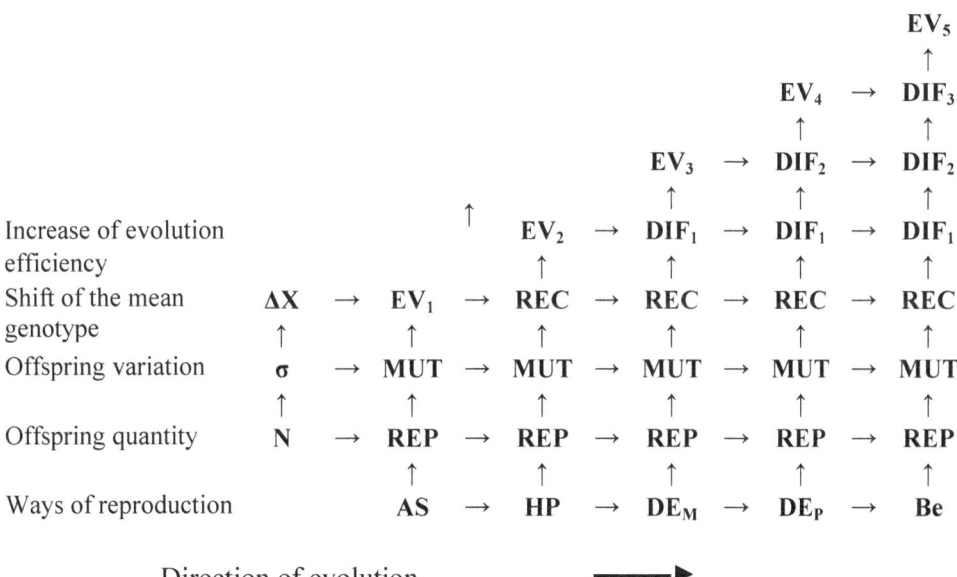

Figure 7.1 Emergence of the main programs and the increase of efficacy of evolution associated with different ways of reproduction (Geodakyan, 2000).
AS—asexual, HP—hermaphrodite, DE_M—dioecious monogamous, DE_P— dioecious polygamous, Be—bees.
Programs: REP—reproduction, MUT—mutagenesis, EV—evolution, REC—recombination, DIF—differentiations.

The next important step was the appearance of the recombination (REC) program, which underlies such biological phenomena as crossing-over, fertilization, or syngamy. The recombination created a new source of diversity, separated from the environment. It gave a foundation for the sexual process and the hermaphrodite (HP) mode of reproduction.

Next in sequence of importance is the program of differentiation (DIF), which has created the phenomena of meiosis, sexual, and other differentiations. This provides a basis for the dioecious forms, castes in social insects, dwarf males in some fish, etc. During evolution, these programs and biological phenomena triggered by them appeared specifically in this order, which reflects constitutive-facultative relations between them. The preceding, more fundamental programs are indispensable for the appearance of the subsequent ones, while the later programs are not indispensable for the earlier ones. If the first two programs had a fundamental nature of innovations, the later ones can be regarded as improvements, which increased efficacy of evolution (see **Figure 7.1**).

Three Main Types of Reproduction

The appearance of *crossing* (fertilization) process was the first important event in evolution of reproduction. Crossing made possible to combine the genetical information of parents. With crossing all types of reproduction, existing in animals and plants was divided by two main types—*asexual* (no crossing) and *sexual* (crossing is present) (**Figure 7.2**). According to this classification *parthenogenesis* should be considered as a form of asexual reproduction.

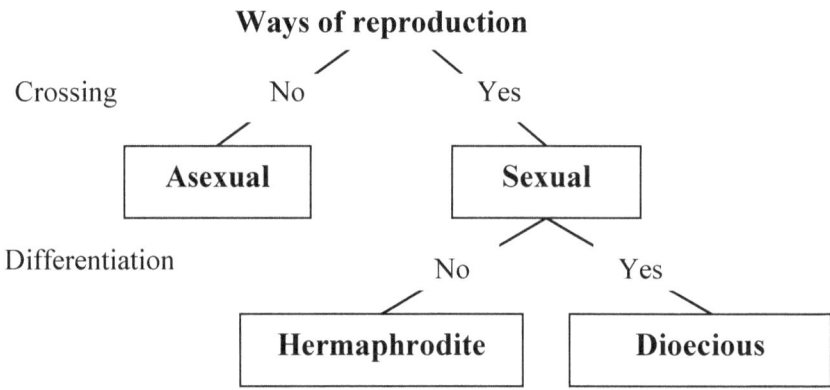

Figure 7.2 Ways of reproduction (Geodakyan, 1983,1989).

Only one parent participates in asexual reproduction and makes similar progeny. Two parents participate in sexual reproduction. The significant difference, however, is not the quantity, so that in the first case two organisms are produced from one, and in the second—"from two-three". Much more important is the qualitative difference, namely that during asexual reproduction there are no new properties, while each time after sexual reproduction, new quality appears. It is very essential. Sexual reproduction requires crossing to create new variants, which is necessary for maintenance of a genetic variety.

The second important event was the appearance of sexual *differentiation*—division into two sexes with prohibition of crossing between the organisms of the same sex. The appearance of sexual differentiation divided sexual types into two main forms: *hermaphrodite* (no differentiation) and *dioecious* (differentiation exist).

Depending on the presence or absence of these phenomena, many existing ways of reproduction can be divided into three main forms: asexual, hermaphrodite, and dioecious (**Figure 7.2**). Some characteristics of those types are listed in **Table 7.1.**

Table 7.1 Main Types of Reproduction and their characteristics

Type of reproduction	Program			Source of diversity*
	reproduction	recombination	evolution	
Asexual	**max**	mid	min	MUT
Hermaphrodite	mid	**max**	mid	MUT+REC
Dioecious	min	mid	**max**	MUT+REC+DIF

* MUT—mutation, REC— recombination, DIF—differentiation.

Advantages of asexual reproduction are—simplicity (no need to find a partner) and high effectiveness (any organism can reproduce alone at any place). Another advantage is that valuable combinations of genes, having appeared as a result of mutations, do not dissipate, and are transferred from generation to generation unchanged.

The disadvantage of asexual reproduction is that mutations are the only source of diversity, which is required for the natural selection. Assume that two good mutations a → A and b → B appear in different individuals of an asexual population. Then there is no way that they can be combined in the genotype of one of their progeny. An individual with genotype AB can appear only if a second mutation B will happen in one of the generations of initial mutant A or vice versa (Fisher, 1930; Müller, 1932).

With appearance of the sexual process, sexual forms lose the advantages of asexual reproduction. Instead, they gain a new, practically inexhaustible source of variety—gene recombination. Another advantage provided by recombination is the possibility to repair the damaged genes. However, evolutionary advantages of recombination are not quite clear, as is noticed that the useful combinations created in one generation, can be destroyed in the next one (Michod, 1987).

The *sexual process* and *sexual differentiation* are different phenomena. The first creates (increases) diversity of genotypes, and this is its evolutionary role; the second, in fact, decreases diversity to one half, and nobody can explain its evolutionary role.

Alteration of Generations and Evolution of Reproduction

Many species can use both asexual as well as sexual ways of reproduction. This is called different *"generations"* of a given species. If they change systematically, the phenomenon is called *alteration of generations.*

The alteration of generations has many similarities in plants and animals. The fertilization process is a border that divides sexual from asexual generation in the development cycle (**Figure 7.3 Appendix C**). A diploid *zygote* appears in the process of fertilization as a result of fusion of haploid *gametes* and sexual generation changes to asexual one. Both asexual and sexual generations can have a haploid as well as diploid set of chromosomes depending on what stage of life cycle meiosis takes place. After meiosis, the number of chromosomes is reduced in half and a diploid set changes to haploid. In the process of evolution, the significance (duration and size) of the haploid phase is diminished and the diploid one is increased (**Figure 7.4** Willie, Detie, 1975).

Figure 7.4
Evolution of asexual (2 n) and sexual (n) generations in plants (from Willie, Detie, 1975). A. Algae (*Oedogonium*). B. Moss. C. Algae (*Ulva*). D. Fern. E. Gymnosperms. F. Flowering plants (Angiosperms).

Alteration of generations and environmental conditions. Alteration of generations depends upon environmental conditions. In favorable conditions, duplication occurs as a rule, asexually (by means of binary fission, budding, vegetative reproduction, or parthenogenesis). Switch from asexual to sexual generation occurs slightly before or after the adverse conditions approach. For example, many water fleas and plant louses use parthenogenesis to reproduce under favorable conditions (usually in summer). Young generation, exclusively females, appears from "summer" eggs with a soft shell. When less suitable conditions approach, they produce a few males which then impregnate females. Impregnated females lay "winter" eggs with a firm shell. These eggs can withstand the long periods of adverse conditions (Cain, 1993). North American aphid *Therioaphis trifolii*, breeds parthenogenetically in normal conditions. When treated with insecticides, it switched to a sexual reproduction, and developed resistance (Blackman, 1981). When yeast cells are supplied with sufficient nutrients they reproduce asexually, but deprivation of nutrients triggers sexual reproduction (Goddard M., 2005).

Another example—experiments of Coen and Tausen on *Rotifera* cited by M. M. Zavadovski (1923). Normally the males were absent. When they moved *Rotifera* from pond into river water or vice versa, they observed the appearance of males after 3-4 days. It is typical, that the direction of moving was irrelevant.

The same situation can be observed in different species and populations of primitive *Crustacean*, which have all the transitions from pure parthenogenesis (without males) through some steps of *seasonal parthenogenesis,* when males appear in unfavorable conditions, and to populations, which have males all the time. It is interesting that the better the environment conditions, the less males in the population. Therefore in small drying ponds, populations with males usually occur, and in large, relatively stable ponds—the parthenogenetic forms prevail (Manuilova, 1965). For example, males were discovered only in 8 out of 25 species of the *Polyphemidae* family, living in the Caspian sea (Morduhaj-Boltovskoj, 1965a).

The evolution of reproduction progressed from *asexual* to *sexual* forms. Primitive forms reproduce only asexually. More complex forms have an alteration of generations. Most progressive species have sexual reproduction only (**Figure 7.3E Appendix C**). Therefore, *alteration of generations* can be considered as a transitional step in the evolution of reproduction.

Evolution of sex determination

Originally created in evolution as a purely reproductive (recombination) event, sex gradually acquired evolutionary functions. Concomitantly, sex determination was consistently undergoing transition from the **gene** level (in hermaphrodite) to the **chromosomal** level (in dioecious forms, beginning probably with fish) and the **genome** level (in the honey bee). The degree of differentiation was increasing in parallel, and sexual dimorphism was undergoing "expansion." It is absent in *asexual* forms; in *hermaphrodites*, sexual dimorphism exists only at the level of primary sex characters (gametes, gonads); in *monogamous dioecious* forms, there is also sexual dimorphism at the **organism** level (secondary sex characters); in *polygamous dioecious* forms, sexual dimorphism appears at the **population** level, including sexual dimorphism in terms of numbers and variance, while in the honey bee (and possibly in other social insects), there is, in addition, sexual dimorphism at the **genome** level (haploidy, diploidy).

Evolution of mechanisms of sex determination in bees. In bees the coexistence of two systems of sex definition is described: haplo-diploid and multiple alleles of a sex gene. Shaskolsky (1971) assumes that this circumstance reflects the evolutionary transformations related with transition from single to family existence. According to the author, it is possible to assume the existence of three stages in sex determination in bees:

1) the most ancient, widespread amongst insects is the XY or XO mechanism of sex determination.

2) the sex is defined by a series of multiple homozygous alleles of a sex gene. In this case only 2–8 % of offspring are males, since their occurrence is defined by a homozygous condition of alleles of a sex gene. At this stage both sexes are diploid.

   Replacement of the first system by the second is related with the termination of single existence and family formation, specialization of individuals in it (a queen, working bees and drones), sharp reduction of the number of breeding females and reduction of the need for males.

3) haplo-diploid system of sex determination. Occurrence of this system is connected, apparently, with the further development of a family, increase in the number of its individuals to several thousand, thus the requirement for drones is limited only to seasonal reproduction.

In bees a new conservative-operative differentiation into casts also appears, which consists of two ecological subsystems. They include drones from other (rich) families bringing in the genetic information (from the environment nearby) and working bees—bringing the information from honey plants (from the environment far away). Therefore, from the theory's point of view, working bees being genetic females perform the function of the second ecological (male) sex, and from the evolutionary standpoint the mode of reproduction in bees appears to be more advanced.

In some salmon fishes, the second differentiation took place at the level of male genome, which was needed for the delivery of environmental information from sea and river into the female sex (see **Ch. 14** for details).

Gamete Type and Sex

With respect to sex, the conflict on conservative and operative evolutionary tendencies was raised at least twice. First, when *isogamy* on the cell level existed, the conflict requirements to their dimensions arose. The operative task was to find another cell. In order to accomplish this task, small size and mobility were required. The conservative task was to preserve the formed zygote, by supplying nutrition and energy resources, and providing protective membranes etc., which is related to a big size. With isogamy, each gamete simultaneously performed both conservative (providing the zygote with resources) and operative (searching for a partner gamete) functions. As isogametes are of medium size, their performance of both functions was mediocre. Differentiation in size allowed the small- and large-sized gametes to optimize the **search** and the **resource supply**, respectively. Hence, the large–small size combination is more advantageous than the medium–medium one, which explains the evolutionary advantage of differentiation. As a result the gametes were differentiated by size and mobility on *ova* and *spermatozoids* (Parker et al., 1972).

The differentiation by the gamete type had the same problem. It was necessary to try recessive genes before including them permanently in the genome. It can be accomplished in autosomes in the homozygous state, and in sex chromosomes only in the hemizygous state in heterogametic set of chromosomes (XY). This is operative tendency. Heterozygous gene combination in autosomes and in homogametic set (XX) realize conservative tendency, because defective recessive genes are not manifested.

Table 7.2 Conservative and operative subsystems on different levels of organization in mammals and birds.

Organization level	Heterogametic males		Heterogametic females	
Organismic				
Cellular	Spermatozoids	Ova	Spermatozoids	Ova
Chromosomal	**XY**	**XX**	**XX**	**XY**

The notion of sex is related to the first differentiation on gamete size. So, we consider **an organism, which produces small mobile gametes to be a male, and the one that produces big gametes—to be a female.** In the process of evolution most species had both operative subsystems (small gametes and heterogametic constitution XY) in males and both conservative subsystems (large gametes and homogametic constitution XX) in females. These are the species with *Drosophila* gamety type (**Table 7.2**).

—"I came from Texas, where men are men and women are women."
—"We have the same arrangement in Kansas".

However, in the evolution of some species the directions of these two differentiations did not match. The conservative ova combined with the operative heterogamety found themselves in females and combination of operative spermatozoids with conservative homogamety—in males. These are species with the *Abraxas* gamety type. So, from the presented point of view, the *Drosophila* type of gamety is consistent, but the *Abraxas* type of gamety is contradictory. This can explain the fact that there are a lot more of *Drosophila* species, assuming that the selection of type was purely stochastic (independent of sex).

Hormonal Sex

> *"The sexes were not two as they are now, but originally three in number; there was man, woman, and the union of the two, having a name corresponding to this double nature, which had once a real existence, but is now lost, and the word 'Androgynous'..."*
>
> *Plato "Symposium"*

Sex determination during an organism's development can occur at the moment of fertilization (gene level), and also can be controlled by internal (hormones) and/or external factors. In humans and the higher animals the significant role is played also by education and training. Sex chromosomes define the sex of a zygote at conception (**Figure 7.5**). In mammals a base or "default" sex is homogametic (XX) female sex. Heterogametic (XY) male sex is a derivative sex. It is initiated by a Y-chromosome transforming "sexless" rudiments of an embryo gonad into testes that produce androgens. In the absence of a Y-chromosome, the same tissues form ovaries, producing estrogen.

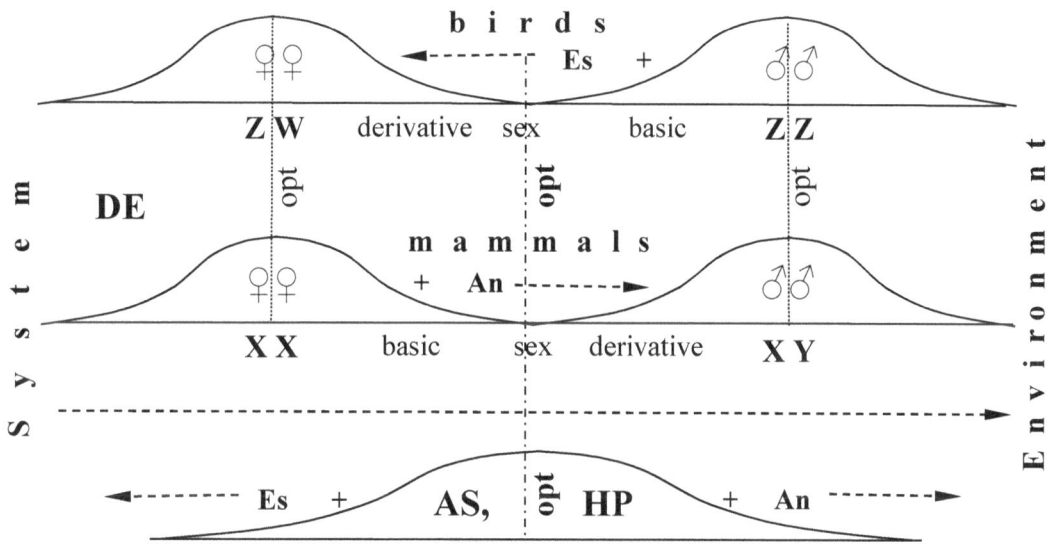

Figure 7.5 The emergence of dioecious (**DE**) forms from asexual (**AS**) and hermaphrodite (**HP**) ones in the phylogeny of mammals and birds. **An** – androgens, **Es** – estrogens.

In birds, the basic sex is also homogametic (ZZ), but a male. Derived sex is heterogametic (ZW)—female. It is formed by W-chromosome, which transforms undifferentiated gonad into ovaries which produce estrogen. Without W-chromosome, the same tissues are converted into testes, producing androgens. In both mammals and birds, the male sex is the "environmental" one, and the female sex is "systemic" sex. Only in mammals, the androgens move males away from females towards the environment, while in birds the estrogens move females away from males and environment. Therefore, sex is more fundamental phenomenon than gamety type, and sex hormones create sexual dimorphism as a vector "system –> environment".

Many sex-related characters (not only primary sex characters) depend from the hormonal environment in a womb of a mother. Animals with the ability to deliver multiple births—rats, gerbils, mice—provide a model to study the influence of sex hormones on embryo's development. Female and male embryos are positioned sequentially in the two-horned uterus of the mother. It has been shown that the females surrounded by two brothers (♂♀♂), are exposed to higher doses of androgens and lower doses of estrogens, than females, located between two sisters (♀♀♀). Females of the first type have more masculine anatomy compare to their more feminine sisters. They achieve sexual maturity later, the longevity and duration of the reproductive period is shorter, the litter sizes are smaller. They are more aggressive to other females and less sexually attractive to males.

Sex chromosomes define sex of a zygote only at gene level, its homo- or hetero-gamety (XX or XY). Further realization of sex in ontogeny, i.e. phenotypic sex, is influenced by sex hormones. While chromosomal sex is **discrete** (intersexual **di**morphism), hormonal sex is **continuous** (inter- and intrasexual **poly**morphism) (**Figure 7.6**).

Sex hormones along with the other features change the harmonics of the voice that occur during puberty. The impact of estrogens lowers the fundamental frequency of the female voice by one third than that of a child. The androgens released at puberty are responsible for the male vocal frequency, an octave lower than that of a child. Testosterone, the hormone that elongates the vocal cords during puberty, is also responsible for other manly features—such as broad shoulders and hairy faces. As indicators of sexual maturity, these traits may play a large role in the process of finding a mate for reproduction. This means that since the beginning of humankind, a man's voice could have functioned as a testosterone advertisement.

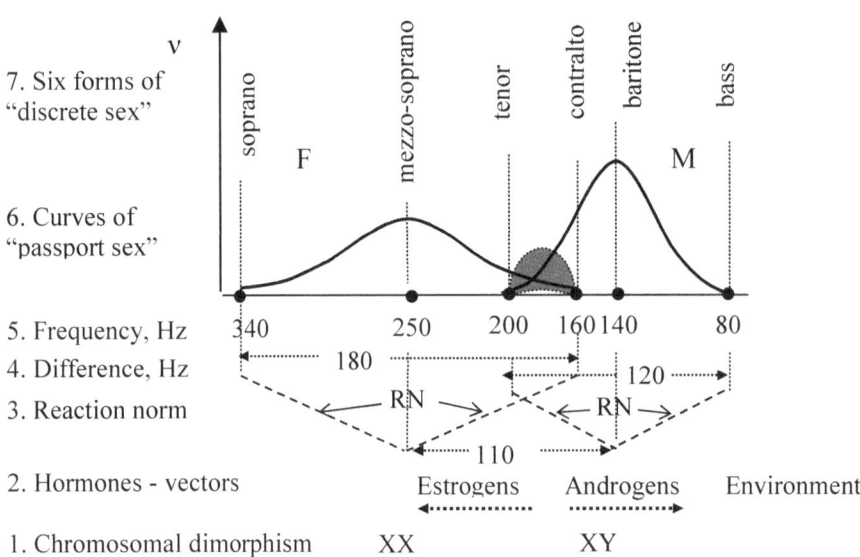

Figure 7.6 Transition from discrete chromosome dimorphism (1), via hormonal hexamorphism of "fractional sex" (7), to continuous polymorphism of passport sex (6). Abscissa: frequencies of male and female voices (Hz) (5), determined by estrogens (E) and androgens (A). Ordinate: frequencies of phenotypes in the population (ν). Dashed curve—the zone of transgression.

To consider the differences caused by sex hormones we shall allocate for simplicity inside each of the two sexes three gradations of a "fractional sex": modal, feminine, and masculine. Since sex hormones define the tone (height) of voice in humans, these forms can be presented as average frequencies of voices (in hertz). For women it is a soprano (340 Hz), mezzo-soprano (250 Hz) and contralto (160 Hz). For men it is a tenor (200 Hz), baritone (140 Hz) and bass (80 Hz).

According to one of Ancient Greece myths, the first people on the Earth were double beings. Some of them were double men and double women; others were half the man and half the woman. These beings were so powerful that they thought they were equal to Gods and started a war. Gods had won, and divided the remaining Titans in half to reduce their power. As a result each of us tries to find the lost half. It is interesting that if we assume the distinction of the divided halves, then Plato's scheme leads to the same six types allocated by us (three masculine and three feminine).

Hormonal Sex and Homosexuality

> *"Each of us when separated, having one side only… is but the indenture of a man, and he is always looking for his other half. Men who are a section of that double nature which was once called Androgynous are lovers of women; … the women who are a section of the woman do not care for men, but have female attachments; the female companions are of this sort. But they who are a section of the male follow the male, and while they are young, being slices of the original man, they hang about men and embrace them …"*
>
> *Plato "Symposium"*

The choice of sexual orientation depends on style, attraction, and tradition. According to the evolutionary theory of sex, male sex is responsible for creation of sexual dimorphism, while female sex—for its' elimination. And behaviors directed towards these goals will facilitate the evolution of the progeny.

It's well-known that sexual dimorphism can determine female choice and preferences. Sexual dimorphism on height can serve as an example of female choice of the partner. On average men are taller than women. Thus women, as a rule, prefer taller men. The most common difference is about 5–10 cm. Too big or reverse differences look unnatural.

Taking into attention that there are active and passive homosexuals, Plato's model leads to the same 6 types (3 male and 3 female). Combination of 6 types of "hormonal sex" gives 15 pairs: 9 hetero- and 6 homo-. Their distinction defines size and a direction of sexual preferences and tastes. As it was already mentioned in **Ch. 4**, the voice height in humans is related with attractiveness for both men and women. Men are attracted to women with more high pitched voices, and women find men with deep voices more attractive and sexual.

If we compare differences in voice frequencies for such pairs (**Figure 7.7**), we can see that homo- pairs do not occupy the last places on the list. Attraction, of course, is determined by many other factors besides voice frequency.

Many scientists have pointed out that homosexuality is not a crime or mental illness requiring punishment or compulsory treatment. It rather needs more understanding and tolerance (Freud, 1995; Kon, 2003). In respect to tolerance certain progress is already been reached, however we still do not understand why homosexuality exists and how it was created. Since homosexuals do not leave posterity, "genes" of homosexuality (if they exist) should be eliminated by natural selection. But despite that, homosexual relationships are discovered in at least 450 species (Bagemihl, 1998).

Genes of homosexuality could be beneficial. It was noticed that in parthenogenetic lizards and humans the genetic factors connected with homosexuality raise fertility of females (Corna et al., 2004). It has been shown, also, that homosexual members of a family can increase the reproductive success by increasing the survival rate of their close relatives (kin-selection on altruistic behavior). In bonobo chimpanzees, homosexual interactions are a form of social connections. It is possible that homosexuality evolved to serve social functions in humans, too (Kirkpatrick, 2000). In humans homosexual behavior frequently occurs in armies, jails, ships, during war-time and starvation. It means that we are dealing with natural adaptive phenomenon to extreme environmental conditions (disturbance of tertiary sex ratio, high population density, and stress).

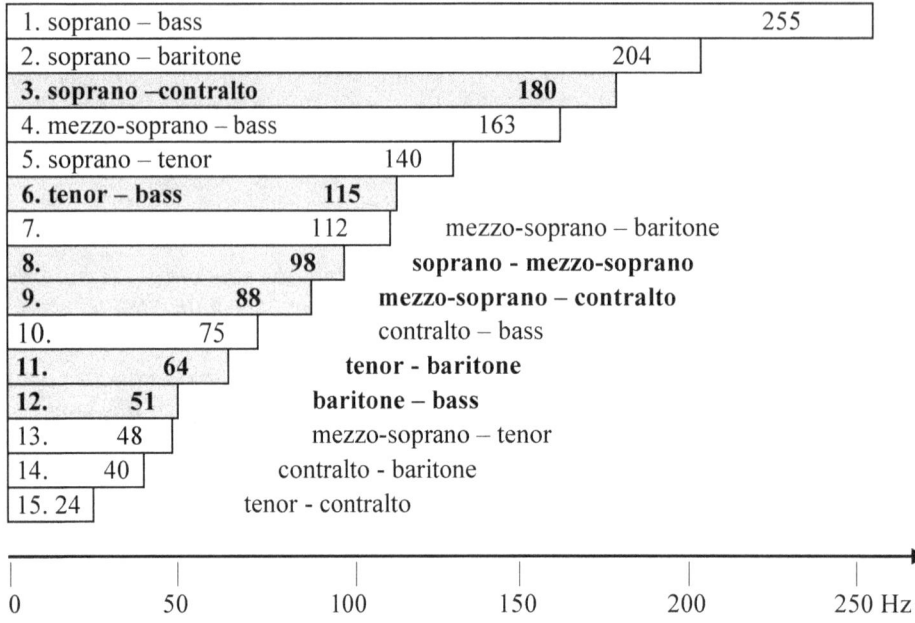

Figure 7.7. Difference in voice frequencies (Hz) for 9 hetero-   and 6 homo-pairs
as a measure of sexual attractiveness.
Average frequency, (Hz): soprano – 340, mezzo-soprano – 248,
tenor – 200, contralto – 160, baritone – 136, bass –85

The evolutionary theory of sex interprets homosexuality as an adaptive regulator of quantity-quality of progeny in extreme environment. It's another binary differentiation: hetero- orientation is a conservative subsystem, homo- —operative one. The ratio homo/hetero is closely related to environmental conditions: it is minimal in the optimal environment and grows in an extreme environment. It is known that severe stress, caused by natural or social conditions at the beginning of pregnancy increases the probability of the birth of gay children of both sexes (Swaab, 2015).

Sex, initially a purely reproductive phenomenon, gradually acquired evolutionary functions and finally became a social-cultural phenomenon closely related to creativity. In humans it has become a cultural phenomenon. Otherwise it's hard to explain close relation of homosexuality with sex (M : F ≈ 2 : 1), left-handedness (L : R ≈ 1.4 : 1) (Lalumiere et al., 2000), high education (homo : hetero ≈ 2 : 1), high percentage of homosexuals amongst notable culture and art workers ("donors" of information), and the influence of sex hormones on physical and mental abilities.

Chapter 8

Quantity, Quality, and Assortment of the Progeny

The product of any manufacturing process can be characterized by three main parameters: **quantity, assortment and average quality.** The same parameters can describe the output of reproduction—offspring of the population.

Asexual and Hermaphrodite Ways of Reproduction

Quantity of offspring in asexual population is directly proportional to the quantity of reproducing individuals (N). Assortment of offspring is directly proportional to the variation of parent generation and the mutation rate. Quantity of offspring in hermaphrodite, as well as in asexual population, is directly proportional to the total quantity of reproducing organisms (mothers). Assortment of offspring besides parent's variation and mutation rate is also directly proportional to the product of multiplication of numbers of mothers and fathers ($\approx N^2 / 2$).

Dioecious Ways of Reproduction

In dioecious population of N individuals quantity, quality, and assortment of the progeny depend upon sex ratio. Maximal quantity of posterity (close to N) is obtained at the minimal tertiary sex ratio (1 male and all others—females), which is probably not reliable. All children will be of one father's type and differ only by mother. Maximal assortment of posterity is obtained at the tertiary sex ratio equal to 1 : 1 ($N^2 / 4$). However, the quantity of offspring will decrease and among them there will be good as well as bad parent combinations.

Selection is required to increase quality, so part of the population should not reproduce. And shift in average quality or rate of average genotype change is proportional from one hand to variation of the population, and from another hand—to the fraction of organisms separated from reproduction. The more individuals get rejected, the more the shift in average quality (**Figure 8.1** C).

Stable environment. Stabilizing selection acting on asexual or sexual populations in stable environment (S) is changing the amount of progeny only (N). Average genotype ($\bar{x}$) remains unchanged (**Figure 8.1**).

Changing environment. If stable environment starts changing (D), the selection influences the population amount as well as the average genotype value. For simplicity, let's consider one trait only changing from the old (0) to new (1) value. The stronger the selection in monoecious forms, the more average genotype is shifted (Δx) and the less amount of progeny is produced (**Figure 8.1** B, C). The evolution rate is proportional to the amount of organisms separated from reproduction (ecological flow of information). Contrary the amount of progeny is proportional to the amount of organisms participating in it (genetical flow of information). So, the conflict exists between the evolution rate and the amount of progeny. Monoecious forms have to choose some intermediate values of these parameters, and can not achieve their maximal values (**Figure 8.1** B).

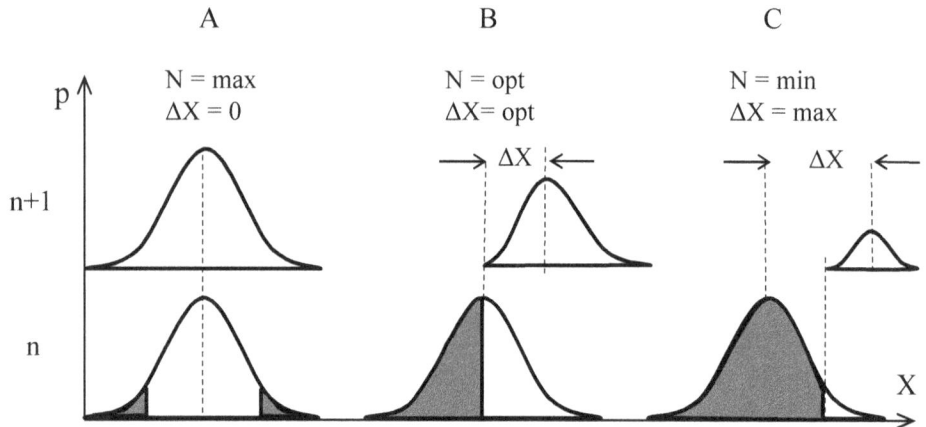

Figure 8.1 Genotypes distribution in generation *n* and *n + 1* (Geodakyan, 1989).
X—genotypes, p—their frequencies in the population. In stable environment (A), average genotype stays unchanged (ΔX = 0). In changing environment (B, C), the more elimination (dark area) monoecious form has, the more average genotype (ΔX) is shifted, but the less amount of progeny is produced (N).

In strictly *monogamous* dioecious population the relation of both sexes to the offspring quantity, assortment and quality is the same and symmetric. Asymmetric relation appears in *panmictic* or *polygamous* population only. Let's consider population consisting of randomly mating individuals. This population has no such limitations as pair or group marriage. Using the ideal gas analogy, we can call this population an "ideal population". "Panmictic" population is very close to an "ideal population". In many cases we can also consider polygamous population an "ideal population".

If it is required to improve or change quality of a dioecious population, it is necessary to create conditions for sexual selection, so that some individuals will not participate in reproduction. For this purpose it is necessary to have a surplus of males competing for females. The competition will lead to that the part of males will not be presented in posterity. Thus, the more surplus of males, the more rigid the selection conditions will be.

Thus, there exists a certain specialization, some kind of a "division of labor" in a role of sexes in reproduction. Their relation to key offspring parameters—quantity and quality—is different. The more females are in the population, the more posterity is produced. On the contrary, the more males, the better selection, and the faster the change of offspring quality.

Such "asymmetry" arises only at the population level. Both parents transfer to each descendant approximately equal quantity of the genetic information. However, considering population as a whole, the new features for that higher level of organization appear.

The potential possibilities for the male to transfer his genetic information is much higher than the possibilities of a female. After all each male basically can become the father of all posterity in a population. Therefore in several conditions rare variants of males unlike rare variants of females can play an essential role in change of an average genotype, and accordingly the population quality.

Ethologists and anthropologists recently re-examined the assumption that monogamy is a "natural" condition. Ethologists now believe that only 1-2% of all species may be monogamous (Angier, 1990). None of the simian species are strictly monogamous; our closest relatives, the chimpanzees, practice a form of group marriage. Among the 849 human societies examined by the anthropologist Murdock (1967), the vast majority (83%) practiced polygyny, men having more than one wife; monogamy was characteristic of only 16% of the societies. Even in societies that are technically monogamous there are often many cases of both tolerated and covert infidelity.

If we would try to formulate briefly the relation of sexes to quantity and quality of posterity, it is possible to say that the quantity of females defines the quantity of posterity, and each female struggles for the quality of posterity. The quantity of males defines population quality, and each male is a "fighter" for quantity of posterity. This simplified formula reflects also to some extent the differences in behavior and "psychology" of the sexes. One vivid example of such different "psychology" can be found in Darwin's writings: "... males of deer hound dogs are attracted to unfamiliar females, whereas females prefer familiar males." In biological categories it means that females in population express a greater degree of the trend of heredity, and males— variability. Such specialization offers species essential advantages.

The process of self-reproduction should provide two opposite tendencies: *heredity* which is conservative and aspires to transmit all parental attributes to the progeny unchanged, and *variability* which is progressive and results in the occurrence of a new attributes.

The ratio between these two tendencies (*evolutionary plasticity*) depends upon environmental conditions. In a stable, constant (optimum) environment, where there is no need for any changes, the quantitative aspect of duplication becomes most important. Therefore, in such conditions, the tertiary sex ratio should go down, reducing evolutionary plasticity of a population. In changing environment, species need to adapt and develop new qualities which would better match the new conditions. In such periods the trend of *variability* would prevail, thus giving basis for evolution.

Evolution is carried out by selection (natural, sexual or artificial). The necessary condition for selection is the presence of a variety of attributes. Selection leads to elimination as well as discrimination of some individuals from duplication. Hence, any change of quality always requires certain loss of population quantity. In extreme environmental conditions, dioecy allows to eliminate or discriminate many male's genotypes, thus achieving a faster rate of selection and maximal shift of quality. At the same time, the amount of progeny (which is proportional to the number of females), and the spectra of genotypes from previous generations (which is represented by female's variation) remain unchanged.

For the best preservation of a species headcount and existing genotypic distribution, it is necessary to reduce elimination to a minimum. On the contrary, for a fast change of old distribution, higher elimination is required. Sex differentiation and panmixy are removing this conflict. Sex differentiation, isolates the tests (and hence also solutions and mistakes) within the male subsystem, and selection and conservation—within the female one. This division allows implementing different solutions of evolutionary problems without risk of preserving unsuccessful decisions. This is a "short-term" advantage of dioecious forms, which overweighs hermaphrodite's double superiority in recombination.

*   *   *

The double superiority of asexual forms over the number of posterity is compensated at sexual forms by their combinatory potential, due to crossing. The double superiority of hermaphrodites over dioecious forms in assortment is very well compensated by flexibility in change of quality, provided by differentiation. Therefore any attempts to understand dioecy as a best way of reproduction are destined to fail. Dioecy is not the best way of reproduction,—it's the inexpensive way of evolution (Geodakian, 1989,1991). Roughly speaking, asexual forms are betting on the offspring **quantity**, hermaphrodites—on offspring **assortment**, and dioecious ones—on offspring **quality** (**Table 8.1**).

Table 8.1 Characteristics of three main reproduction types (Geodakyan, 2000).

Type of reproduction	Efficiency		
	quantity	assortment	quality
Asexual	**max**	mid	min
Hermaphrodite	mid	**max**	mid
Dioecious	min	mid	**max**

Chapter 9

Two Sexes—Two Streams of Information

*Each male constitutes an experiment in which
a different set of genes is tested against the environment*

F. Hapgood "Why Males Exist" (1979)

The population of individuals, evolving in the changing environment, requires high stability. It is logical to assume that this stability is provided with its division into two parts (two sexes). In other words, differentiation of sexes is a reflection of the same general *principle of conjugated subsystems* (see **Appendix B**). It means that one of the sexes should specialize on the conservative aspect and represent the system's "core" and the other—on operative aspect, thus become the system's "shell". Then, division into two sexes becomes favorable to population information contact with the environment. Such contact is carried out through two streams of the information: generative (transfer of the genetic information from generation to generation, from the past to the future) and ecological (the information from environment, from the present to the future). Two sexes participate differently in them. The specialization on generative stream demands more stability and resistance, and specialization on the ecological stream—more sensitivity and mobility. There appears a question, which of the sexes specialize on the genetic, and which—on the ecological information stream? Before answering this question, it is necessary to find out what is the essence of sexual distinctions.

There are two fundamental differences between males and females: channel "cross-section" of the information transfer with posterity and norm of reaction.

Information Channel "Cross-section" to the Progeny

The basic distinction of males and females consists that potential male opportunities in transfer of the genetic information is much greater, than that of a female. Father and mother transfer each to descendant approximately identical amount of the genetic information, but the quantity of progeny that males can produce is much more compared to female. One male basically can transfer the information to the entire proceeding generation of a population, the females cannot do that. In terms of the information theory, the throughput or "cross-section" of a male's information channel with offspring is much greater when compared to a female's (**Fig. 9.1**).

A woman can produce 400-500 eggs during her life cycle. In real life she has much less offspring—two registered records are 69 and 32.
Guinness Record Book, 1980, p. 29

The vast majority of species produce a lot more small gametes compare to large ones. For example a woman in a lifespan can produce about 500 ova. At the same time man has hundreds of millions of spermatozoa in each ejaculate.

Figure 9.1

Participation of males and females in reproduction.

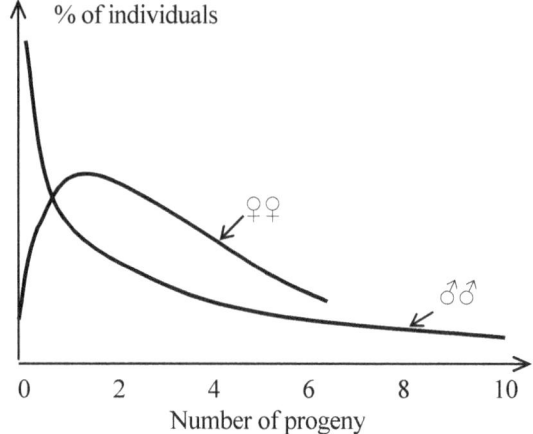

A man can make billions of spermatozoids and theoretically can become the father of all children on Earth.

Sultan Morocco Moulay Ismail fathered more than 1,000 children.
Guinness Record Book, 1980, p. 29

This results in hundreds of billions spermatozoa during life time, which comes to more than a billion sperm cells per 1 egg. We can say that for each spermatozoid successfully fertilizing the egg, there are more than a billion of rejected ones. Besides, any male can impregnate many females, but females are deprived from such opportunity. For example, a cow can leave a maximum of 15 calves, but the bull—tens of thousands. The ratio of information streams to the progeny is approximately 1 : 1000. Assuming the population sex ratio of 1 : 1, it means that inevitably, about 1000 males will be discharged from reproduction.

Bateman's principle

This conclusion was verified experimentally. Using *Drosophila melanogaster* with chromosomal markers, Bateman (1948) has counted up number of pairings in a population broken on several groups consisting from 5 males and 5 females. It appeared that in these conditions, 21% of males and only 4% of females did not participate in crossings.

Bateman showed that there are basic distinctions in sexual strategy of males and the females, related with various "cost" of their contribution to reproduction. Female contribution is almost always much higher, than the contribution of the male (even if there is no parental care) and consequently, the female decides which males will leave descendants. All species-specific features of mating behavior are based upon that principle.

There are also other observations. For example, LeBoeuf (1974) estimated that more than two thirds of elephant seal (*Mirounga*) males in any given season fail to reproduce. The higher male's social-hierarchical ranking, achieved in fights, the more often he approaches females for mating. According to other research, 85% of sea elephants females were impregnated by only a 4% of males (Le Boeuf, Peterson, 1969).

In laboratory conditions, similar results were attained with house mice: dominant or more aggressive males were the fathers of the greater number of descendants, than less aggressive (Levine, 1958). Less than 10% of black grouse (*Tetrao tetrix*) males participated in more than 70% of all pairings (Wiley, 1973). This means that rare male variants, contrary to rare female variants, can play the basic role in transfer of managing information from the environment to the progeny. So, males define the amount of change of the average character value.

Channel "cross-section" of the information transfer to the progeny depends upon the mating system. For monogamy it's the same for both sexes. For polygyny even if the ratio is only $1\male : 3\female$, the difference in volumes of information is almost ten-fold (male transfers 3 times more information and females receive the same information 3 times).

The "Coolidge Effect"

There is another effect that can play an important role in more even distribution of information to the offspring. The males of most mammalian species have a definite urge towards seeking variety in their sexual partners. It is the so called "Coolidge Effect" (Bermant, 1976). If a male rat is introduced to a female rat in a cage, a remarkably high copulation rate will be observed at first. Then, progressively, the male will tire of that particular female and even though there is no apparent change in her receptivity, he eventually reaches a point where he has little apparent libido. However, if the original female is then removed and a fresh one supplied, the male is immediately restored to his former vigor.

The President and Mrs. Coolidge were visiting a government farm in Kentucky and after arrival were taken off on separate tours. When Mrs. Coolidge passed the chicken pens she paused to ask her guide how often the rooster could be expected to perform his duty. 'Dozens of times a day' was her guide's reply. She was most impressed by this and said, 'please tell that to the President.' When the President was informed of the rooster's performance he was initially dumbfounded. Then a thought occurred to him. 'Was this with the same hen each time?' he inquired. 'Oh no, Mr. President, a different one each time' was his host's reply. The President nodded slowly, smiled and said, "Tell that to Mrs. Coolidge!'

Bermant (1976)

Rams and bulls are unmistakably resistant to repeating sex with the same female (Beamer, et al., 1969). Thus for breeding purposes it is unnecessary for a farmer to have more than one male to service all his sheep and cows.

D. Symons (1979) has pointed out that the term "indiscriminate" has been applied inaccurately to male sexuality. Male animals do not choose their mates randomly; they identify and reject those that they have already had sex with. In the case of rams and bulls it is notoriously difficult to fool them that a female is unfamiliar. Attempts to disguise an old partner by covering her face and body or masking her vaginal odors with other smells are usually unsuccessful.

Data illustrating the Coolidge effect in human beings have been reported by Wilson (1981). During review 63% of women chose "more sex with their spouse or steady partner" (compared with 38% of men) while only 18% of women chose "more partners" (compared with 37% of men). Partner variety therefore is of greater interest to men than to women.

Variation of the Sexes

The population mortality curve should be in close contact with the harmful environmental factors. If the environment starts changing, the population will get this information and be able to adapt. This means that population always has to pay for the information in the form of elimination of organisms most sensitive to the given factor. The "payment" is directly proportional to the amount of information received and is related to population's *phenotypic variation*. If phenotypic variation is insufficient, in a stable environment there is no elimination, and hence no information contact with the environment. In this case the next unexpected change of environmental conditions can catch population unaware and extinguish it. On the other hand, if phenotypic variation is too big, the payment for the new information is too high, which is wasteful. So, there should be some optimal value of *phenotypic variation*, which provides the necessary information with minimal payment for it.

Let's divide a phenotypic distribution curve of the population into three parts, as shown in **Figure 9.2A**. In well-known old environmental conditions, the organisms from the central part of a population *(G)* are most adapted to them. They live under comfort conditions and are maximally involved in the transmission of genetic (specific) information to their progeny. This part can be named the "genetic" part of the population. The peripheral parts E_1 and E_2 are in discomfort, suffering from alternative values of the same factor. For example, if part E_1 suffers from cold, then part E_2 will suffer from heat. Therefore their reproduction is limited. It means, that even in populations of the animals living in tropics, monkeys for example, some animals are dying not only from heat, but also from cold, while in populations of penguins or polar bears, some animals perish from heat.

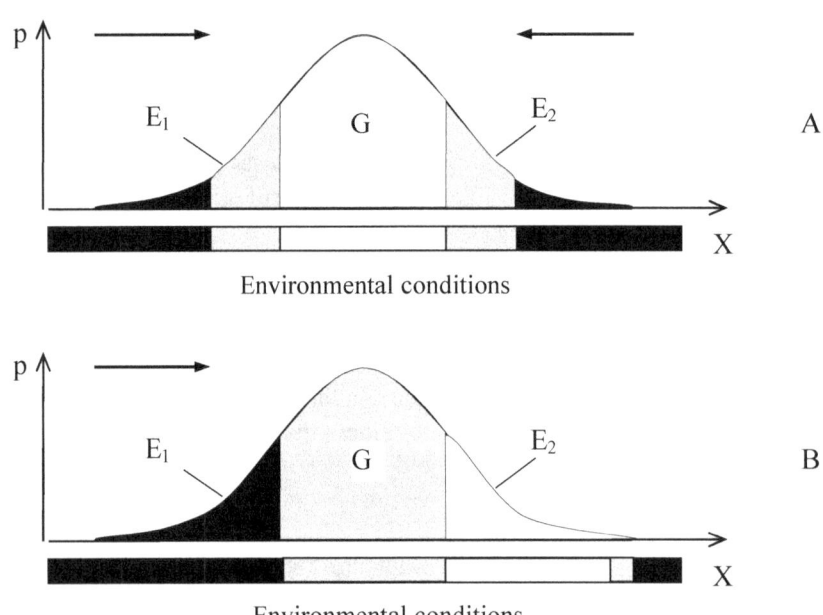

Figure 9.2 Relationship of population and environment.
Zones of: □—comfort; ▒—discomfort; ■—elimination.
A—under stabilizing, B—under directional selection (→ —direction of the selection);
X—generalized character (resistance to environmental factor), p—probability (or concentration) of the character in population, G—genetical part of the population,
E_1, E_2—ecological parts.

When the environmental conditions change, one of the peripheral parts (E_1) gets shifted into the elimination zone and perishes. Part G moves into discomfort zone, and part E_2—into comfort zone accordingly (**Figure 9.2B**). Individuals of E_2 class who usually suffered from discomfort, having got at last, in comfortable conditions, are breeding and "withdrawing" population from the coming harmful factor of environment.

Information value of individuals on the ends of the genotypic distribution curve (areas E_1 and E_2) is maximal, since they can cause the maximum shift of genotypes in the following generation (**Figure 9.3**). The further genotype from mode of a distribution, the more original it is, and the higher its information value. However there are only few most original individuals and their contribution to posterity's gene pool is insignificant. Therefore the real contribution will be defined by product of a degree of originality on concentration of individuals with the given value of a trait, and also on cross-section of their communication channel with posterity. As the channel cross-section of males with posterity is much wider than females, their contribution also will be higher.

Consequently, the informational roles of different population parts differ. The "genetic", past oriented, central part *(G)* "works" more for storing the genetic information, as it realizes more of a conservative trend. The peripheral "ecological" parts E_1 and E_2 are both maladapted to the old environmental conditions. Their reaction to the new conditions is different—organisms that belong to E_1 class are perished, while organisms that belong to E_2 class become most adapted. One can say that parts E_1 and E_2 are "working" more towards **alteration**. They realize mainly the relations with the environment and are more oriented towards the future. In other words, distinct specialization of the parts exists in the population, "division of labor" of some kind—performing relationship with preceding generations (*inheritance*) and with the environment (*variability*). Are male and female sexes equally represented in these specialized classes of the population (E and G) or is the sex ratio in them different from 1 : 1 proportion?

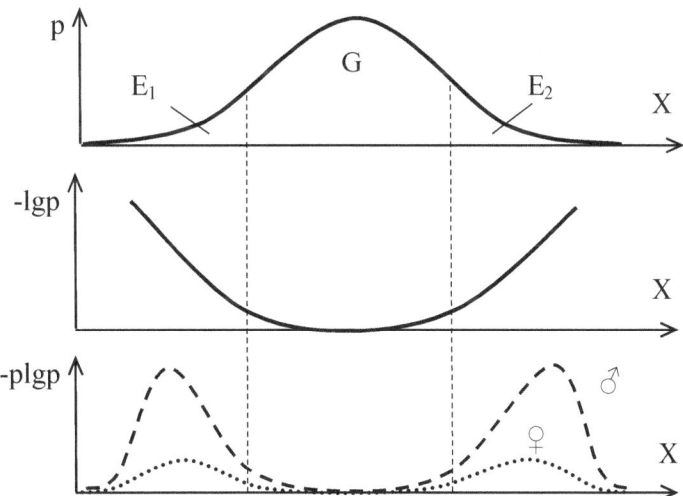

Figure 9.3 Distribution of phenotypes (p), their information value (-lgp) and
information efficiency (-plgp) in a population (Geodakyan, 1983b).
X—generalized attribute (or a phenotype), p—probability of its presence in
a population, G—genetic part of a population, E_1, E_2 —ecological parts.

It was noted already that males are first who die from all damaging environmental factors. The male sex is a biologically weaker sex. If we draw a mortality curve for each sex separately, the male curve should contact with the front of the harmful environmental factor. Hence, either curves are shifted, so that the male curve is positioned between the front of the harmful factor and a female curve, or male's curve should have a greater variation. The first decision is not satisfactory, because rescuing females from the given harmful factor of environment (let's say, cold) exposes them to an alternative factor (in this case, heat). This leaves us with the second decision—**male's curve should have greater phenotypic variation than female's curve (Figure 9.4)**. This prediction is the corner stone of the new theory. This conclusion is confirmed by the data shown in **Ch. 2** (see **Greater Male Variability hypothesis**).

MALES SHOULD HAVE WIDER PHENOTYPIC VARIATION COMPARE TO FEMALES

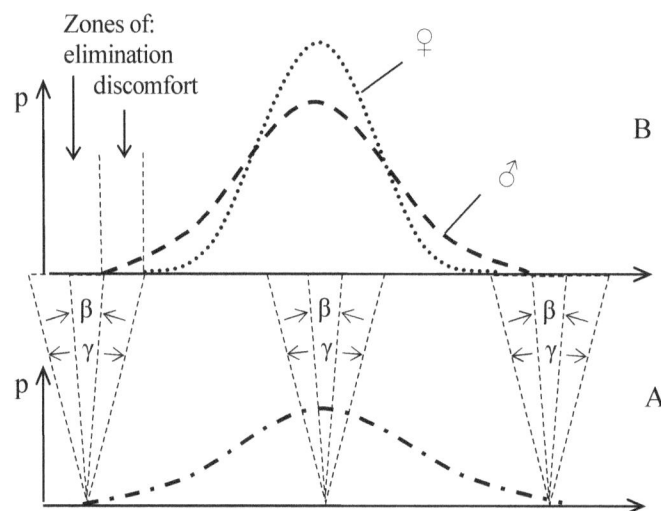

Figure 9.4 Distribution of male (- - -) and female (. . .) genotypes (A) and phenotypes (B) in population. β and γ—male and female reaction norms accordingly (Geodakyan, 1983b).

In order for males to receive the information from the environment, they should have more sensitivity (mortality) to all harmful environmental factors. This is achieved due to the greater variation of male attributes, in other words on all attributes, males should be more various than females. Wider phenotypic variation of males allows population "to pay" for the new information mostly by male individuals. At the same time males can leave much more descendants compared to females. Males receive and transmit ecological information from the environment to posterity more effectively, than females. On the other hand the picture of genotypic distribution in the population is more fully transferred to posterity by females.

The genetic stream of information (from generation to generation) is transferred mainly by females, and the ecological stream (from environment to posterity)—by males. Wider phenotypic variation of males can be a consequence of their wider genotypic variation. It can also reflect wider reaction norm of females which allows them to leave zones of elimination and elimination discomfort. Wider genotypic variation of males can be due to their higher mutation rate. Also, more additive inheritance of parental characters by female offspring can decrease their variation compare to males.

Higher Mutation Rate of Males

It is known that spermatogenesis has more cell divisions than ovogenesis, and errors in DNA replication and repair are the main source of mutations in molecular evolution. On this basis, it was suggested that the mutation frequency in sex chromosomes is greater than in autosomes and that, at least in the evolution of mammals, males serve as mutation generators (note that a higher level of spontaneous and induced mutagenesis in both homo- and heterogametic males, as compared with females, has been repeatedly demonstrated for Drosophila, silkworm, and mammals, including humans (Kerkis, 1975). Searle (1972) found that male mice have higher level of point mutations compared to females.

By comparative analysis of nucleotide substitutions in human and mouse genes located on autosomes and sex chromosomes, it was demonstrated that males do provide the main source of mutations for molecular evolution, and that the ratio of gene evolution rates is Y : A : X = 2.2 : 1 : 0.6, which conforms to the theoretical expectation 2 : 1 : 2/3 (Miyata et al., 1987). A similar method was used in another study to compare Y/X ratios of replacement rates of nucleotide sequences in synonymous genes of man, orangutan, baboon, and squirrel monkey. It has been shown that Y-genes diverge faster and are more divergent than X-genes; in other words, also in higher primates, males are at the forefront of molecular evolution (Shimmin et al., 1993).

Inheritance of Parental Attributes
Badr and Spickett (1965) found distinctions between male and female mice in a relative role of additive and nonadditive hereditary factors in determination of adrenal glands weight. The share of an additive genetic variety in phenotypic variability of weight of adrenal glands was 40% for males and 60% for females.

According to L. Schuler et. al. (1976) relative weight of adrenal glands, thymus, sex glands and a hypophysis of female mice hybrids practically matched with a half-sum of values of these characters for the parents. Male hybrids had the deviation of average weight in a direction of smaller value for adrenal glands, thymus, sex glands, and greater—for a hypophysis. Female mice hybrids also had more additive inheritance of the genes responsible for motoric activity (Borodin et. al., 1976).

X chromosome inactivation. Females are mosaics of X-linked gene expression with approximately half of the cells with paternally derived X chromosomes and the other half of the cells with maternally derived X chromosomes (Amos-Landgraf et al., 2006; Check, 2005; Gunter, 2005). Therefore, one would expect that the effects of each X-linked gene will be averaged in females, and thus likely to be less extreme as the full effect of a particular gene variant would require the female to be homozygote for this variant. In contrast, males are fully dependent of the one X chromosome copy they have (Zechner et al., 2001).

Wider Reaction Norm of Females

A man marries a woman thinking she will never change
and a woman marries a man thinking he will change.
And they are both wrong.
Atlanta Journal Constitution "The Vent".

Different reaction norm for males and females can be responsible for their different phenotypic variation. The reaction norm of female individuals should be wider than the appropriate reaction norm of males. In other words, a male's phenotype is more "rigidly" related to genotype than a female's. In ontogeny, the environment has more influence upon realization of a female genotype into a phenotype (**Figure 9.2** and **9.5**).

There is a certain zone of "ontogenetic discomfort" in the "space of opportunities". In this zone individuals do not perish, but experience inconveniences and difficulties, suffer from lack of adaptation (freezing, starving and so forth). The wide reaction norm allows female individuals to leave this zone and survive, while males stay in it and perish. Female individuals have bigger "space of abilities" due to their wide reaction norm compared to males. Therefore the sex differentiation can be treated also as a specialization on mutually additional qualities of perfection (female sex) and innovation, progressiveness (male one) in their relationship with the environment. The adaptation of female individuals to existing environmental conditions is more perfect, than males. On the contrary, the presence of various deviations (imperfections) from the norm (mode) of a population, give males more advantage in the adaptation to changed conditions of environment in the future.

Figure 9.5

Change of the wing size and form for Drosophila males and females depending on the temperature of the environment (Shmalgausen, 1969).

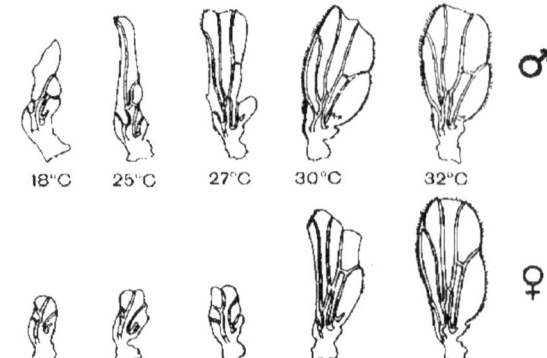

The narrow reaction norm of males provides them wide phenotypic variation. It allows mostly males to be the subject of elimination and discrimination. And high channel cross-section with the offspring maximizes replication of selected, the most suitable, male genotypes.

Contrary, wide reaction norm of females narrows their variation because it will allow extreme genotypes to "escape" zones of "discomfort" and to come closer to phenotypic norm (mode) of a population (**Figure 9.4**).

High ontogenetic plasticity provides female sex high stability in phylogeny. Hence, the male sex is more flexible in the phylogeny, and the female sex is more flexible in the ontogeny. Such, at first sight paradoxal, role distribution in phylogeny and ontogeny actually is consecutive and logical. It realizes uniform idea of specialization of sexes on conservative and operative tasks of evolution. Slightly exaggerating, it is possible to tell, that informational relationship of a population with the environment is based upon the elimination of males and the "education" (ontogenetic shift) of females.

> **" ... If someone will combine all males of the population into one team and all females into another, then males will be champions in most of the "individual events", while females will win most of the "team competitions".**
> Geodakian, 1966

The hypothesis of broader hereditary reaction norm in females, as compared to males, allows to make predictions, that can be experimentally verified. For example, in males the share of "hereditary component" must be larger and the "environmental" one—smaller than in females. If this hypothesis is correct, the intrapair differences between monozygous (identical) female twins must be greater than between the male ones. Contrary, interpair differences must be greater for male twins. between monozygous (identical) female twins must be greater than between the male ones. At the same time in dizygous (fraternal) twins like in common siblings, everything must be vice versa. Further, phenotypic variation in a pure line can be rather wider for females, and in a polymorphic (wild) populations it should be wider for males.

Most of the data shown in **Ch. 2** support the hypothesis of reaction norm. The most revealing research was performed on twins. In full accordance with the theory, female identical twins had wider variation, while in fraternal twins the variation was wider between males (Vandenberg et al., 1962; Nikityuk, 1977; Chovanova et al., 1980; Loat et al., 2004). More direct and indirect evidence can be presented in favor of the hypothesis as to the wide reaction norm in females. For example, greater conformism of females well known to psychologists has not been adequately interpreted untill now (Harper et al., 1965; McCoby, 1966; Kon, 1967).

The hypothesis of wide reaction norm of a female extends previous idea of the evolutionary concept of sex differentiation on ontogeny, combining phylogenetic and ontogenetic laws of dioecy and a population's relationship with the environment. It allows to explain differential mortality of the sexes, and also other ontogenetic phenomena such as polymery, penetrance, and expressivity. It can also explain many other features of sexes, including psychological.

The resulted reasoning allows constructing a sequence of phenomena leading from different reaction norm of the sexes to population sexual dimorphism: the wider the reaction norm, the more the plasticity in ontogeny, the less the phenotypic variation and death rate. The less the death rate is, the better preservation of genotypic variation, and the less the plasticity in phylogeny, which leads to appearance of sexual dimorphism.

Clay and Marble

The evolution goes by a method of "trial and error". It is more favorable to try on part of the system, but not on the whole system. For this purpose it is enough to divide the system into two subsystems and provide their dichronous evolution so that one of them (experimental) should start and finish evolution earlier, than the basic, more valuable subsystem.

Division of the population into two sexes and their "specialization" on quality and quantity of posterity, leads to the situation that any information from the environment is initially received and transformed by males. Such division into a stable core and labile shell allows population to distinguish short-term changes of environment (for example unusually cold winter or especially hot summer) from long-term regular changes in the same direction (say, approach of the glacial age).

A. Aleksakhin and A. Tkachenko (1977) sought to test the proposition that females represent constant memory of the population, storing the information "from the past", and transmitting it to the offspring, while males represent operative population memory, by collecting and transmitting the information "from the present". They suggested that the ability of the child to process information should depend on the experience of past generations on the maternal side, as well as on the personal experience of his father. Mother's personal experience and the experience of the ancestors of the father should have much lesser effect on the ability of the child. As an indicator of individual experience they have chosen the age of the parents. After processing the biographical information about famous people, they compared the age of the parents with the abilities of their children. The likelihood of talent appearance was not influenced by the age of the mother, and grew rapidly with age of the father (**Fig. 9.6**).

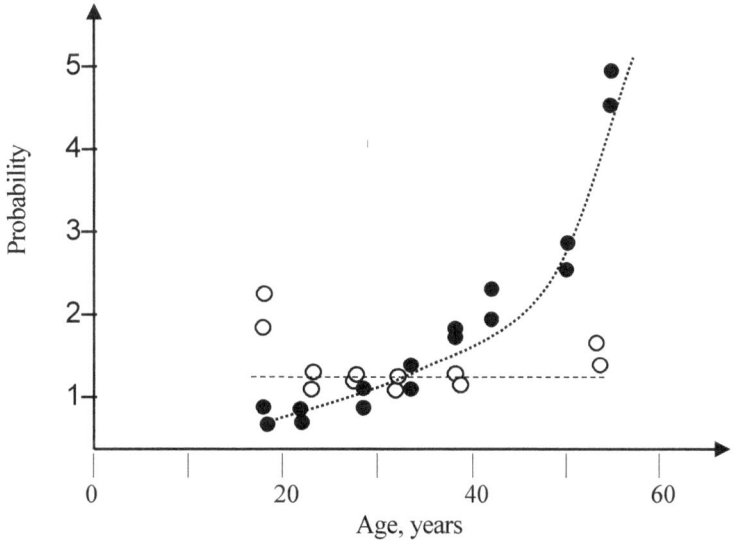

Figure 9.6 Dependence of the relative likelihood of the talent from the age of the father (●) and mother (○) (from Aleksakhin, Tkachenko, 1977 with changes).

Authors think that such pattern may be due to the fact that ova are formed simultaneously with the mother's body during fetal development and, therefore, use only the information received from ancestors. During life time of the female no new eggs are formed. Contrary, production of sperm begins with the onset of puberty and lasts almost entire life. Thus, the male gametes are constantly renewed.

It is possible to imagine males as the vanguard of the population advanced far forward, towards harmful factors of environment. Between this vanguard and "core"—"the gold fund" of the population—females there is a certain distance necessary for verification and selection. Evolutionary inertness, delay of females is a payment for their perfection. And, on the contrary, progressiveness of males is a reflection of their imperfection. Thus, we can answer "yes" to a question, whether males are necessary. They are necessary mainly for the adaptation to the changed environmental conditions.

Differentiations into two sexes are based on the same, main for evolving systems, specialization principle: on **preservation** (P) and **change** (C) of the system. First, the conservative and operative aspects of evolution are two of its main indispensable conditions. If one of them is absent, there is no evolution: the system either disappears, or is stable. Secondly, their ratio, C/P, characterizes evolutionary plasticity of the system. Thirdly, these conditions are **alternative**: the more C, the less P, and vise versa, because they supplement each other to one: P + C = 1. Therefore without specialization into subsystems, the system should choose certain intermediate optimum C/P, while the specialization allows maximizing both parameters. This is the evolutionary advantage of all conjugated differentiations.

*   *   *

As mentioned earlier, the question "what does sex provide?" includes two independent questions: 1) "what kind of advantage does crossing provide?" and 2) "what does sex differentiation provide?" The answer of the new theory: crossing provides combinatory potential and assortment of genotypes. In other words it serves a genetic function. Sex differentiation provides effective information contact with the environment and change of an average genotype that is carried out as an ecological function.

Chapter 10

Sex Ratio

Tertiary sex ratio determines the proportion between maintenance and variation tendencies as well as the species evolutionary flexibility. The population needs different evolutionary flexibility at different stages of evolution, and also in different environmental conditions (and consequently there exists a definite optimal tertiary sex ratio value for each of these conditions). And this value is not necessarily equal to 1 : 1.

The Relationship Between Sex Ratio and Environmental Conditions

In stable environmental conditions low evolutionary plasticity is required. Rapidly changing environment requires higher plasticity. So, optimum value of tertiary sex ratio should be low in the first case and high in second. Therefore tertiary sex ratio is a very important parameter of dioecious population which determines its evolutionary plasticity and is closely related to the environmental conditions. From the other side the environment changes the tertiary sex ratio through the differential mortality of the sexes. Unstable environment leads to higher male mortality and lower values of tertiary sex ratio. In severe conditions the population can loose males completely. So, we have a conflicting situation, the more the environmental conditions change, the less males is available, and the more males required for adaptation.

The only way not to lose males is to have a mechanism that allows preservation or even increases the tertiary sex ratio in extreme environmental conditions. This can only be accomplished by an increase of the secondary sex ratio. So, in extreme environmental conditions when tertiary sex ratio decreases, secondary sex ratio should increase. Both male mortality and male birth-rate increases which means that "turnover" of males increase. According to new concept secondary sex ratio is also a variable dependent on the environment, rather than a constant specific for a species, as it was believed. In a stable environment the secondary sex ratio is at its' optimum level. It increases in changing extreme environmental conditions, when more males are required (Geodakian, 1978).

The Relationship Between Secondary and Tertiary Sex Ratio

In 1965 the hypothesis that besides the direct relation, there exists a **negative feedback** between the secondary sex ratio and the tertiary one was proposed (**Figure 10.1**). So, in many species the secondary sex ratio may be regulatory related to the tertiary sex ratio, controlling the sex ratio in the population when its optimum is disturbed (Geodakian, 1965 b).

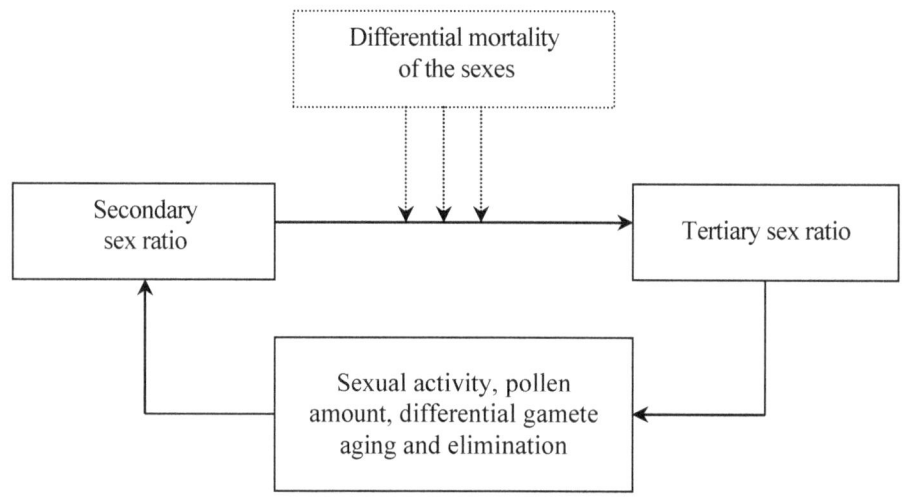

Figure 10.1 The scheme of negative feedback, regulating population sex ratio (Geodakian, 1965 b).

Several negative feedback mechanisms were proposed (see **Chapter 16**).

Ecological Rule of Sex Differentiation

Any dioecious population is characterized by three main parameters: sex ratio (ratio of males to females), variation (the ratio of diversity values within each sex), and sexual dimorphism (difference between average trait values for males and females). By assigning the conservative tasks to females, and the operative tasks—to males, the theory links these population parameters to environmental conditions and evolutionary plasticity of a species.

In a stable (optimal) environment, when no changes are required, the conservative trends are strong and evolutionary plasticity is minimal. In a changing (extreme) environment, when high plasticity is needed, the operative trends grow. In some species, such as lower *Crustaceans*, these transitions are made by switching from one type of reproduction to another (for example, parthenogenesis in optimal conditions, and dioecious in extreme ones). Most dioecious species have gradual regulation: in optimal conditions the main parameters decrease (birth-rate of males drops, their variation narrows, and sexual dimorphism decreases), and in extreme conditions—increase (*"The Ecological rule of sex differentiation"*).

SEX RATIO, VARIATION AND SEXUAL DIMORPHISM ARE VARIABLES,
CONNECTED WITH THE ENVIRONMENT AND THE EVOLUTIONARY FLEXIBILITY
OF A SPECIES. THE HIGHER THESE CHARACTERISTICS ARE, THE HIGHER IS THE
POPULATION FLEXIBILITY IN PHYLOGENY AND VISE VERSA.

IN STABLE CONDITIONS (THE OPTIMUM ENVIRONMENT) THEY
SHOULD DECREASE, AND IN CHANGEABLE CONDITIONS
(THE EXTREME ENVIRONMENT)—GROW

Freeman et al. (1976) noticed that male plants are usually found in relatively harsh environments, whereas females predominate in more favorable habitats.

According to the ecological rule, the main characteristics should increase under any environmental or social cataclysms (earthquakes, war, starvation, migrations). Therefore these parameters can serve as indicators of the ecological niche condition. When ecological niche "goes away", the male birth-rate, variation, and sexual dimorphism increase abruptly and irreversibly. This is an SOS signal. If this is an isolated tribe and only boys are born, then after some time only men will be left. One can say that genetical death comes earlier than physical one.

One unique type of buffalo is living on the Mindanao Island (Philippines). This isolated herd is destined to become extinct, because the number of males greatly exceeds the number of females. In Russia the birth-rate of boys in the Aral Sea region and Karakalpakia for the period 1980–1990 increased by 5% (A. Ergashev). This is a reaction of the population to the destruction of the Aral Sea. In one Chechen hospital, on the day of referendum (March 23, 2003), 13 boys and 4 girls were born ($325\male : 100\female$!) (normal sex ratio 106–$108\male : 100\female$). There was a report of one American Indian tribe living in harsh conditions, the number of male newborns was so high, that in time there were no females left.

Increased Male Mortality

The effect of differential death rate of sexes consists of two components (**Figure 10.2**). The first component, explained well by the theory of imbalance of genes, is the contribution of gamety type. The second component—is a contribution from specialization of sexes at the population level. A male sex and heterogametic constitution (XY) are responsible for the operative task, while female sex and homogametic constitution (XX) – for the conservative one. The directions of these two differentiations do not coincide in some species (for example, for birds).

The population effect arises in panmictic, freely crossed population only, and is depends, probably, upon a degree of polygamy. Therefore, to explain various death rates of sexes one needs to take into account a gamety type as well as mono- or polygamy of a given species. For a monogamous species the population effects are minimal or close to zero and the increased death rate is observed for heterogametic sex (for males of Drosophila type and females of Abraxas type). The facts are well described by the theory of gene imbalance. For the polygamous species of the Drosophila type, the population effect is imposed upon the effect from gamety thus strengthening it. This explains the maximal difference in death rate observed for such species (McArthur, Baillie, 1932). For polygamous species of Abraxas type, the population effect is directed against the effect of gamety: heterogametic constitution reduces the life span of females, while the populations effect—the life span of males. Therefore it is possible to expect, that the difference in death rate for such species will be less expressed, than for the polygamous species of the Drosophila type with the same degree of polygamy. The consequence from abovementioned statements is that monogamous species of the Abraxas type should have higher female mortality, which actually was observed (Lek, 1957).

Therefore, the new approach explains the observed pattern of differential mortality of sexes well. It is possible to explain even small discrepancies that seem not to fit in the common scheme. For example, for doves (Cole, Kirkpatrick, 1915), though total death rate is a little bit higher for females, but there are more males among victims within the first four weeks. In order for the new information from the environment to get to the male subsystem it is quite enough. Strictly speaking, it is important that male mortality should "precede" the female one. More intensive early death rate of a male is shown, for example for humans (**Figure 3.3 Chapter 3**, Gavrilov, Gavrilova, 1991). At an adult age man's death rate exceeds the female one (**Figure 3.4 Chapter 3**). It partly can be connected by that all "new" illnesses, illnesses of the "century" or "civilization" (arteriosclerosis, hypertension, cancer, AIDS, coronary diseases and schizophrenia), as a rule, are illnesses with the prevalence of males.

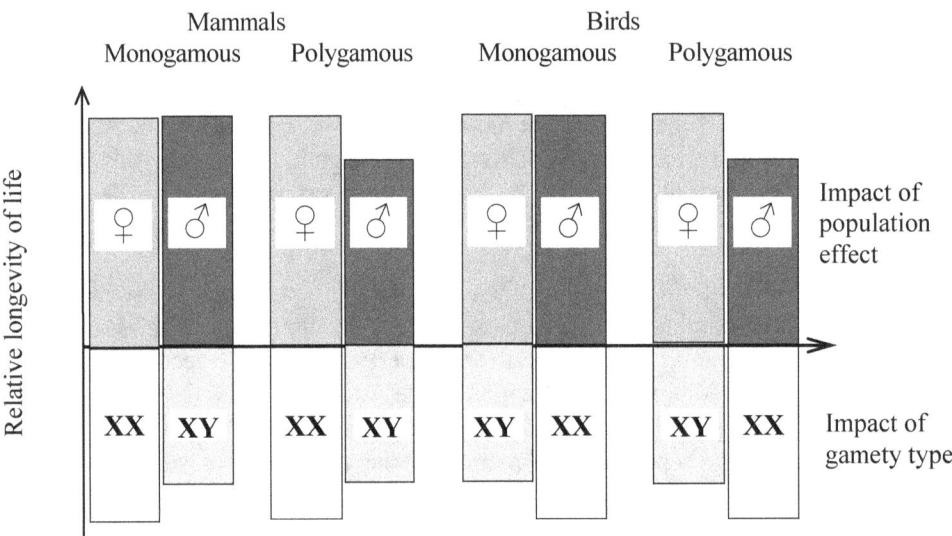

Figure 10.2 The impact of gender specialization at the population level (the top part of figure) and the contribution of gamete type (the bottom part of figure) in differential death rate of sexes (Geodakyan, 1983b).
Population effect arises only in freely crossed (polygamous) population, and is not valid for monogamous species. For polygamous species the higher death rate of males is common.

Existing theories—*chromosomal imbalance* and *metabolic*—consider differential mortality of sexes as a passive consequence of chromosomal constitution or level of metabolism. They explain the mechanism of a phenomenon and substitute the evolutionary problem by genetical (gene imbalance theory) or physiological one (metabolic theory). New theory considers increased mortality of male sex as an active feature, increasing evolutionary stability of the population. New approach clarifies evolutionary meaning of differential mortality relating it with different reaction norm of the sexes.

Chapter 11

Sexual Dimorphism and Character Evolution

I n this chapter the evolution of asexual, hermaphrodite and dioecious populations will be compared. We will also discuss how the sexual dimorphism appears and how it is regulated.

A population living in a stable environment is subject to stabilizing selection only. The average genotype stays unchanged. If stable environment starts changing, the population moves into a new state, which is optimal for new conditions. The evolution of traits will eventually stop after the new optimal value is established.

Evolution in Monoecious Forms

Stable environment. Stabilizing selection acting on asexual or hermaphrodite population in a stable environment (S) is changing the amount of progeny only (N). Average genotype ($\bar{x}$) remains unchanged (**Figure 8.1 Ch.8**).

Changing environment. If stable environment starts changing (D), the selection influences the population amount as well as the average genotype value (**Figure 8.1** B, C). For simplicity, let's consider one trait only changing from the old (0) to the new (1) value.

When environment is changing the monoecious population increases its variation and mutation rate. The evolution of the character begins—new mutations randomly appear and spread within the population (**Figure 11.1** phase E_1). When all organisms have a trait and its evolution is finished, the stable stage (s_2) starts.

After many generations if the environment is changing back and the character is no longer needed it disappears (phase E_2). In this phase the variation of the character is increasing again. When the character is gone, in the next stable post evolutionary stage (s_3) the variation goes back to normal due to stabilizing selection.

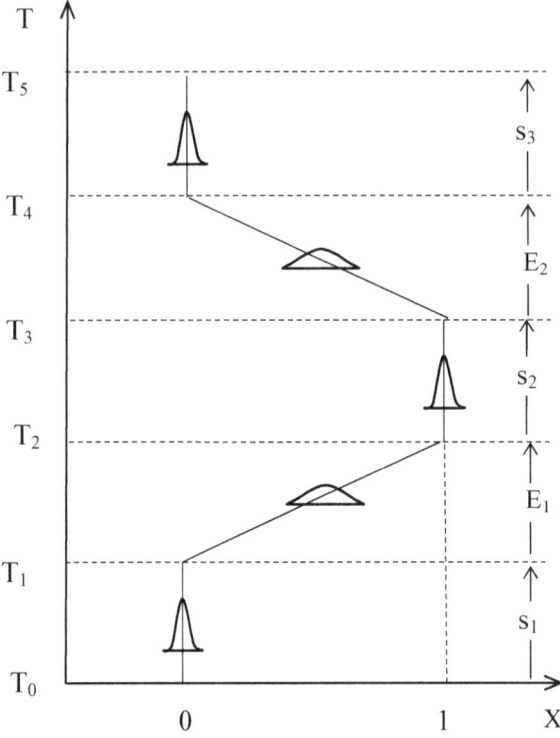

Figure 11.1 Evolution of the trait $(0 \rightarrow 1)$ in monoecious forms (Geodakyan, 2000).

Abscissa: X—mean population genotype for a given trait,
(0)—preevolutionary, (1)—post evolutionary.
Ordinate: T—time of phylogeny, T_1 and T_2—beginning and end
of trait evolution. E—stage of trait evolution;
s_1—preevolutionary, s_2—post evolutionary phase of the trait stable state.
Small distribution curves show the magnitude of genotype variance
in the population during different phases.

Dichronous Evolution

Sex differentiation resolves conflict between the **evolution rate** and the **amount of progeny**. It is based on two main differences between sexes: wide *"channel cross-section"* of the male sex with the progeny and wide *reaction norm* of the female sex.

Stable environment. Different reaction norm of males and females allows transforming their similar genotypes into different phenotypes, maximally adjusted to the environment (**Figure 11.2 f**). The phenotypical distribution of males before selection approximately follows initial genotypic one, because of a narrow reaction norm (**Figure 11.2 g**). Distribution of females' phenotypes is more adaptive to the environmental pressure and can have higher deviation from genotypic one. In a stable environment this narrows the phenotypic distribution of the females. High ontogenetic plasticity allows females to leave selection zones, move to a comfort zone and more fully preserve the past genotype spectra. Males stay in dangerous areas and undergo selection. The influence of the environment is either in the elimination and discrimination from reproduction, or in modification but with participation in reproduction. Due to the different reaction norm, the elimination and discrimination mainly involve male sex and modification—female one.

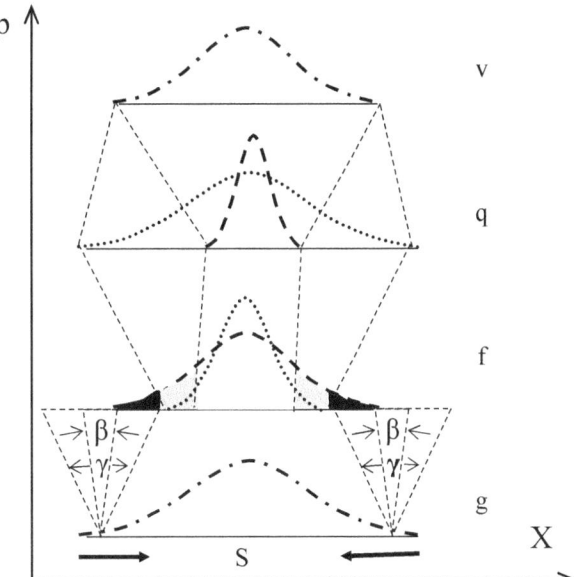

Figure 11.2 Transformation of genetic information in one generation (n) in stable (S) environment (Geodakyan, 1989,1991).

X—genotypes or phenotypes, p—their frequencies in population. Dashed lines—male sex, reaction norm—β; dotted lines—female sex, reaction norm—γ. ■—eliminated by natural selection (mainly male sex). ▓—rejected by sexual selection (male sex only). Distributions: g—genotypes received from generation n-1 (zygotes); f—phenotypes, realized from g (before and after selection); q—genotypes transmitted to generation n+1; v—genotypes received from generation n (zygotes).

In a stable environment (S) all transformations of the genetic information involve variation of the sexes, but does not change the modal values of traits (**Figure 11.2**). Therefore sexual dimorphism is absent. Only difference in variation occurs, which disappears in the next generation.

Changing environment. In a changing environment (when the ecological niche is "moving") phenotypical distribution of males before selection approximately mimics initial genotypic distribution (**Figure 11.3 g**). A high reaction norm of females lead to a shift in their phenotypical distribution and to an appearance of temporary, *phenotipical sexual dimorphism* (**Figure 11.3 f**). Females leave the zones of discomfort and selection thus preserving the spectra of current genotypes. Males stay in dangerous areas and undergo selection. After selection the fraction of males decreases and their genotypic variation narrows (**Figure 11.3 q**). In a changing environment (*D*) the modal values are also involved: the reaction norm creates temporary, phenotypical sexual dimorphism, selection—genotypic (**Figure 11.3 f,q**). Males are getting new ecological information. It is important, whether this difference remains, or gets leveled at fertilization? The existence of the *reciprocal effects* and Y-chromosome which is never transferred from father to daughter suggests that at least a portion of genetic information stays in the male subsystem and is not transferred into female one.

Because of selection the number of fathers, as a rule, is lower than the number of mothers. But wider channel "cross-section" of the information transfer allows males to fertilize many females and preserve the amount of progeny. The quantity of genetic information transmitted by both sexes is approximately the same, because each child has one father and one mother. But the information itself is different. Females transmit a wide spectra of information about the past (permanent, phylogenetic memory of a species). Males transmit narrow spectra of more selective ecological information about the present (temporary, onthogenetic memory of a species).

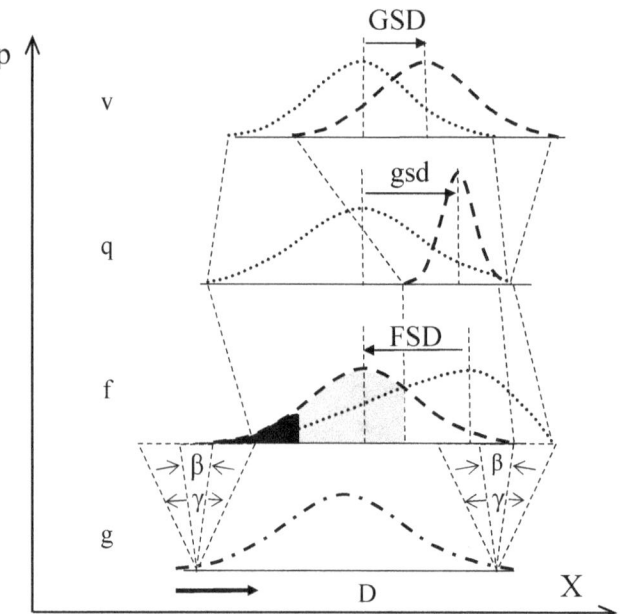

Figure 11.3 Transformation of genetic information in one generation (n) in changing (D) environment (Geodakyan, 1989,1991).

X—genotypes or phenotypes, p—their frequencies in population. Dashed lines—male sex, reaction norm—β; dotted lines—female sex, reaction norm—γ. ▪—eliminated by natural selection (mainly male sex). ▨—rejected by sexual selection (male sex only).

Distributions: g—genotypes received from generation n-1 (zygotes); f—phenotypes, realized from g (before and after selection); q—genotypes transmitted to generation n+1; v—genotypes received from generation n (zygotes). PSD—phenotypic sexual dimorphism (temporary), GSD—genotypic (temporary), gsd—genotypic (real).

So, in a changing environment, different reaction norms and channels to the progeny create *genotypic sexual dimorphism* in the first generation (**Figure 11.3 v**). *Genotypic sexual dimorphism* (GSD) then increases in the subsequent generations.

Sexual Dimorphism in Phylogeny

Appearance of a new trait. On phylogeny time scale when a stable environment starts changing, only males' trait variation changes and evolution begins (**Figure 11.4**). Evolution trajectory separates into male and female branches, divergence of the trait between sexes appears—appearance and growth of *genotypic sexual dimorphism*. This is *divergent phase*—(polarization), in which the variation and evolution rate in the males is higher. The duration of the divergent phase or *sexual dichrony* (SDC) equals the time difference between the sexes (**Figure 11.4**). This temporary "distance" is necessary to test the new traits in males. However, the divergence of sexes cannot continue indefinitely, otherwise it would result in reproductive isolation. A mechanism of GSD relaxation is turned on; and outflow of information from males to females; that is starts their evolution.

This is the second, *parallel* phase of the trait evolution, when both sexes evolve at a similar rate. A steady state regimen is established for *genotypic sexual dimorphism*, which continues to the end of the phase.

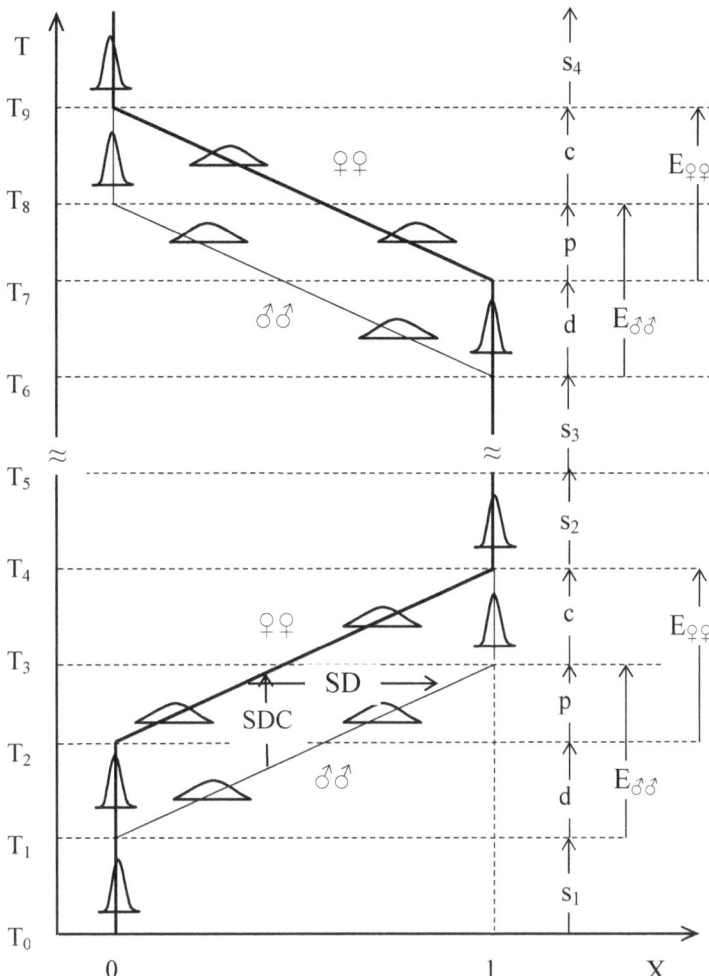

Figure 11.4. Dichronous evolution of appearance $(0 \to 1)$ and elimination $(0 \to 1)$ of traits
in males ($\male\male$) and females ($\female\female$) (Geodakyan, 2000).
T_1–T_3—beginning-end of trait evolution in $\male\male$; T_2–T_4—in $\female\female$; Phases of evolution:
d—divergent, p—parallel, c—convergent; $E_{\male\male}$ and $E_{\female\female}$—evolution of trait in $\male\male$ and $\female\female$.
SD—sexual dimorphism, SDC—sexual dichrony. For other designations, see **Figure 11.2.**

The third part of evolution is *convergent*, when only the female sex is evolving. This phase begins when the environmental differential is no longer acting on the male sex, but the *genotypic sexual dimorphism* continues to act on the female sex. As a result, *genotypic sexual dimorphism* decreases and disappears; i.e., the trait dimorphic in the course of evolution again becomes uniform and stable. This ends the trait's evolution. So, the phases of character's evolution are shifted in time: for males they begin and end earlier compare to females.

Since evolution of a character always starts with widening and ends with narrowing of the genetic variation, in the *divergent phase* the variation will be higher in males, in the *convergent phase*— will be higher in females, and will be equal in the *parallel phase*. Therefore, one can determine the direction of the character's evolution by the sexual dimorphism and the phase of evolution—by the variation of a trait.

So, sexual dimorphism on any character is related to the character's evolution: it appears when the evolution starts, exists when it is going, and disappears when it ends. Sexual dimorphism is related to the dichronous evolution of the sexes, with the formation of the "distance" between them on chronological and morphological axes.

The theory predicts the existence of "*sexual dichronism*". Together with the *sexual dimorphism* they are two dimensions of the more common phenomenon of "*dichronomorphism*". "*Sexual dichronism*" can be found in paleontology as a difference in depth levels where the trait can be found in two sexes. The trait in males can be found deeper underground (earlier appearance of the trait) compare to females.

In monomorph stages one can distinguish three phases: *preevolutionary* (s_1), *post evolutionary* (s_2), and *intermediate* (s_3). Preevolutionary stage belongs to stable condition before evolution, the other two—after the evolution ends. Post evolutionary stage immediately follows the *convergent* phase; the intermediate phase is much far away in the future (**Figure 11.4**).

Higher phylogenetic variability of males and ontogenetic (environmental) variability in females was shown in experiments of D. Zhukov and coworkers. They selected rats on opposite values of the trait (high and low speed of learning) for more than 50 generations. Two diverging lines (KHA and KLA) were formed. The difference between males on the number of avoidance responses in the shuttle chamber (92.7 versus 1.4) was significantly higher than among females (84.3 and 25.8). On the other hand, within each line the variations between males ($85 \div 98$ and $0 \div 5$) were lower compare to females ($62 \div 97$ and $9 \div 68$) (Zhukov, 2007, p. 371) (**Fig. 11.5**).

Figure. 11.5. Higher phylogenetic variability of males and ontogenetic variability in females (from Zhukov, 2007 with modifications).
Abscissa: number of avoidance reactions
Ordinate: phylogenetic time (number of generations).

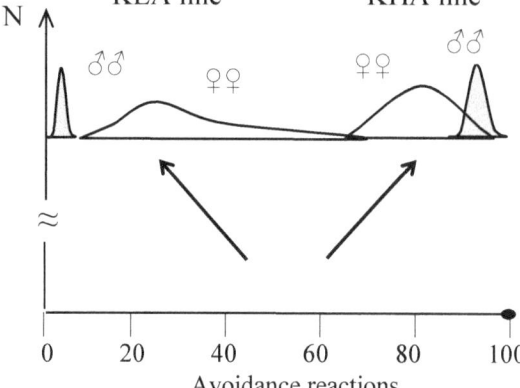

Elimination of a trait. When the character is no longer needed it disappears, first in males and later in females (**Figure 11.4** $T_6 - T_9$).

Forms of Sexual Dimorphism

Sexual dimorphism on different characters appears as a result of the *differentiation* of sexes. The evolutionary theory of sex considers sexual dimorphism not to be a uniform phenomenon but to consist of the main *reproductive sexual dimorphism* (RSD) and earlier unknown *evolutionary sexual dimorphism* (ESD) predicted by the theory and appearing at the beginning as the consequence of the *modification* changes of female *sexual dimorphism* (MSD), which then replace *selectional sexual dimorphism* (SSD) in the male sex (**Table 11.1 Appendix C**).

Reproductive sexual dimorphism is a permanent, constitutive, baseline dimorphism with respect to primary and secondary sexual traits different in the male and female sex and having direct relation to reproduction (gametes, gonads, genitals, androgen-estrogen ratio, and all traits determined by them: *reaction norm*, mammary glands, beard in man, lion's mane, heel of cock, etc.). These are fundamental species-specific traits, the genes of which, according to the theory, should be present in autosomes; i.e., they are common for the two sexes. Since there cannot be any genotypic sexual dimorphism for common genes, the reproductive sexual dimorphism is hormonal; i.e., **phenotypic**. It appears in embryogenesis and remains constant in ontogeny and during phylogeny. Its function is to set up programs for the two sexes.

Modification sexual dimorphism (MSD) is a temporary or facultative (during ontogeny) dimorphism that originates as a result of changes in females because they have a broader reaction norm and an increased phenotypic plasticity. It precedes evolution of any adaptive trait, and the wider the reaction norm for the trait, the higher the associated sexual dimorphism. The purpose of modification sexual dimorphism is to protect the female sex from selection during the time of dichrony, until new genes appear that have been tested in the male genome. An example of such dimorphism is found in adaptations of females living in the Arctic and having thick layer of subcutaneous fat, short legs, high mineralization of skeleton not only in comparison with "their own" men, but also with females of the control group (Alekseeva, 1975).

Any population that exists in the stable environment for a long time has only reproductive sexual dimorphism. Evolutionary sexual dimorphism equals zero, and modification sexual dimorphism is only of a variational nature, since the phenotypic variance of males in a stabilizing environment is greater than that of the females.

When the stable environment starts changing, the selection influence predominantly the males. Due to limited capacity of the information channel between male and female subsystems and also due to its' selectiveness, preventing the mixing of all genetic information, there exists some female sex inertness (shift), stability or lag behind. This reflects in later appearance or development of the characters in phylogeny and consequently in ontogeny. *Genotypic sexual dimorphism* (GSD) appears and increases from generation to generation as a result of the different inertia of the sexes.

Evolutionary sexual dimorphism (ESD) appears in the evolution of any trait representing a "distance" between sexes in conjunction with any selection (natural, sexual, artificial) as a result of the anticipatory change of the male genome. Therefore, it is *genotypic* in nature. The purpose of ESD is to create dichrony for effective evolution. The ESD vector (from the female form of the trait to the male one) corresponds to the direction of trait evolution. The female form of the trait indicates the past state, while the male form indicates the future state.

Sexual dimorphism is closely related to the evolution of the character: it must be absent or minimal for stable, non-evolving characters and maximal for characters which are in "evolutionary march", it must be more pronounced for phylogenetically young (evolving) characters.

Sexual dimorphism must be closely related to the reproductive structure of population: in strict *monogamous* species it should be minimal, since they use sex differentiation at the organism level, rather than at the population one. In *panmictic* and *polygamous* species more completely using advantages of differentiation it must rise with the level of polygamy.

If **new** information (I_n) has already entered the male sex but did not yet enter the female, or **old** information (I_o) has already been lost by the male sex but is still retained in the female, their sum total equals *evolutionary sexual dimorphism*. Therefore, information contained in the male genome is $I_m = I_c + I_n$, while in the female genome, it is $I_f = I_c + I_o$, where I_c is **common** information present in both cases. It should be pointed out that when two populations interbreed (races or ethnic groups), the common information undergoes mixing after the first cross, while the new and the old information remain segregated throughout the period of sexual dichrony.

This view easily explains differences of interspecies, interracial or interethnic reciprocal hybrids associated with the direction of crosses, since in reciprocal hybrids only I_c is identical, while I_n and I_o are received by them from different forms (e.g., the mule and the hinny). If offspring received identical genetic information from the father and mother, then no reciprocal effects would have been present. As has already been mentioned, reproductive sexual dimorphism defines the hormonal status and sexual dimorphism in terms of the reaction norm. But specifically, why is the reaction norm narrow in the males? And is it the males that have operative specialization? Can it be the other way around?

Sexual Dimorphism and Mating Systems

Then sexual dimorphism must be closely related to the reproductive structure of a population: in strict *monogamous* species it should be minimal, since they use sex differentiation at the organism level, rather than at the population one. In *panmictic* and *polygamous* species more completely using advantages of differentiation it must rise with the level of polygamy.

Inversion of sexual dimorphism in polyandry

Polyandry; i.e., the phenomenon when the female is mating with several males; i.e., has a higher reproductive index (the number of realized gametes) as compared with male, is found in invertebrates, fish, birds, and mammals. This is often associated with the **inversion** of sexual dimorphism (females are larger than males and have bright color, males make the nest, hatch the eggs, and take care about the litter, there is no fight for the female, etc.). In the case of *polygyny*, the picture is reversed. This implies that the direction of sexual dimorphism and also of asynchrony and the ratio of rates of sex evolution depend upon the direction of polygamy or the ratio of the reproductive indices of sexes. In *monogamous* species, the indices of males and females are identical; since the number of fathers and mothers is equal.

On the other hand, in order for sexual dimorphism with respect to the reaction norm to precede the evolution of any trait, the width of the reaction norm should be determined by sex hormones (according to theory, sex hormones are the substances, which regulate the "distance" between the system and the environment: androgens makes this distance shorter, while estrogens increase it). Moreover, it is known that the Y chromosome triggers the synthesis of testosterone, and its concentration determines sexual dimorphism. One can construct causal relationship between the direction of *evolutionary sexual dimorphism* i.e. ratio of evolution rate for males and females (E_M/E_F) with their reproductive indexes (P_M/P_F):

$$P_M/P_F \sim N_{mothers}/N_{fathers} \sim T_M/T_F \sim RN_F/RN_M \sim S_M/S_F \sim E_M/E_F \sim ESD \qquad [4]$$

where N is the number of individuals in the population; T_M/T_F —testosterone levels; RN_F, RN_M —reaction norms; S_M, S_F —selection coefficients; $\sim$ —sign of proportionality.

Consequently, the more **polygamous** sex always receives the role of evolutionary **"vanguard"**, while the **monogamous** sex is always behind. The higher the polygamy index, the higher may be *the evolutionary sexual dimorphism* (in strictly monogamous forms, *ESD = 0*, and therefore sexual dimorphism is minimal).

All three forms of mating relations may be present in panmictic population: monogamy-full sibs are born (FS); polygyny-paternal half-sibs (PhS), and polyandry-maternal half-sibs (MhS). Their concentrations in the population are defined by an equilibrium *[PhS] ↔ [FS] ↔ [MhS]* and are controlled by environmental conditions: Under optimal conditions, the equilibrium is shifted to the right, thereby increasing the evolutionary stability of the population. Under the extreme conditions, it is shifted to the left, thereby increasing the evolutionary plasticity of the population.

Polygyny occurs widely in nature, and polyandry is more exotic. This can be explained by a potentially greater reproductive opportunity of the male sex (due to the larger number of gametes). In fact, **poly**andry as such does not exist. Strictly speaking it should be called **oligo**andry, since the opportunities of the female sex in this respect are limited.

Chapter 12

Sexual Dimorphism—Rules

Most theories consider sexual dimorphism closely related to sex. For example Darwin considered it a consequence of sexual selection. Therefore these theories were unable to explain sexual dimorphism on characters, not related to sexual selection, sexual dimorphism in plants, or sexual dimorphism on congenital malformations.

The evolutionary theory of sex links sexual dimorphism with the evolution of the character. Therefore it is a consequence of any selection as a result of asynchronous evolution of the sexes. In this chapter we will describe different forms of sexual dimorphism and rules that can be formulated based of this new approach.

Phylogenetic Rule of Sexual Dimorphism

Taking into account known irreversibility of evolution processes (*Dollo's law*) males can be considered as evolutionary "vanguard" of population, and sexual dimorphism for a character—as a vector indicating the evolutionary trends of this character. It is directed from the female's norm in the population for the given character to the male's norm ("phylogenetic rule of sexual dimorphism"). Therefore the characters which often appear and are more pronounced in females ought to be of the "atavistic" nature, and those more often manifested in males—of the "futuristic" one (search). In other words, males conduct evolutionary search and trial and therefore include both innovations (progressive things) and mistakes (imperfections), while females fulfill the selection and fixation of already tested solutions (perfection).

If there exist a *population sexual dimorphism* on the character that occurs in both sexes, theory can link it with the character evolution. *"Phylogenetic rule of sexual dimorphism"* can be formulated: if there exists genotypic populational sexual dimorphism for some character, so one can distinguish two forms of the character—male and female (either on frequency of occurrence (penetrance) or degree of expressivity), then the evolution of the character goes from female to male form.

PHYLOGENETIC RULE OF SEXUAL DIMORPHISM

IF THERE IS A GENOTIPIC POPULATIONAL SEXUAL DIMORPHISM
ON ANY CHARACTER, THEN THE EVOLUTION OF THE CHARACTER
GOES FROM FEMALE TO MALE FORM

Genotypic populational sexual dimorphism can serve as a "compass", showing the direction of the trait evolution. Variation of the trait points to the phase of evolution. If males have higher variation—the phase is divergent, equal variation means parallel phase, and higher female trait variation—the convergent phase (*"Phylogenetic rule of variation of the sexes"*).

PHYLOGENETIC RULE OF VARIATION OF THE SEXES

IF THE VARIATION OF A TRAIT IS LARGER IN MALES —
THE PHASE OF ITS EVOLUTION IS DIVERGENT; IF VARIATIONS
ARE EQUAL —THE PHASE IS PARALLEL; IF VARIATION
IS HIGHER IN FEMALES — THE PHASE IS CONVERGENT

The phylogenetic rule of sexual dimorphism was successfully checked up on a large group (173 species) of lower *Crustacean* (Geodakian, Smirnov, 1968) (see **Chapter 14**).

Sexual Size Dimorphism

In the evolution of sizes in two branches of the animal world a different trend was noted. In vertebrates more often enlargement of sizes took place, while in insects they diminished. *The phylogenetic rule of sexual dimorphism* allows prediction of the main direction of sexual dimorphism in such groups: bigger male should be found in large vertebrate species, and smaller males—in small groups of insects and *Arachnoidea*.

Wingspans of one of the largest known flying insect Meganeura (300 million years ago) was more than 75 cm. Modern Dragonflies have wingspans 1–11 cm.

The same tendencies should be observed also inside smaller taxa, say in mammals females of small forms are larger than males, while large forms has larger males. For example African savanna elephant males weight up to 6.5 ton, but females up to 3.5 ton only. Small forms—some bats, flying squirrels, spotted hyenas, dwarf mongooses, rabbits and others frequently has larger females. The same trend can be followed within one order. Data on primates (**Figure 12.1**) fully proves predicted by the theory relationships between sexual dimorphism and animal size and body mass, and polygamy.

Figure 12.1

Relationship between sexual dimorphism and body weight for some primates (Clutton-Brock, Harvey, 1977).
○—monogamous species,
●—polygamous species.

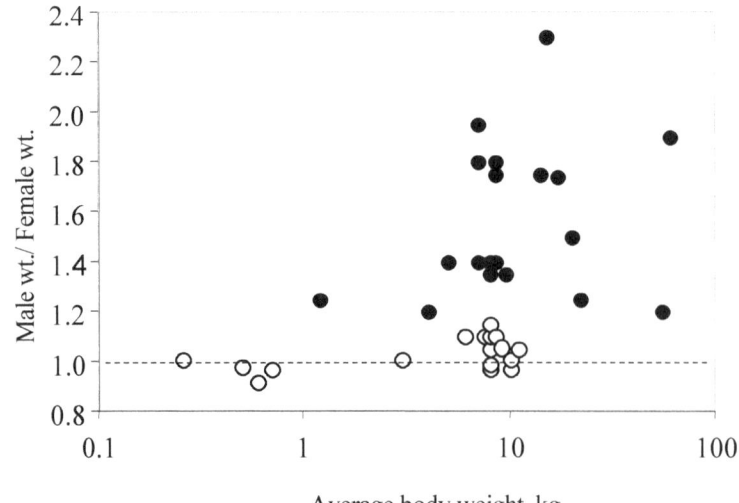

Dogs undergo intensive selection in different directions to produce new breeds. Dog's ancestors (wolf, fox, jackal) seem to have average size as compared to large (mastiff, St. Bernard weighing up to 70 kg) and small (toy-terrier up to 400g, Chihuahua from 1 to 3 kg.) dogs. The theory predicts larger males in large breeds and larger females in small ones.

Semi-aquatic species derived, probably, from "land-dwelling" should have increased in size during their adaptation. In 4 species studied males are bigger than females (Panteleyev, 2003; Corbet, 1977), and there are paleontology data that the muskrat (*Ondatra zibethicus*) and water vole (*Arvicola amphibius*) were derived from smaller ancestors (Aghajanian, 1993, 2001). Contrary, for tree-climbing species—squirrel (*Sciurus vulgaris*) and Siberian chipmunk (*Tamias sibiricus*) males are smaller than females (Panteleyev, 2003; Degn, 1973; Vericad-Corominas, 1970).

Divergent and convergent evolution

Under any divergence or disruptive selection the divergence of males should be more advanced compare to females. So, appearance of bimodal male distribution together with the monomodal female distribution (Altuhov, Varnavskaja, 1983) can be a sign of a started disruptive selection. The same reason why the ecological rules of Bregman and Allen should be more contrast in males. It can be suggested that the length of extremities, ears and tails should be relatively shorter for males from arctic populations, compare to females. Males from tropical populations should have longer extremities.

Interesting adaptations to hot climate of deserts was developed in *Pterocletida*—water transportation for fledglings in wetted feathers. Changed abdomen feathers make it possible to take 3–4 times more water than other birds do. It has been shown that male feathering keeps 1.5 times more water than the female one.

The processes that contribute to sexual dimorphism may also lead to speciation and morphological differences among related species, as argued originally by Darwin. Where sexes are separate and dimorphism is well-developed, males of related animal species (both vertebrate and invertebrate) are often strikingly different from each other, while females may be virtually indistinguishable. A similar pattern may exist in plants: it is frequently the males (of dioecious taxa) or the male portions of the flower (in co-sexual flowers) that apparently have diversified (Willson, 1991).

The *phylogenetic rule of sexual dimorphism* can be easily checked up on species whose evolution is well known (elephant, horse, and camel). It is known, e.g., that in elephants took place enlargement of size, development of the trunk, original specialization of the tooth system. Consequently sexual dimorphism for all these characters can be predicted. Many of these predictions were confirmed.

Horse and donkey diverge ~4-8 thousand years ago from common ancestor tarpan (*Equus ferus ferus*). During evolution horse become longer and higher. The tail is formed from long hairs. Donkey has long ears, big head, and narrow hooves. The tail has brush of hairs. Both their hybrids—mule (donkey father) and hinny (horse father) look similar to their fathers.

This rule is also valid for plants. e.g., in poplar female specimens have more elongated leaves, the male ones more rounded ones. Leaves of gingko female tree have even edges and are smaller, of male ones larger and cut. As known, poplar phylogenetic ancestors had narrow (like willows) leaves, while gingko ancestors the uncut ones.

During evolution process the gametophyte reduction took place (**Figure 7.4 Chapter 7**). In complete accordance with the theory, the reduction of male's gametophyte went further than female's: female gametophyte has 8 cells, while male one—only 3 (Willie, Detje, 1975).

The *phylogenetic rule of sexual dimorphism* can be applied to domestic animals. Their artificial evolution (selection) was directed by humans and followed their goals. So, all economically valuable characters should be more advanced in males. This prediction of the theory is fully proved. In all cases if the trait is present in both sexes phenotype, it is more expressed in males. Males (especially after castration) give much more meat with higher quality, with more effective utilization of feed, dynamics of weight increase compare to females. In wool production sheep and wether are more productive. They sometimes give 1.5–2 times more wool than eves. Male horses are more valuable for use as sports as well as working animals. Same can be said about camels and deers. In fur production males have higher quality furs. In silk production, males have 20–25% higher productivity compare to females. In plants, male hemp plants produce 1.5 times more fiber than female plants.

Another example is provided by chickens subject to artificial selection for either high- or low-copulation frequency. After twenty generations there was a striking change in mating frequency among cockerels, but no change in the propensity for hens to copulate (Halliday, Arnold, 1987; Cheng, Siegel, 1990).

So, *phylogenetic rule of sexual dimorphism* is confirmed on the group of economically valuable characters.

Phylogenetic rule of sexual dimorphism predicts the relationship between sexual dimorphism and the phenomenon of heterosis. Considering heterosis as merging of new achievements acquired by two species during divergent evolution, we can assume that, the contribution of the father in heterosis should exceed contribution of the mother. Also, the effect of heterosis should be more pronounced in sons than daughters. This theoretical prediction was confirmed (Levine, Cormody, 1967).

On convergent evolution, males should have the same value of a trait, but females are yet different. This pattern can be observed in one-dimensional niche, where one environmental factor dominates: cold in polar region or hot in the desert.

Parasitic forms. It's known that the evolution of parasitic forms was going in two different directions. From one hand, such species developed the whole complex of traits that provide host search and mutual coexistence. From the other hand, many characters, organs, and even systems of organs underwent reduction (Zimmer, 2001). The *phylogenetic rule* predicts that males should lead in both cases; therefore on disappearing characters males should be more primitive. For example, a female of the marine Green Spoon worm (*Bonellia viridis*) has retained all the functions, while the male, have lost them, and become a small (1-3 mm) parasite living on her body or in the genital tract.

Ontogenetic Rule of Sexual Dimorphism

Let's take a look at three phenomena, *phylogeny*, *ontogeny*, and *sexual dimorphism*. Long time ago scientists noted the existence of the relationship between ontogeny and phylogeny. According to *the Haeckel-Müller biogenetic law* (theory of recapitulation) in developing from embryo to adult, each organism goes through stages resembling successive **adult** stages in the evolution of their remote ancestors. According to K. Baer's laws of embryology (1828) general characteristics of the group to which an embryo belongs develop before special characteristics. Embryos will resemble each other before attaining characteristics differentiating them as part of a specific family, genus or species, but embryos are not the same as the final forms of lower organisms. Despite the notion that E. Haeckel's point of view is not acknowledged by modern biology, some relationship between ontogeny and phylogeny can not be denied. Evolutionary changes must be expressed in ontogeny (Gould, 1977).

Phylogenetic rule of sexual dimorphism establishes an association between the phenomena of phylogeny and sexual dimorphism. Therefore we have three phenomena—phylogeny, ontogeny, and sexual dimorphism, connected by two principles (the *Haeckel-Müller law* and the phylogenetic rule). Consequently, the existence of a third principle can be predicted and formulated which establishes a direct association between phenomena of sexual dimorphism and ontogeny (**Figure 12.2**).

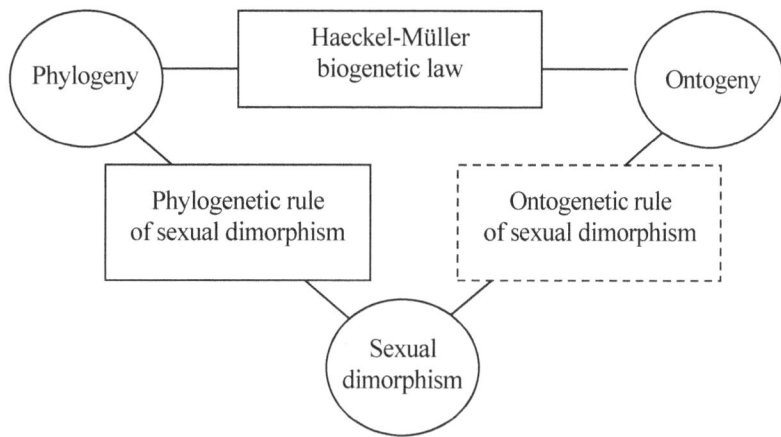

Figure 12.2. Principles associating the phenomena of phylogeny, ontogeny, and sexual dimorphism (Geodakyan, 1983).

The new principle can be called the *"Ontogenetic rule of sexual dimorphism."* If there is populational sexual dimorphism according to a certain trait, then during ontogeny (with age), this trait changes, as a rule, from the female form to the male, i.e., the female form of a trait is more characteristic of the initial, juvenile stage (the stages of childhood, growth, and formation), while the male form is more characteristic of the definitive stage (mature, adult). In other words, female forms of traits should, as a rule weaken with age, while male are intensified.

ONTOGENETIC RULE OF SEXUAL DIMORPHISM

IF THERE IS POPULATIONAL SEXUAL DIMORPHISM ACCORDING TO A CERTAIN TRAIT, THEN DURING ONTOGENY (WITH AGE), THIS TRAIT CHANGES, AS A RULE, FROM THE FEMALE FORM TO THE MALE.

Dynamics of Sexual Dimorphism in Ontogeny

Each of phylogenetic phases discussed in **Ch. 11** (**Figure 11.5**), can be projected into ontogeny and get their 6 related types of sexual dimorphism in individual development (**Figure 12.3**) (Geodakian, 1994). Dichronism will manifest itself in ontogeny as a delay in character development in females. This means that female form of a dimorphic character will dominate at the beginning of ontogeny and male one—at the end.

Such interpretation of age dynamics of sexual dimorphism is confirmed, for example, by well-known sequence of appearance of horns in deer and antelope males and females during ontogeny and phylogeny. One can see clear relationship between the extent of horn development in a species and the age of appearance of horns in males and females. The more pronounced are horns in a species, the earlier they appear in life first in males and then in females. This age delay of character development in females is nothing but ontogenetic projection of phylogenetic sexual dichronism.

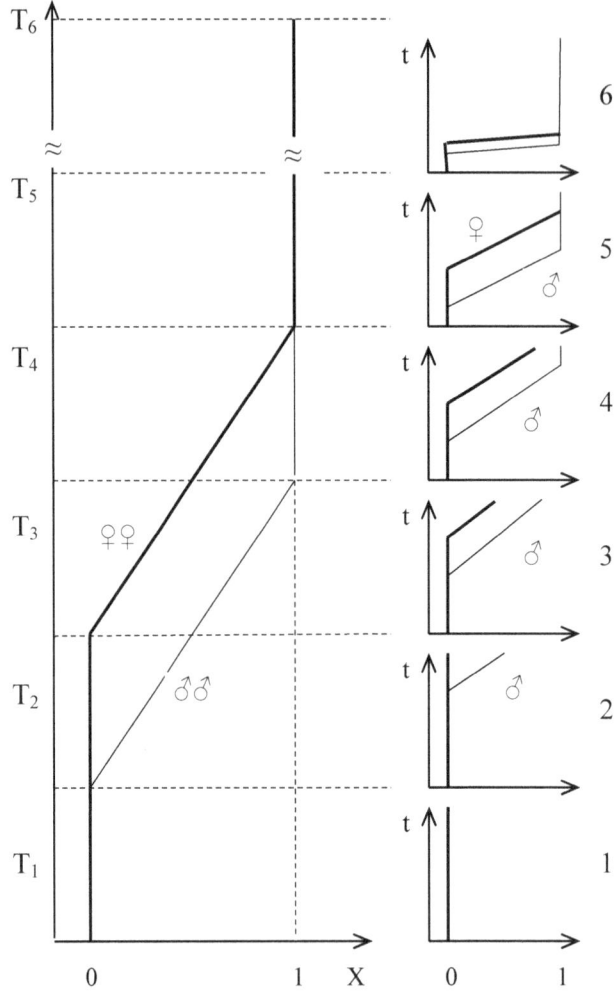

Figure 12.3 Phases of a character evolution (0→1) in phylogeny and 6 related
types of sexual dimorphism in ontogeny (Geodakyan, 1991).
Abscissa: X — mean population genotype for a given trait, T - time of phylogeny.
t - time of ontogeny.
1 — Ontogeny of preevolutionary phase. Character is constant and the same
for both sexes. Sexual dimorphism is absent.
2 — Ontogeny of divergent phase. Character is constant in females and starts changing
in males at the end of life. Sexual dimorphism appears in old age and only grows.
3 — Ontogeny of parallel phase. Character is changing in males and later in females.
Sexual dimorphism appears in earlier age (compare to 2). It grows first, then becomes constant.
4 — Ontogeny of convergent phase. Sexual dimorphism appears in earlier age (compare to 3).
Contrary to 3, it grows first, then becomes constant, then decreases, but not disappears.
5 — Ontogeny of post evolutionary phase. The whole picture is shifted to the beginning
of the ontogeny. At the end of ontogeny sexual dimorphism disappears.
6 — Ontogeny of intermediate phase. The dichronism "loop" is compressed into
embryogenesis. In postnatal ontogeny sexual dimorphism is absent.

C. Darwin (1953) paid attention to more close relationship between female sex and earlier stages of ontogeny. He wrote: "In all animal kingdoms, if males and females are different on appearance, the changes involve males with very few exemptions, while female usually stays similar to young animals of the species and to other members of the group". Anthropologists also noted the proximity between female and children types (more fragile bones, less expressed eyebrow arches, less body hair etc.) (Ginzburg, 1963; Roginskij, Levin, 1955; Aksyanova, 2011).

In order to test the new rule, it is necessary to compare the direction of sexual dimorphism according to different traits with the dynamics of these traits during ontogeny. Such data is presented in **Table 12.1 (Appendix C)** for several anthropological traits (relative length of legs, forearm, fingers, head index, tooth arch, epicanthus, aquiline nose, erythrocyte concentration in the blood, pulse frequency, brain asymmetry, norm and time of reaction, olfaction, and perception of bitter taste of phenylthiourea), for which information could be found concerning both sexual dimorphism and ontogenetic dynamics. It is seen from the table that the proposed theory does actually exist.

Teratological Rule of Sexual Dimorphism

On species-specific (and higher ranks of a generality) traits (multicellularity, warm-bloodedness, the number of organs, the plan and a basic structure of a body etc.) sexual dimorphism in norm is absent. It is observed only in the field of pathology and expressed in different frequency of occurrence of congenital developmental anomalies at males and females.

The idea of classification of congenital anomalies on "atavistic" (returns or stops of development) and "futuristic" (search of new ways) allows recognizing in some cases of sexual dimorphism the abovementioned common tendencies predicted by the theory. For example, among 2000 newborns with one kidney there were twice as much boys, while among 4000 newborns with three kidneys there were 2.5-fold more girls. Is it accidental? Or is it a reflection of a well-known evolutionary trend of oligomerization of multiple organs? We shall remind, that lancelets and sea worms (old predecessors of mammals) have a pair of the specialized secretory organs—metanephridia in each segment of a body. So, it is possible, in the certain sense to consider the occurrence of the three kidneys as an "atavistic" tendency, and one kidney—as "futuristic" one. The same picture is observed among newborns with above norm number of ribs, vertebra, teeth and other organs that underwent reduction during evolution (oligomerization)—among them there are more girls. Contrary, there are more boys among newborns that have less than normal amount of such organs.

Another pathology—congenital dislocation of a hip occurs at girls in 4-5 times more often, than at boys. We shall note that children with this defect can crawl and climb on trees better, than normal children. Anencephaly can be two times more often is found at girls (WHO reports, 1966). Darwin mentioned above permitted standard muscles, which 1.5 times more often are found out in corpses of men, than women. He also listed the data on frequency of occurrence newborns with 6-th finger. Here also the number of boys exceeds the number of girls 2 times (Darwin, 1953).

With reference to anomalies of development it is possible to formulate a *"Teratological rule of sexual dimorphism"*: **if for any attribute at the given stage of evolution sexual dimorphism is absent, but it existed previously, at earlier stages of evolution, it can be found out as a "relict" in an "asymmetric" sex ratio in pathology. Thus deviations from norm in an "atavistic" direction will appear more often at females, and in "futuristic"—at males.**

TERATOLOGICAL RULE OF SEXUAL DIMORPHISM

FEMALES SHOULD HAVE THE ANOMALIES OF DEVELOPMENT
THAT ARE OF "ATAVISTIC" NATURE MORE OFTEN,
WHILE MALES SHOULD PREDOMINANTLY HAVE THE
ANOMALIES OF "FUTURISTIC" NATURE (SEARCH).

The rule can be verified by comparing sex ratio of persons with congenital anomalies with various stages of phylo- and ontogenetic development. Many congenital heart defects have a pronounced sexual dimorphism. Application of the *teratological rule of sexual dimorphism* enabled to explain the entire spectrum of congenital heart malformations observed in children of different sexes (see **Chapter 15**).

"The Rule of Correspondence"

Past and Future Forms of Characters in Phylogeny and Ontogeny

Let's go back to the three principles (the Haeckel-Müller law, phylogenetic and ontogenetic principles of sexual dimorphism), which associate three phenomena (phylogeny, ontogeny and sexual dimorphism) with one another in pairs (**Figure 12.2**). It is possible to attempt to combine all three specific principles into one general principle. For this purpose, we shall introduce the concept of two forms of a trait, associated with a vector of time, into each of the three phenomena. We shall select and differentiate the following: During phylogeny of a trait, "atavistic" and "futuristic" forms; during ontogeny of a trait, its "juvenile" (young) and "definitive" (mature) forms; and for populational sexual dimorphism, its "female" and "male" forms. Then, the generalized principle associating the phenomena of phylogeny, ontogeny, and sexual dimorphism can be formulated as "principle of correspondence" between the atavistic, juvenile, and female forms of a trait on the one hand, and between the futuristic, definitive, and male forms on the other.

Mutations, Dominance and Sexual Dimorphism

The "principle of correspondence" can also be extended to other phenomena systematically associated with phylogeny and ontogeny (evolution) and for which past and future forms can be separated. For example, the phenomena of *mutation* (the phylogenetic process of gene origin), *dominance* (the ontogenetic process of gene manifestation), *heterosis* and *reciprocal effects* all permit two forms of a trait to be distinguished: a past and a future. The association between phenomena of phylogeny, ontogeny, mutation, dominance, and sexual dimorphism is also indicated by known facts such as the higher degree of spontaneous mutations in the male sex; greater additive inheritance of parental traits by female offspring, which means greater dominant inheritance by male offspring (Borodin et al., 1976; Schuler, 1976); known autosome genes manifested in the female genome as recessive traits, but in the male as dominant and intensified during ontogeny, such as the gene for horned-hornless in sheep or the gene causing baldness in humans; dominance of the paternal form over the maternal according to evolving (new) traits (the "paternal effect"), et al.

In 1930 R. Fisher showed that recessive mutations outnumber the dominant ones, and they usually lower the fitness of the organism. He hypothesized that if there are two alleles of the gene, the more favorable for the specie's changes, will evolve in the direction of dominance (Fisher, 1931). J. Haldane (1935) supported the hypothesis about the evolution of dominance.

The concept that traits acquired later during phylogeny are dominant was stated independently by D. D. Romashov and A. S. Serebrovskii (Malinowski, 1970). Such an association and associations between phenomena of phylo- ontogeny, mutation, and dominance were discovered during the extensive experiments of V. S. Andreev et al., on plants. Interpreting recessive mutations (in a homozygote) as interruptions in the development of a trait at a particular stage of its formation, thus revealing the results of the operation of previous, earlier genes, and the significantly rarer dominant mutations as the addition of a new link to the existing chain of development, they showed that during ontogeny traits always dominate, the development of which has reached later stages of phylogeny, i.e., younger traits (Arkatov et al., 1976; Ratkin et al., 1977,1980). Since, in such an interpretation, a recessive mutation is manifested as a dominant trait if its carrier interbreeds with an earlier form, such mutations are more logically called not recessive, but "retrospective". Dominant mutations are, correspondingly, "prospective". In fact, mutations can also be in haploid forms, but the phenomenon of dominance-recession is associated with a diploid state.

This can be understood from more general considerations. In the process of progressive evolution the volume of genetic information has grown steadily. If at fertilization mother and father contribute the DNA molecules of different length, then the new molecule has two possibilities: either it will have a length of the shorter chain (i.e. unpaired links are not being completed), or—the longer one (unpaired links are completed). Obviously, in the first case, useful evolutionary traits are not likely to survive. The second option, conversely, promotes progress, because the evolutionary more advanced form dominates.

This conclusion is valid only for the *progressive* evolution, when a new feature (longer molecule) appears. When the trait gets lost (*regressive* evolution), the old form of the molecule will be longer. This means that dominance is determined, above all, by the amount of information, and the form that have more information will always dominate, because missing information can not dominate over the existing one.

An association between *heterosis* and *dominance* has been indicated repeatedly: the overdominance hypothesis of G. Shull and E. East, the hypothesis of favorable dominant factors (C. Davenport). The association between *heterosis* and *phylogeny* was given attention by D. D. Romashov and A. S. Serebrovskii (Malinowski, 1970), we have shown the association between *heterosis* and *sexual dimorphism* (Geodakyan, 1981), etc. Consequently, different investigators, even long ago, had focused attention on the associations between phenomena of phylogeny, ontogeny, sex, mutation, dominance, heterosis, et al. In order to reveal a general principle, considering the phenomena enumerated above as a unified system, we shall again distinguish the two forms of traits associated with the vector of time in each phenomenon: the **past** and **future** forms (**Table 12.2**). Then, the expanded and generalized principle of correspondence can be formulated in the following way. If there is a system of interrelated phenomena in which forms oriented in time (past and present) can be distinguished, then a specific correspondence (a closer association) exists between all past forms on one hand, and between future forms on the other.

THE RULE OF CORRESPONDENCE

IF THERE EXISTS A SYSTEM OF INTERRELATED PHENOMENA
IN WHICH FORMS ORIENTED IN TIME (PAST AND PRESENT) CAN BE
DISTINGUISHED, THEN A SPECIFIC CORRESPONDENCE (A CLOSER
ASSOCIATION) EXISTS BETWEEN ALL PAST FORMS ON ONE HAND,
AND BETWEEN FUTURE FORMS ON THE OTHER.

Table 12.2 Past and Future Forms of Traits in Different Phenomena (Geodakyan, 1983, 1984).

Phenomenon	Form of trait	
	Past	Future
Phylogeny	Atavistic	Futuristic
Ontogeny	Juvenile	Definitive
Sexual dimorphism	Female	Male
Dominance	Recessive	Dominant
Mutation	Retrospective	Prospective
Heterosis	Parental	Hybrid
Reciprocal differences	Maternal effect	Paternal effect

It is interesting to note that any of the phenomena listed in **Table 12.2** can serve as a "compass" indicating the direction of the evolution of a given trait (Geodakyan, 1965a, 1981; Malinowski, 1970; Ratkin et al., 1980).

Chapter 13

Parent of Origin Effects

In some cases phenotype of a progeny depends on the genetic contribution of the father or mother (so-called "parent-of-origin effects"). If there exist two varieties of an organism, A and B, the classical Mendel's genetics presumes that the hybrid AB phenotype will be equivalent regardless of whether a female of type A mates with a male of type B or vice versa. On the contrary, when ♀A x ♂B ≠ ♀B x ♂A, we deal with parental effects (Gray, 1972). For example, crossing of a female tiger (*Panthera tigris*) with the male lion (*Panthera leo*) gives a hybrid called liger. Liger is bigger than each of the parents and often is up to 3 m long and weight more than 350 kg (Morison et al., 2001). The posterity received from return crossing (a female lion with the male tiger) usually smaller than each of the parents.

Genomic Imprinting

Placenta—the "Male" Organ?

The main idea of dichronism that from mothers we receive old genetic information (about the past), and from fathers—the "latest evolution news" (about the present) allows to explain the mysterious discovery of two groups of English scientists (Surani et al., 1984; McGrath et al., 1984). They combined genes from two eggs or two spermatozoids in one zygote in order to receive progeny from one parent only (♀+♀, ♂+♂). Only control embryos, which had genes of both mother and father (♀+♂), were developed normally. Unisex embryos died in the end. The ♀+♀ embryo at first developed normally, but did not form membranes and placenta, and perished. In a case of ♂+♂ embryo, on the contrary, the placenta was more than normal, but instead of an embryo, an amorphous lump of cells was developed. So, development of extra-embryonic membranes and placenta is determined by male's genes (instead of female's), and development of the "common" embryo—female genes (instead of male's and female's)! After all an embryo has equal amount of parental genes, and grows in maternal "environment". The phenomenon have received the name "*genomic imprinting*", and its biological meaning is not clear till now. Scientists explain the mechanism of "*genomic imprinting*", by a different expression of paternal and maternal autosomal genes.

The evolutionary theory of sex explains this phenomenon to that an embryo is an evolutionary old system, and membranes and placenta—new ones: they have appeared in higher mammals. The embryo receives old information from mother, and new (about placental development) —from the father. Thus it's possible that genes of a placenta either not yet arrived in female genome, or have already arrived, but are not yet verified and therefore cannot work.

Hydatidiform Mole

A relatively common human disease termed hydatidiform mole is characterized by excessive trofoblast (placental) tissue growth and absence of a fetus. Hydatidiform moles are usually characterized by either lack of a maternal genome or by a 2 : 1 paternal/maternal genome ratio due to fertilization by multiple sperm.

The similar phenomenon has been discovered at floral plants. The endosperm is the second fertilization product in flowering plant reproduction and develops after the fusion of the two polar nuclei (sister nuclei to the egg) with a sperm nucleus. It's generally considered to be involved in nutrition of the embryo similar to placenta in mammals. Normal development of endosperm usually requires a parental genomic ratio of 2 ♀ : 1 ♂ and deviations from this ratio often are associated with seed abortion. Maternal excess (ratio of more than 2 ♀ : 1 ♂) generally is correlated with inhibited proliferation of the endosperm, and paternal excess generally is correlated with overgrowth (Haig, Westoby, 1991).

Theories of Genomic Imprinting

Prevention of parthenogenesis. One of the consequences of genomic imprinting is prevention of parthenogenesis. Then it is possible to expect that it should not occur in groups reproducing by parthenogenesis.

The parental conflict hypothesis. The kinship–parental antagonism hypothesis (Haig, Westoby, 1989) explains genomic imprinting as a result of the conflict of interests between mother and the father for quantity of the resources transferred from mother to fetus. For plants it is expressed that endosperm (responsible for the accumulation of resources) is a possible place of genomic imprinting manifestation. Similarly, at animals the placenta also acts as the main "battleground" and paternal genes should promote embryonic growth by increasing the extraction of nutrients during pregnancy, whereas maternal genes should inhibit growth.

"The Paternal Effect"

"—"Really, you don't recognize me? Most people say
I look just like my father."
—"I look like my father, too" said the chairman impatiently.
—"What is it you want, Comrade?"
"The Golden Calf" (I. Il'f, E. Petrov, 1930)

Because phenotype of one sex lucks the character, one can judge genotypic sexual dimorphism by reciprocal effects. It follows that while with "old" stable characters the genetic contribution of father to the offspring is somewhat smaller than that of mother due to maternal effect stipulated by cytoplasm inheritance and uterine development, with "new", evolving characters paternal contribution must exceed the maternal one. This may result in the compensation of maternal effect for such characters and even in the initiation of opposite "paternal effect". In other words, when "new" characters are transmitted paternal characters must to some extent dominate over the maternal ones.

Under divergent evolution there is always a stage when males as vanguard are already bimodal, but females as rearguard are yet monomodal. Then both hybrids from two different males with a common female on additive inheritance (♀+♂)/2 will be closer to the paternal species. Under convergent evolution of two species males are already the same on a new character, but females are still different, and both hybrids will be closer to the maternal species. Since evolution in multidimensional niches (tropics) and all selection is divergent, "paternal effect" is observed more frequently then maternal one. Convergent evolution occurs in one-dimensional niche, where one environmental factor dominates: cold in polar region or hot in the desert.

PHYLOGENETIC RULE OF RECIPROCAL EFFECTS

RECIPROCAL HYBRIDS HAVE DOMINANCE OF PATERNAL FORM ON DIVERGENT CHARACTERS, AND MATERNAL—ON CONVERGENT ONES.

Reciprocal "paternal" effect allows distinguishing the character that undergoes evolution from the stable one. The direction of character evolution can be determined based on *genotypic sexual dimorphism* and *heterosis*. Considering heterosis as a sum of new evolutionary achievements acquired through divergence it can be suggested that paternal contribution to heterosis must exceed the maternal one. In the light of new views, it becomes clear why in heterosis we, as a rule, observe increasing of the characters useful for the human, but not for the species itself. In addition, this phenomenon is independent of species undergoing heterosis. Heterosis gives nothing to the species and can be even unhealthy. But since selection can be considered as human-forced artificial evolution for those species, the direction of this evolution and the direction of heterosis are in accordance with human's interest, but not with the interest of the particular species.

Considering evolution of the character in phylogeny as some kind of an abstract "movement", one can speak about a "distance" between male and female sex on this character.

Suppose that an initial form diversified in phylogeny into two different forms (breed, line or race) by this character. Then, according to *"phylogenetic rule of sexual dimorphism,"* we can expect that males from both forms (A and B) should be more advanced compare to females. So, one can speak about the "distances" between breeds according to the trait (how fare they are gone from each other) and between males and females within each breed (**Figure 13.1**). It's possible to distinguish the "impact of breed" and "impact of sex" in hybridization. The effect of heterosis can tell about the "distance" between the breeds and sexual dimorphism—about the "distance" between the sexes. The direction ("maternal" or "paternal" effect) and value of reciprocal effect can tell about divergent or convergent evolution of the trait. Therefore appears a possibility to explain more completely the reciprocal effects, which are nothing but the algebraic sum of maternal and paternal effects.

Figure 13.1

Estimate of reciprocal effects.

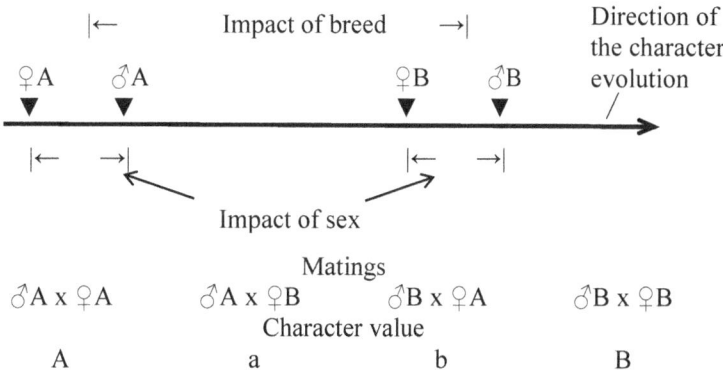

The following formula can be used for the measurement of reciprocal effects (r):

$$r = (b - a) / (B - A) \qquad\qquad [9]$$

where A and B —are values of the character for the initial forms; a —the same for hybrid ♂A x ♀B; b —for reciprocal hybrid ♂B x ♀A.

Then positive value of r ($r > 0$) will correspond to "paternal" effect, negative ($r < 0$)—to "maternal". Absolute value of r ($|r|$) will provide relative measure of the effect in units (B - A).

Existence of "paternal effect" for the given character as well as population sexual dimorphism point to the evolution of the character. However, unlike population sexual dimorphism, which requires the presence of the character in both sexes, the "paternal effect" may be for the characters specific for one sex only (primary and secondary sex characters among them). In other words, the "paternal effect" gives principal possibility to reveal genotypical sexual dimorphism.

In agricultural animals and plants, "evolutionary new" characters are evidently all economically valuable characters. **Table 13.1 (Appendix C)** shows the results of studies by different authors on inheritance of some characters by reciprocal hybrids in hens, pigs and cattle. Roberts and Card (1933) studied the inheritance of brooding instinct in reciprocal hybrids of *Leghorn* and *Cornish* breeds. Due to industrial incubator breeding and selection for egg yield the white Leghorn breed has practically lost the brooding instinct. Among *Cornish* and other breeds the brooding instinct was preserved completely. This feature is important, since it is well known that hens with this instinct have as a rule lower egg yield. The frequency and rate of hatching was higher when fathers were from the Cornish breed. The same results were obtained by Morley and Smith (1954), breeding *Leghorn* and *Australorp* (has brooding instinct) and Saeki et al. (1956) on *Leghorn* and *Nagoya* (has brooding instinct). As one can see, on brooding instinct clear "paternal effect" was observed (r = 0.45; 0.38 and 0.50 accordingly).

Another trait—sexual maturity (age of the first egg laying) was studied on *Rhode Island* breed that has both early and late maturing lines (Warren, 1934). On early maturing the "paternal effect" was observed (r = 0.59). Same results were obtained by Australian and Japanese scientists (Morley, Smith, 1954; Saeki et al., 1956) on hybrids from mating *Australorp* and *Nagoya* (late maturing breeds) with early maturing *Leghorn* breed.

Data on egg-laying capacity were obtained by Warren (1942) for reciprocal hybrids of different breeds. Clear "paternal effect" was observed in four of five hybrids. Similar data obtained by Nox on *Leghorn* and *Rhode Island* breeds from 1946 to 1956 also confirm the existence of "paternal effect" (r = 1.14).

Dobrinina (1958) conducted breeding *Leghorn* with *Moskovskaya*, observed the "paternal effect" on egg-laying capacity and live weight and "maternal effect"—on the weight of eggs. As expected, daughters of *Leghorn* father had higher egg-laying capacity, and daughters of *Moskovskaya* father's—higher live weight.

Large father's contribution to the egg yield of daughters was explained by the fact that in hens the female is heterogametic and the male—homogametic, therefore its single X-chromosome the hen receives from its father (Morley, Smith, 1954). If so, it should be expected that in mammals everything must be vice versa, since their males are heterogametic, i.e. greater maternal contribution must be observed, notwithstanding the fact, whether an "old" or "new" character is inherited. According to evolutionary theory of sex, disregarding the gamete pattern of sexes, in all cases the "paternal effect" for evolving (selected) characters must exist.

Aslanian (1962) investigated inheritance of vertebra number and some parameters of digestive system on two contrast breeds of pigs—*Swedish Landras* and *Large white* and their reciprocal hybrids. *Swedish landras* is the breed of meat-bacon selection. During 50 years, the body of these pigs became longer by 14 cm. The average number of vertebrae increased by 1.2. Simultaneously the efficiency of food consumption increased due to the lengthening of their small intestine. The large white breed is that of universal meat-fat selection.

Later Aleksandrov (1966) investigated inheritance of vertebra number using X-ray. As seen from the **Table 13.1** pronounced "paternal effect" for inheritance of vertebra number is observed in reciprocal hybrids. The inheritance of various characteristics of the digestive system reveals a "paternal effect" only with respect to the average length of the small intestine and esophagus. Maternal effect was observed for other characteristics (average weight of the embryos, digestive system and it's various parts, and the large intestine length). The "paternal effect" was found also on growth dynamics of piglets (r = 1.8). So, the "paternal effect" was observed precisely on characters for which the selection of Landrases took place: the vertebra number (selection for long body), the length of small intestine (selection for best food utilization), and growth dynamics (selection for early maturity). It should be noted that the "maternal" effect was observed for the weight of newborn piglets.

In three breeds of cattle, "paternal effect" was observed for milk and fat production (Fohrman et al., 1954).

It is of interest that small maternal effect is observed for egg size in hens and percentage of milk fat in cows. Both characters are economically valuable and were selected during domestication. Comparing the number and size of eggs in cultivated hen breeds with those of their wild ancestors, one can see that in the first place the egg number was increased significantly. The egg size remained approximately the same.

Modern hens bear on 126–200 eggs in a year (a record—1515 of eggs for 8 years).

Similarly, comparing fat percentage and milk yield in cultivated cattle breeds with their wild ancestors which produced small amount of milk with high fat content, we shall see that due to selection the amount of milk was strongly increased, but percentage of fat was even reduced.

Wild ancestors of hens produced 5–14 eggs per year.
Akimuskin I., 1973, p. 207.

Existence of "paternal effect" in mammals shows that the interpretation of this phenomenon in hens because of females' heterogametic constitution is insufficient. It explains the chromosome mechanism of this phenomenon in birds but is inapplicable to mammals. This phenomenon is much broader; it is related not to heterogametic pattern, but to sex and is closely connected with evolutionary transformations of populations. In mammals, it is realized probably through other mechanisms, different from heterogametic one.

"Paternal effect" provides explanation for the century-old mystery of parental "asymmetry" for reciprocal crossings which classical genetics was unable to explain. Reciprocal hybrids of horse and donkey appeared ~8-10 thousand years ago, but it's still unclear why mule and hinny both look like their fathers (**Figure 13.2**).

Horse and donkey diverged from common ancestor tarpan (or Przewalski's Horse). More old ancestors were of the size of a cat. After that horse become longer, higher and faster. The tail is formed from long hairs. Donkey has long ears, big head, and narrow hooves. The tail has brush of hairs. Genes of these new characters should be initially in males only.

Figure 13.2 Mule and hinny both look like their fathers.

So, we can consider that father's impact on heterosis is more than mother's and dominance of paternal form on characters that undergo evolution or selection really exist. Male and female sex both have genetic information on secondary sexual characters (and may be on primary ones also). "Paternal effect" as well as genotypic sexual dimorphism can serve as an evidence of character evolution. In a practical aspect "paternal effect" permits qualitative prediction of hybridization results and correct selection of parents at crossings.

Genomic imprinting is a particular case of more common phenomenon of parent genomes asymmetry. This asymmetry can manifest itself in the form of presence – absence of genes, as well as in the form of their expression – suppression.

Chapter 14

Applications of the Theory

 nly some examples showing how the evolutionary theory of sex "works" are presented in this chapter. Theory application in pathology is described in **Chapter 15**, and to selection of plants—in **Chapter 16**.

Problems of Evolution and Speciation

Evolution of Lower Crustacean

The *phylogenetic rule of sexual dimorphism* was successfully tested on a large group (173 species) of lower Crustacean (*Chidoridae* family) (Geodakian, Smirnov, 1968). Males in all cases were smaller than females. According to the rule, we can assume that this group has a common evolution trend of size reduction. In fact, it is well-known that morphologically more primitive forms of Crustacean are larger. Inside the group we can pick out arrays of forms with sequential size and number of extremities reduction, and specialization according to phylogenetic preemptivity and smaller male size.

On several traits the males from morphologically more primitive sections preceded those features that appear in subsequent sections (morphologically more developed). We can conclude that the evolutionary trend of characters, defined on one hand by morphological criteria and on the other hand, based on sexual dimorphism rule, is identical. This conclusion allows us to apply the new rule for future research. You can make some assumptions about the biological sense of the changes. With decreasing size, the relative surface increases, which contributes to gas exchange and increases buoyancy. S. Chetverikov (1915) suggested that the evolution of invertebrates with an exoskeleton in the direction of downsizing is related to the skeleton's mechanical properties.

The *phylogenetic rule* was used to answer some questions concerning the evolution of *Chydoridae*. After an analysis of three characters (body length, number of anal spinules and basal spines of claws) with marked sexual dimorphism, a new place for the *Leydigia* group was proposed (Geodakian, Smirnov, 1968).

Selection of Spermatozoids (scrotum—"refrigerator" or also a "thermometer"?)

> *"The original reason that a scrotum evolved has long been a subject of debate among reproductive biologists and still has no universally accepted answer."*
>
> *Left testis is usually suspended below the right one, so that they will not compress each other between the hips"*
> Saladin (2004).

The testes of elephants and hedgehogs, as well as whales and seals are placed internally, just as they are in birds and most other animals. In adult humans and many other mammals the testes hang externally in the scrotum. The testes start out in the abdomen but during the early years of life descend downwards, so that by puberty they reside outside of the body. For a long time scientists attempted to understand why such a fragile design is needed. It is considered, that testes are placed outside of the body to keep the temperature low (the blood that reaches the testes is $1.5°$ to $2.5°C$ cooler than the core body temperature (Saladin, 2004)) for the spermatogenesis. However, it's obvious that the sperm can be produced by internal testes at the body temperature (Freeman, 1990). It seems that testes of birds should first of all be brought out of the abdominal cavity, as the body temperature of birds is several degrees higher than that of mammals.

With the development of internal fertilization, the necessity of a great number of gametes disappears. But only a female line is subjected to reduction, passing from millions of ova in fishes to just several ones in mammals. The number of spermatozoids remained on the level of hundred millions, in spite of evident possibilities for reduction and a certain trend of evolution to lose everything no longer needed. The maintenance of the "population" system in spermatogenesis and the ecological specialization of males make one to assume the existence of the natural selection of spermatozoids. For animals with a little posterity and a seldom change of generations it should be advantageous.

It is necessary for the realization of such a selection that selected genes should be represented in the phenotype of the spermatozoids. It can be assumed, that these are primarily the genes of stability for the most fundamental factors of the environment (temperature and others). If this is true, Mendel's distribution of phenotypes must be disturbed by the crossing of a heterozygous male with a homozygous female. Such disturbances were found in mice by the transmission of the taillessness gene from a heterozygous male to his offspring. In a reciprocal crossing this phenomenon was absent. The length of the tail is connected with an ambient temperature. According to Allan's rule, the more northern the habitat of animals, the shorter their tails. In rodents kept under extreme temperatures their tail length varied accordingly.

The second necessary condition for the realization of an adequate selection of spermatozoids is the availability of their information contact with environment. In case of external fertilization the spermatozoa were in direct temperature and chemical contact with the environment. After the transition to the internal fertilization the chemical contact with the environment was lost. Among warm-blooded animals only surface parts of the body, keep a temperature contact with the environment. Perhaps this compelled evolution to bring testis of mammals out of the abdominal cavity, while the ovaries remain inside the body.

In very large underground and water animals, not subjected to great temperature fluctuations, as well as in birds, having homogametic males, the testes are in the abdominal cavity. It allows us to conclude that selection of spermatozoa is possible only in the heterogametic sex and is realized only in spermatozoa carrying Y-chromosome. First, the Y-chromosome is the "ecological" chromosome, which realizes contact with the environment. Second, appearance of the male with a required genotype is more efficient for adaptive transformation of the population than of the female, since the male produces more offspring. One more indirect argument in favor of the fact that selection proceeds among Y-carrying spermatozoa is the increased sexual activity of males of rare genotypes. And it is known that sexual activity of males is regulated namely by the Y-chromosome.

The sperm-training hypothesis. A. Zahavi and S. Freeman (Freeman, 1990) proposed that by keeping the testes relatively cool in the scrotum the sperm are trained to endure environmental hardship, so that when they finally find themselves in the hostile environment of the vagina they perform more competitively.

Why Three Sexes?

The majority of dioecious species have two sexes. Among vertebrata only at some fishes it is possible to find two types of males. None of the species has two kinds of females. It is considered that two kinds of males result from selection on various strategies of mating.

The interesting riddle is represented by three sexes mating system at salmon fishes. A pacific salmon is a unique group of monocyclic kinds of fishes. They breed only once and perish after spawning so the age of the first reproduction is equal to the age of sexual maturity and life span. For example, spawning population of red salmon consists of fishes at the age from 3 to 7 years. One can easily distinguish groups of three different sizes: females and two kinds of males (large, long-living, more homozygous than females and small, short-living, more heterozygotic than a female).

Small males are young fishes. In intact herds their quantity is less than one percent. They grow quickly and ripen early, usually in 3 years. The large males are slow growing, late ripening, 5–7 year old fishes. Females have intermediate sizes and age of 4–5 years. The degree of heterozygosity on genes coding synthesis of isoenzymes is maximal at small males, minimal at large ones, and is intermediate at females (Altukhov, 1983, 1994; Varnavska, 1983).

The observed picture can be explained by the way of life of salmons. Salmons breed in the rivers, and then move down to the sea (ocean), where they grow, develop and, having reached sexual maturity, come back to the rivers for spawning. Therefore they need ecological information from two different environments—the "sea" and the "river" which is provided with two kinds of males (an ecological sex). According to the evolutionary theory of sex, the male sex is the evolutionary vanguard; therefore one kind of females and two kinds of males says that there is a disruptive selection and disruptive dichronous evolution which will lead sooner or later to disintegration into two species. So, this is an initial stage of speciation, when males are already bimodal, and females are still monomodal.

The Evolution of Ontogeny

"Women live longer than men, especially widows."
G. Clemenceau.

Total genetic information of the population transferred from generation to generation $(\sum)$, is comprised from **genetic** part (G), transferred through **gametes** and zygotes and **cultural** (C), transferred through **nonzygote "shunt"**:

$$\sum = G + C \qquad\qquad [4]$$

Primitive forms have **zygote** channel between generations only $(C = 0)$. For example, butterfly lay eggs and die after that leaving the progeny nothing but genes. Therefore children get the information about expedient behavior genetically in the form of inborn instincts. Such a way of transferring of behavioral programs—"on any occurrence in life", is, of course, very rigid. Sometimes parents can transfer the nonzygote information to descendants, without meeting with them—in absentia. For example, warps-equestrians provide food to the future larvae by laying eggs in the paralyzed body of a victim. Blest (1959) has found that the post reproductive period of moths with aposematic coloration is markedly longer than that of moths with cryptic coloration. It means that duration of the post reproductive period of ontogeny is of adaptive value and is controlled by intergroup selection. It is useful for the population when adult individuals incapable of reproduction being caught by predators make them not attack the fertile individuals with the least loss for the species.

More flexible and effective way is to teach children in ontogeny. For this the overlap of generations is necessary. It arises and grows for more advanced forms, which have an opportunity of transfer ontogenetic information (experience and "knowledge") to descendants by training and not through heredity. The training process requires time, so with growth of volume of the cultural information, the overlap of generations grows accordingly. There appears long "old age" (usually donors of the information), "childhood" is increased (acceptors of the information) and ontogeny as a whole is extended. So, for humans up to 4-5 generations can live at the same time simultaneously! The transition to anthroposphere complicates the organization even more: races, the nations, languages, social hierarchy, and professional specializations. And with occurrence of culture the complication gets avalanche-like character: religions, temples, trades, libraries, sciences, arts. The personal contact between adjacent generations becomes insufficient to transmit increased volumes of the cultural information. Other channels suitable for communication with **non-adjacent** generations are created (books for example).

There are five important moments to be noted in the ontogeny of mammals: conception, birth, appearance and loss of the reproductive capacity, death. The ontogeny is thus divided into four stages: prenatal development (embryonic), postnatal-prepubertal development (growth), reproductive, post reproductive (terminal). The sum of last three stages makes life longevity (**Table 14.1**). The duration of life and its stages are the specific characters developed in the course of phylogeny. They are of great adaptive value and are controlled by the intergroup natural selection.

Table 14.1 Longevity and the period of intra-uterine development, growth and development at some primates (Bunak et al., 1941)

Species	Stages			
	prenatal, weeks	growth, years	reproductive, years	life, years
Lemurs	Up to 20	2–3	10–15	15–20
Old World Monkeys	Up to 24	7	11–20	20–30
Hominids	33–39	8–12	20–30	30–40
Homo Sapiens	40	20	45–60	70–80

Various stages of ontogeny play different roles in receiving, realization and transmission of genetic and ontogenetic information. The genetic component is passed at the reproductive stage, received at the moment of conception and realized (with decreasing intensity, it seems) throughout the ontogeny. The ontogenetic component is connected with postnatal life. With age the share of incoming information decreases and that of outgoing—increases. Since it requires time to receive, realize and transmit information (e.g., it takes 20 days for the zygote to grow into the newborn mouse or 660 days to grow into an elephant), the increase of the scope of this information in phylogeny is followed by enlargement of corresponding stages of ontogeny (**Table 14.1**). This, in particular, can explain the known correlations between the life span and a) the animal mass (large forms live usually longer than the small ones); b) encephalization quotient (brain mass per body mass ratio)—the higher the index the longer the life; c) the length of intrauterine development, period of growth and reproductive period: the longer these periods the longer the life (the growth period, e.g., makes up some 20% of the life duration), and others (Malinovskij, 1962; Comfort, 1964; Korchagin et al., 1973).

From these correlations one can judge the evolution of the duration of ontogeny and its stages. Thus, e.g., knowing that man has a maximum life span, growth and reproductive periods among the mammals, and maximum duration of prenatal life among the mammals of comparable mass, one can think that both the duration of the ontogeny as a whole and that of its stages were enlarged in the course of evolution.

If we compare the evolutionary tendency towards an enlargement of ontogeny and all its stages in man with the *"phylogenetic rule of sexual dimorphism"* we can conclude that both ontogeny as a whole and all its stages should be longer in males than in females.

The average duration of intrauterine life of boys is longer than that of girls. Despite this the girls are born more mature than boys (by 3–4 weeks) as shown by X-ray studies of the bone age (Harrison et al., 1964). Therefore we can think that at the moment of birth the difference in the duration of intrauterine life between the sexes is roughly a month. After birth the difference begins to grow progressively: girls begin walking on an average 2–3 months earlier than boys, and talking 4–6 months earlier. By the time of puberty this difference amounts to nearly 2 years (Harrison et al., 1964; Kolesov, Selverova, 1978). The next stage—reproductive—is also longer in men. In women the reproductive stage lasts 35–45 years (from the age of 13 to 45–55 years old), while in men it lasts 45–55 years (from 15 to 60–70 years old), i.e. here the difference is no less than 10 years (Davidovskij, 1966).

So, first three stages of ontogeny in accordance with the conclusion of the theory—are clearly longer in the males. As if to the average life span, contrary to the theory, it is longer for the female gender. Thus a strange picture develops: Women age earlier and men die earlier. What is happening?

> **"…at the age 30-60 years for one unmarried man there are two to six unmarried women."**
>
> N. I. Kozlov, 1999.

The duration of life, just as any other character, is determined by the genotype and environment. Let us imagine the following idealized situations: (1) If the genotypic diversity in population is eliminated (i.e. the clone of genetically identical individuals is taken) and unfavorable environmental factors are removed (i.e. optimal conditions are created for a complete realization of the single genotype) the life span of all the individuals in such a clone will be similar, and the death rate curve will be of a rectangular shape (variation $\sigma = 0$ **Figure 14.1 A**). (2) If instead of the clone a heterogeneous population is taken in an optimum environment which permits realization of all the genotypes, some variance in the life span will appear which is conditioned only by genotypic variability (variation σ_{gen} **Figure 14.1 B**). (3) Now if the clone again will be taken, but placed into a real environment, the variance in the life span will be conditioned by the environment (variation σ_{env} **Figure 14.1 C**). (4) If a genetically heterogeneous population is placed into real environment, the overall variation will include both, the genotypic and environmental components ($\sigma = \sigma_{gen} + \sigma_{env}$) **Figure 14.1 D**). (5) Now let us imagine a situation when the death rate is determined only incidentally (by the environment). Then the genotype will make no contribution to the life span, and the latter will depend neither on age, nor on health. In this case the death rate curve will have an exponential shape (**Figure 14.1 E**).

Curves (B), (C), and (D) (as all the real cases met in life) are intermediate between the extreme types (A) (the death rate is totally controlled by the genotype) and (E) (it is under total control of the environment), therefore they can be presented as a superposition of these extreme types. Consequently, as the population gets "emancipated" from the environment, the pattern of its death rate approximates the rectangular type and moves away from the exponential type, and vice versa. Hence, the more optimum is the environment, the nearer to the rectangular type is the death rate pattern of the population; the more extreme it is—the closer to the exponential type.

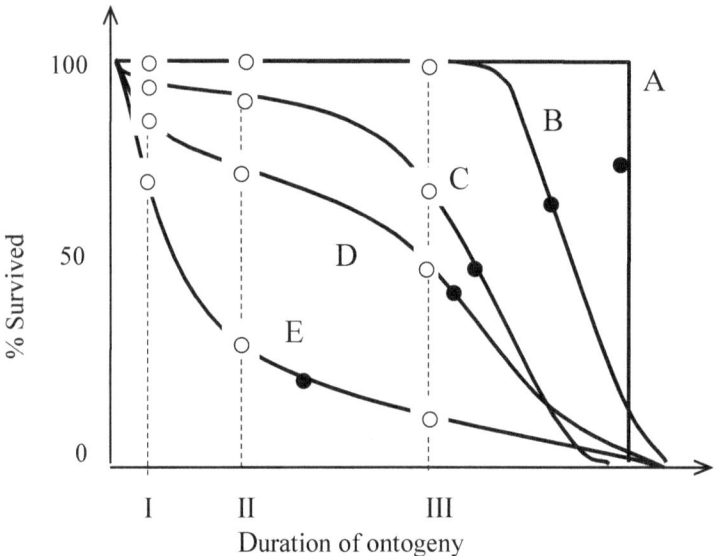

Figure 14.1 Death rate curves determined to varying degree by genotype and environment.
A—clone in ideal environment; B—heterogeneous population in ideal environment;
C— clone in real environment; D—heterogeneous population in real environment;
E—death rate is determined only stochastically. I—average duration of intrauterine life;
average age of appearance (II) and loss (III) of reproductive capacity, ● —average life
span, ○ —average duration of ontogeny stages. (Geodakjan, 1982)

The transition of the population from the optimum environment to the extreme one affects the duration of ontogeny and its stages. It is evident that mortality strongly affects the average life span and the post reproductive period and has practically no effect on the average duration of periods of intrauterine development, growth, and reproduction (**Figure 14.1**). This means that a decrease in the average life span in extreme environmental conditions is the result of the environmental component of mortality and not the genotypic. Thus, the application of the law of sexual dimorphism is a basis for the hypothesis that genotypic average life span of men must be greater than that of women. And if it was possible to eliminate the effect of environment completely (to place the population in an ideally optimal environment), then men might possibly live longer than women.

Comparing the groups differently "emancipated" from the environment, such as men and women of one population, blacks and whites from one country, the same country in the course of history, etc. (Comfort 1964), one can see that the pattern of the death curve changes regularly in the direction indicated by the theory. The more is the group "emancipated" from the environment, the nearer is its mortality curve to the rectangular type and vice versa (see **Figure 14.2 A**). Greater "emancipation" of females as compared to males results from a wider hereditary reaction norm of the women. Greater "emancipation" of white population in the

U.S.A. compared to the African-American one, as well as of modern population as compared with those of the former years is conditioned by social and economical factors (nutrition, medical service, etc.) (**Figure 14.2 B**).

Let's remind that according to new theory, in extreme conditions of environment, the genotypic variation of attributes in a population increases. It concerns also the duration of ontogeny and its stages. The short- and long-living male genotypes are appearing. First immediately increase infant and early mortality, and second will show up as long-living individuals after a 100 years. This interpretation permits also to understand the seemingly paradoxical fact that the phenomenon of long living is encountered in the populations living under far from optimum conditions (Comfort 1964, Davidovskij 1966).

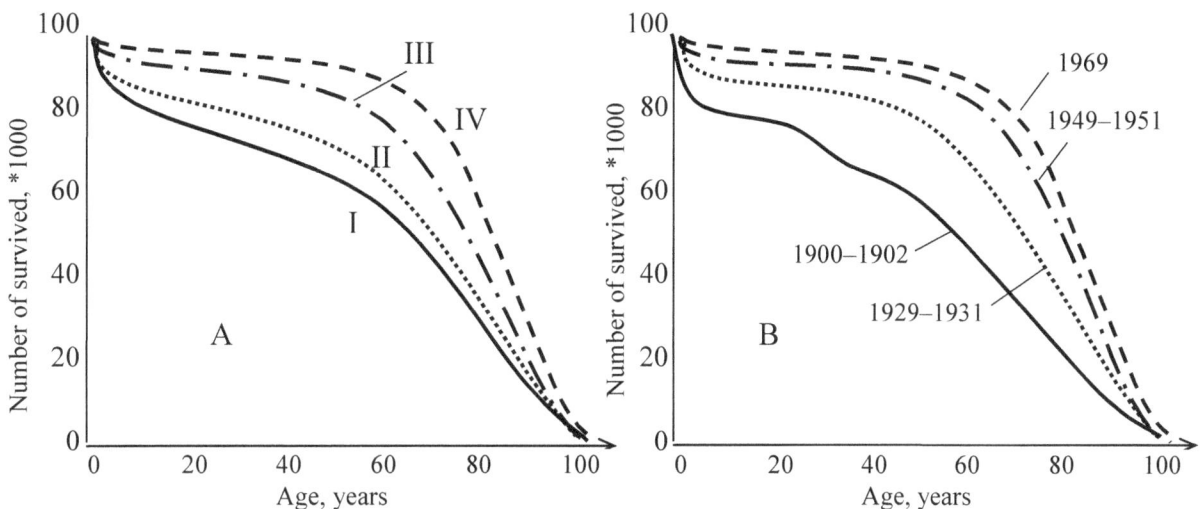

Figure 14.2 Death rate curves (USA) (Comfort, 1964).
A: Number of survived per 100 000 (USA, 1939–1941): I—African-American men;
II— African-American women; III—white men; IV— white women.
B: life span in the USA in different years.

If the phenomenon of longevity is really conditioned by an increased variation of the life span under extreme environmental conditions, it should be closely related with stress as a transmitter of ecological information in animals. It should be accompanied by an increased death rate (the infants one first of all), by an increased variation of other characters, by the increased secondary sex ratio (birth of boys) and by a rise of sexual dimorphism. It becomes clear why the average life span of women is longer, while the "champions" of longevity are men.

The populations in which long-living individuals are found should have the death rate curve nearer to the exponential type, characteristic of extreme conditions. Contrary, the death rate curve in an optimum environment should be closer to the rectangular type, where the long living is not observed. O. M. Pavlovskii (1985) showed that described tendencies are observed in real conditions.

The Evolution of Human Dentition

Theory of sex has been applied to the analysis of the evolution of human dentition. It's known that mammalian predecessors had all teeth of conical shape with one cusp. Conical crown fits conical root with the root canal (Romer, Parsons, 1992). G. Manashev (2005), analyzing sexual dimorphism for different groups of teeth, proposed that the multi-rooted teeth were formed in the process of evolution by merging of simple conical teeth with the fusion of crown, root, and tooth cavity.

Sexual Dimorphism in Anatomy and Physiology

Consider the features that are found in both sexes, but are distributed in a population with different frequency and degree of expression. These are quantitative traits: height, weight, dimensions and proportions, many morphophysiological and etologo-psychological characters. Sexual dimorphism on such characters can be measured as the ratio of their average values. This populational sexual dimorphism serves as a "compass" of a trait's evolution.

The *phylogenetic* and *ontogenetic* rules of sexual dimorphism can be used as tools for research: if the sexual dimorphism is known, it is possible to predict phylogeny and ontogeny of the trait, or, conversely, knowing the phylogeny and ontogeny, it is possible to predict the presence and direction of sexual dimorphism.

During evolution humans increased height and developed bipedal locomotion. The skin lost hairs and developed the ability to intensive sweating. Scientists assume that loss of hairs is connected with disposal of skin parasites, and sweating has improved thermoregulation. It is possible to assume that on all of these characters men should be more advanced. Indeed, on average, men are taller than women, their bones are thicker and stronger, and the muscular system is more developed. Women more often have a congenital hip dislocation (4 to 5 times), varicose, osteoporosis, and bones fractures. Some but not all of these effects can be associated with the action of hormones, child-bearing, and breast-feeding.

Men release heat through the sweat glands faster than women (Wade et al., 2007). Children have a reduced ability to sweat (per gland (Bar-Or, 1980; Falk et al., 1992) and per sq. m. of skin (Araki et al., 1979; Falk et al., 1992)) and as a result quickly overheat. The transition to an adult pattern of sweating occurs during puberty.

Men have much more hair compare to women. At first glance it contradicts the made prediction. However, women have more total number of hair bulbs not only on a head, but also on a body. Not all the hair bulbs are active, some of them are functioning, and others are dormant. Men have higher percentage of the sprouted hair. In women hair grow faster, than in men, and drop out less often. Most intensive hair growth occurs in young men at the age of 15–30 years, and in the period of 50–60 years their growth is sharply slowed down. Men more often grow bald. Still Hippocrates has noticed that eunuchs never grow bald.

In humans, the ratio of the right temporal plane to the left one is almost always less than unity, i.e. the left plane is longer than the right. However, most of the brain specimens, which indicated an inverse relationship, were from women (Springer, Deutsch, 1989; Wada et al., 1975). On the basis of "*phylogenetic rule of sexual dimorphism*," we can say that during evolution this ratio decreases. And applying the "*ontogenetic rule*," it's possible to predict that this ratio should decrease with age. Indeed, in infants it is 0.61, and in adults— 0.55 (Wada et al., 1975).

Another example. It is known that the relative size of a *corpus callosum* strongly increases in ontogeny (Springer, Deitch, 1989; Trevarten, 1974.). Hence, according to "*ontogenetic rule*" the sexual dimorphism on its size should exist: in men it should be bigger than in women. And it means that it should increase in phylogeny.

Vision

It is well known, that phylogenetic predecessors of humans had eyes located laterally (on the sides of the body). Their visual fields were not crossed. Each eye was connected only to an opposite hemisphere of a brain—contralaterally. During the evolution process, some vertebrata including our ancestors developed three-dimensional sight. As a result they had eyes moved on the frontal side. The left and right visual fields were overlapped and new ipsilateral connections (left eye—left hemisphere, right eye—right one) emerged.

Based on *"Phylogenetic rule of sexual dimorphism"* it is possible to predict that males will have evolutionary more advanced ipsi-communications compare to females. And because spatial-visual abilities and 3-dimentional imagination are closely related to stereoscopic vision and ipsi-communications it becomes clear why men have advantage on these abilities. Having applied *"Ontogenetic rule sexual dimorphism"* it is possible to predict increase of visual (and other) ipsi-communications and improvement of spatial-visual abilities with age.

Table 14.3 has data on some mammals concerning ratio of ipsi- and contra-lateral fibres in an optic nerve. One can see that in the process of transition from animals with lateral direction of visual axes to animals with frontal orientation of visual axes the share of ipsi-fibers grows (Blinkov, Glezer, 1964). The appearance of ipsi-lateral communications provides hit of the visual information from both eyes in one hemisphere for comparison and reception of a three-dimensional picture—perception of depth. It was shown, that the most important requirement for the realization of three-dimensional sight is the difference between images on a retina of two eyes (Bishop, 1981). So, ipsi-lateral communications are evolutionary younger (have appeared later) than contra-lateral ones. Such conclusion is fair, probably, not only for visual conducting pathways, but also for all: motor, somato-sensoric and acoustical pathways.

Table 14.3 Amount of crossed and non crossed fibers in the optic nerve of some mammals (Blinkov, Glezer, 1964).

Species	The ratio of direct to crossed fibers	Author
Sheep	1 : 9	Nihterlein, Goldbi, 1944
Horse	1 : 8	Dexler, 1897
Dog	1 : 4.5	Rogalski, Rimashevski, 1945
Possum	1 : 4	Bodian, 1937
Guinea pig	1 : 3	Gess, 1958
Cat	1 : 3	Chang, Cheng, 1961
Ferret	1 : 3	Jefferson, 1940
Macaca	1 : 1.5	Clark, 1942
Human	1 : 2 1 : 2 1 : 1.5 1 : 1	Cahal, 1899 Rogalski, 1946 Santa, 1942 Clark, 1942

Olfaction

Having applied the same rules to an olfactory receptor of the person, it is possible to come to another conclusion. It's known that sense of smell is worsened and undergoes reduction in human phylogeny. Therefore, it is possible to predict the existence of sexual dimorphism on a quantity of olfactory fibers: women should have more fibers compare to men. Or men should have higher degree of atrophy (or its speed) compare to women. With age sense of smell also should be worsened.

Table 14.4 shows an age dynamics of human olfactory nerve atrophy. One can see that the amount of fibers in an olfactory nerve decreases steadily (Smith, 1942). Thus, the tendencies predicted by the theory really exist.

Table 14.4 Age dynamics of an atrophy of fibers of an olfactory nerve in humans (Blinkov, Glezer, 1964; Smith, 1942).

Age (years)	Number of atrophied fibers
0–15	8
16–30	20
31–45	33
46–60	57
61–75	68
76–91	73

Hearing and Voice

As stated in **Chapter 4**, the evolution of hearing in humans was going in the direction of increasing sensitivity in the low and at the same time decreasing it in the high frequency range. Therefore applying *"Phylogenetic rule of sexual dimorphism",* we can predict better hearing sensitivity of women (or worse, of men) at high frequencies, and applying *"Ontogenetic rule of sexual dimorphism"*—hearing impairment, especially hearing loss at high frequencies with age.

Hearing of the newborn girls was much more sensitive than boys, especially in the region from 1 to 4 kHz which is important for speech recognition (Cassidy, Ditty, 2001). Girls hear better than boys, especially in the area above 2 kHz (Corso, 1959; Sato, 1991). With age, hearing sensitivity decreases, extending from high to low frequencies. Children can hear sounds with a frequency up to 30 kHz. For teenagers (up to twenty years) the susceptibility is reduced to 20 kHz, and at sixty years, up to 12 kHz. On average women can hear better than men. They also distinguish sounds at high frequencies very well. In men compare to women hearing loss occurs at an earlier age and to a greater extent (Corso, 1959; Karlsmose et al., 1959). For example, 15-year-old boys are losing hearing 70% more often then girls (Sorri, Rantakallio, 1985).

Voice. The presence of sexual dimorphism suggests that the voice in humans is an evolving trait. Applying the *"Phylogenetic rule of sexual dimorphism,"* we can predict that the voice frequency decreases. Wide range (variance) of women voices (~180 Hz) and narrow range of men voices (~120 Hz), the theory of sex considers as a converged phase of voice evolution **Fig. 14.3**. In this case, the voice evolution in men has already going to an end, and in women it is still continues. So, the bass and contralto are evolutionarily young voices, and the soprano and tenor are vanishing.

Applying *"Ontogenetic rule of sexual dimorphism"* we can assume that voice frequency should be higher during childhood and decrease with age. The effect should be stronger in men than in women.

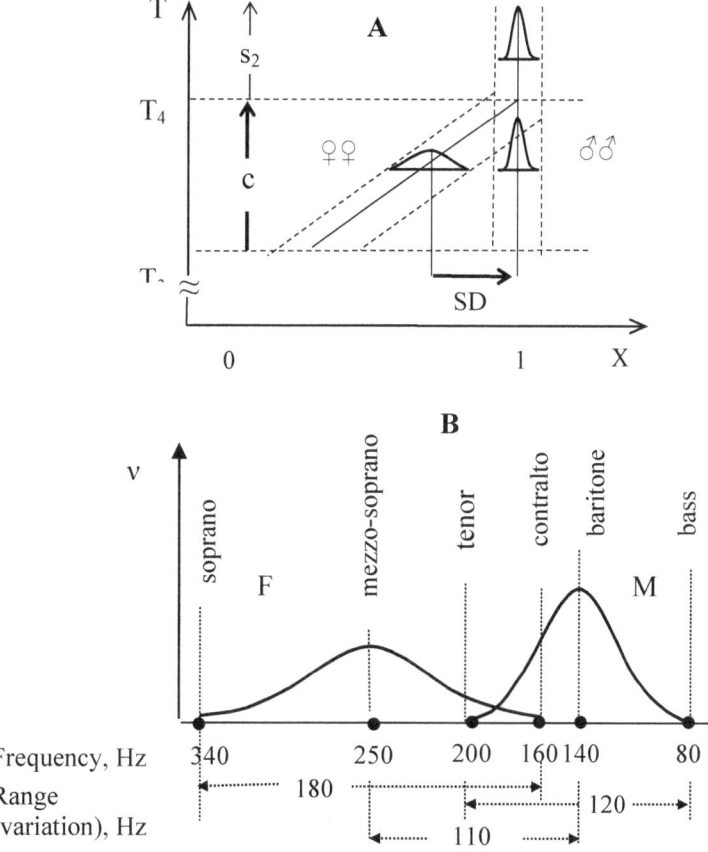

Figure 14.3. Theoretically predicted pattern of convergent evolution phase (upper part of **Fig. 11.5**) (A) and sexual dimorphism in the frequency of voice in ontogeny (B).
Abscissa: frequency of female (F) and male (M) voices, Hz.
Ordinate: frequency of phenotypes in the population (ν).

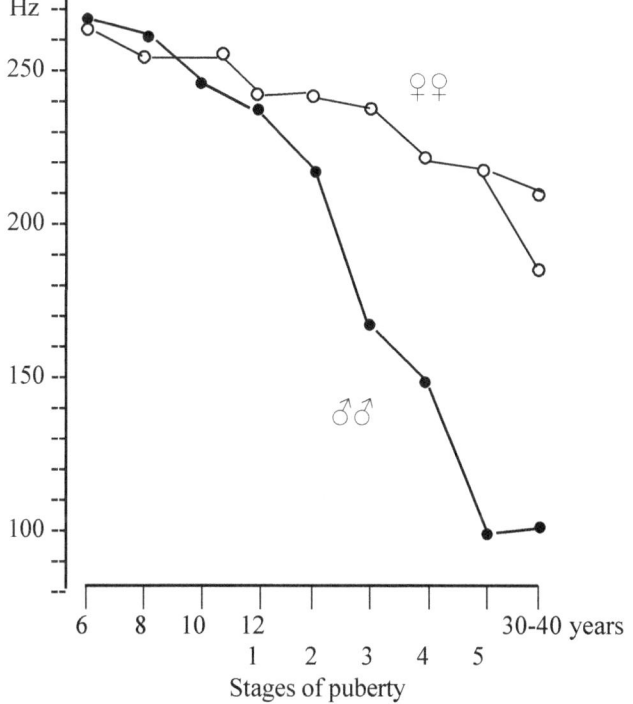

Figure 14.4. Change in primary voice frequency (Hz) depending on the age and stage of puberty (from Vuorenkoski et. al, 1978).
Abscissa: age.
Ordinate: the fundamental frequency of voice (Hz). For women the value for soprano and alto are given separately.

Vuorenkoski et al. (1978) measured the change in the fundamental frequency of voice depending on the age of 374 subjects starting from 6 years. From 8 to 10 years, the frequency was reduced from 259 to 247 Hz in boys but not girls (253 Hz). During puberty there was a much stronger reduction in the voice frequency for boys (up to 100 Hz), compared to 213 Hz in girls (**Figure 14.4**). Eguchi and Hirch also found that from 3 to 7 years for both sexes a fundamental voice frequency decreases from 298 to 262 Hz and further declines after 10 years (Eguchi, Hirch, 1969).

Sexual Dimorphism in Psychology

Evolutionary approach based on the notion that the features of social behavior as well as the features of the morphological structure are determined by adaptation of organisms to the environment, is becoming more frequently used in the social and psychological research (Bagrunov 1981, Kochetova, 2010). Based on an evolutionary approach, the researchers have analyzed psychological differences between the sexes in education (Yeremeeva, Hrizman, 1998), the psychology of creativity (Zhuravleva, Soboleva, 2005), the difference in status preferences (Kochetova, 2010) and the desire for power and control (Trofimova, 2011, 2012). Some of these works use the findings of the evolutionary theory of sex.

The sex differentiation is a specialization on genetic and ecological streams of information. At the same time it's also a specialization on endo- and ekzo-communications: females are specialized on intrapopulation relations, while males—on environmental ones. Therefore it becomes possible to understand sexual dimorphism on verbal, spatial-visual and other psychological abilities. Language, verbal abilities is a main communication tool between members of the population; therefore they are better developed in women. On the other hand, the spatial-visual abilities are more connected to the environment by external communications (protection, hunting, fighting enemies), and are more advanced in men.

Another demonstration of such kind of psychological sexual dimorphism features of handwriting in men and women on which psychologists and forensic experts determine a person's gender. Among the distinctive features of female handwritings are: correct, uniform, precise, beautiful, standard and symmetrical. The following features related to man's handwriting are: abnormal, irregular, sprawling, ugly, erroneous, individually-original, letters "t" and " i " without dashes and dots. That is the women's handwritings are more perfect, and are closer to the standards of training (requirements of environment), than men's (Kirsanov, Rogozin, 1973; Young, 1931). In another special study the same amount of information (the same time of training) increased the *intelligent quotient* (IQ) of boys by 1.5 units, and girls—by 4.5 (Geodakian, 1984b).

Note that most of the verbal tests take into account the perfection of execution. It is fluency of speech, speed of reading, spelling etc. On the contrary, the spatial-visual tests often require a search. These are mental manipulations with geometric shapes—rotations, making the whole figure from the parts, or folding the volume figures from the plane sweep.

It would be interesting to compare the results of males and females on the verbal tests based on the search, and the spatial-visual tests related on training. We should be getting the opposite results: the advantage of women on spatial, and men—on verbal tests. The better men's ability to find verbal associations or solve crossword puzzles can serve as a confirmation of this prediction. For both tasks verbal search is required.

Sexual Dimorphism in Ethology

During evolution in zones of discomfort and elimination there is a selection in different directions. Females are selected based on "adaptability" and "learning ability". Males were selected based on "ingenuity", "resourcefulness" and "creativity" in the widest meaning of these words. For example, males and females with identical genotype will behave differently in zones of temperature discomfort. Females will adapt physiologically by increasing a layer of hypodermic fat. Man will change their behavior—will invent a fur coat, fire, find a cave. They will die if unable to do that. So the different reaction norm is all that required for the appearance of ethological sexual dimorphism.

Ethological features of females are caused by their bigger participation in a genetic stream of information. These features promote stabilizing evolution tendencies, and are aimed at the preservation of old, already mastered and familiar. They can be characterized as refinements of solutions to the old problems. Flexible "compliance" (to adapt, survive and leave offspring) characterizes females' relations with the environment. Therefore, females are often deterred, "avoid" harmful environmental factors, more malleable and susceptible to their modifying influences and able to learn faster. They are more conformal (Con, 1967) and more susceptible to group pressure (Ward et al., 1988).

Male's behavior is determined by their involvement in an ecological stream of information. They are aimed at change and finding new solutions. The relationship of males with the environment is "uncompromising" (inappropriate genotypes die, while the adapted ones can leave offspring). They are willing to engage in close contact with various environmental factors in order to assimilate them, are often involved in more risky, "exploratory" behavior. Modifying effect of environment on them is limited, therefore they are worse trained and less conformal. Behavioral features of males can be treated as resource display in search of new ways, as the innovation of trailblazers focused on the future.

At any age the area, explored by boys, was 1.5–2 times bigger the girl's area despite the fact that the parents put more restrictions on boys in terms of where they allowed to go.

Hart, 1978

Therefore, a little exaggerating and schematizing, it is possible to say that males prefer and better solve new problems, which can be solved somehow (maximal requirements for innovation and minimal—for perfection). Females, on the contrary, prefer and better solve routine problems, which should be solved with perfection (minimal requirements for innovation and maximal—for perfection). Although there are few studies on sex differences in business, all authors note that women are less prone to entrepreneurial activity (Davidsson, 1995; Crant, 1996; Veciana et al., 2005; Routmaa et al., 2004).

In the historical (evolutionary) aspect of mastering any new skill it is possible to allocate two phases: an initial phase of search, finding of the new decision, and a phase of perfection. Males should have advantage in the first phase when a problem is still new, unfamiliar (external). Females will have advantage in the second phase, when a task is not so new (internal). Imaginary chart of perfection of the decision (on an ordinate axis) versus chronological number of the repeated decision making (training, search for the decision and its perfection) (on abscissa axis) for males and females is presented on **Figure 14.5**.

Man in Italy invented knitting in XIII century. For several centuries it was solely men's business. Now it became mostly women's task.

In VII-X cent. Japan, among men has become fashionable to write poems in Chinese, however women remained faithful to the tradition of the national folk poetry and never wrote Chinese poetry.

Tyugashev, Popkova, 2003

All professions, kinds of sports, games, and hobbies initially were mastered by men, and then women followed. Women are now frequently participating in sports that were previously thought to be only for men, such as ice hockey and boxing. Such an approach explains many known facts very well. For example, men's prevalence amongst composers, chess players and inventors. Thus, in the field of music among composers men were always numerically dominated, while among the performers since the mid XVIII century until today the number of women is roughly equal to the number of men. Such a pattern is easy to fit into the proposed scheme: in fact innovation is very important for the work of the composer, while in the performing arts, apparently it is mostly perfection.

Figure 14.5
Hypothetical curves of training for males (♂♂) and females (♀♀) during repeated problem solving (Geodakyan, 1986c, 1989a).
N—chronological number of the task solved;
C—perfection value (speed or quality) of the solution;
a—phase of search; b—phase of perfection.

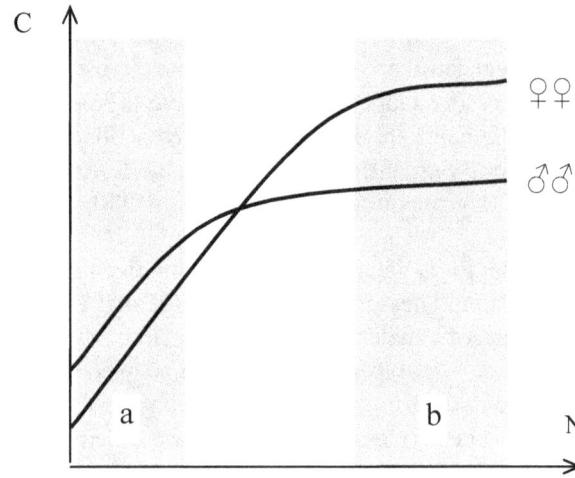

The proposed treatment allows for experimental verification. For example, we can predict that with repeated solution of the problem which includes search as well as training, at first, until success is determined by searching for and finding new solutions, better results should be in males, but in the end, when the solution is already known and perfection becomes the limiting factor—in females.

In the experiments of the Australian scientist Landauer (1981) the examinee held a finger on the central button and after a signal should press one of 8 buttons located around it. Two intervals of time were measured: between submission of a signal and removal of a finger from the central button and between removal of a finger and pressing of the necessary button. In such arrangement the performance of the first part of the task does not require any search. It can be done automatically, reflectively. The second part of the task demands search. The distinct sexual dimorphism matching theoretically expected was found: women were faster in the first part of the task, and men—faster in the second stage. It becomes clear, why women can handle work on the conveyor more easily, while men cannot keep up, make more mistakes, cannot maintain rhythm for a long time and more often are exposed to psychological frustration.

The attitude towards the opposite sex
Since females are responsible for the quality of the offspring, they are more inclined to prefer monogamy and "familiar" males. Males, on the contrary, tend to keep the maximum number of offspring, and so are more inclined to polygamy, and prefer "unknown" females (Geodakyan, 1978). A striking example of such a different "psychological" behavior was described by C. Darwin, who wrote that "male deer hounds prefer unfamiliar females, while females prefer males to which they were accustomed." See also the *Coolidge Effect* (**Ch. 9**).

Sexual Dimorphism in Anthropology

Paleoanthropologists have long paid attention to the fact that the magnitude of sexual dimorphism is likely to reflect the degree of homogeneity of the population (Velikanova 1975, Usupov, 1986). Magnitude of sexual dimorphism can be used as a test for the instability of the genetic structure of the population and a measure of the discrepancy between adaptive capacity of human populations and the rate of environmental change (Solovenchuk, Bondarenko, 1989). Male part of the population is the first to respond to changing external conditions. Higher physiological lability of males compare to females is recognized by many anthropologists (Eveleth, 1960; Stini, 1969).

Anthropological data can be used to check the validity of the evolutionary theory. Ancient humans occupied caves Tabun and Skhul (Karmal mountain, Israel) in the period 500 000 to 40 000 years ago. In the museum of anthropology of Moscow State University (Russia) there are woman and man skulls from Tabun and Skhul caves. The first has still distinct Neanderthal traits (slated lower jaw, arcs, low skull fornix), the second has distinct Cromagnon-like features. According to one theory, *Homo sapiens* and *Neanderthals* coexisted together, and according to another—humans evolved from *Neanderthals*.

Using the technique of generalized photo portrait, a distinct sexual dimorphism has been found in the Turkmen's population: there is just one type of female portrait and two types of male portraits (Pavlovsky, 1980). Similar phenomenon has been found in craniology of Bashkirs: there was a monomodal distribution of characters for female skulls (1 type) and tetra modal distribution for male skulls (Yusupov, 1986). Female type was close to Ugro-Finnish type (geographically these are North-Western neighbors of modern Bashkirs), and male sculls—close to Altai, Kazakh and others (East and South-East neighbors). Same pattern has been identified based on dermatoglyphics: when women had a form of just one adjacent ethnic group and men had a form characteristic of another adjacent ethnic group. In Udmurt population women had Volga-Vyatka region type (North-Western) of dermatoglyphics, while men—of East-Siberian type (Dolinova, 1989). Also Kavgasova noted resemblance in dermatoglyphics between Bulgarian and Turkish men, while women were close to Lithuanian type (**Table 14.4**).

Table 14.4 Phenotypic sexual dimorphism in different populations.

Population	Type of Phenotype	
	Women	Men
Bashkirs	1 type (Ugro-Finnish)	4 types (Altai, Kazakh etc.)
Bulgars	1 type ("Lithuanian")	1 type ("Turkish")
Turkmen	1 type	2 types
Udmurts	1 type ("North-Western")	1 type ("East-Siberian")
Japanese	1 type	1 type

These paradoxes get natural explanation from the theory. Transmission of genetic information from parents to their progeny can occur through four channels: mother → daughter, mother → son, father → daughter and father → son. Common part of information, which is the same for both sexes, is transmitted stochastically through all four channels. Therefore it is mixed fast and is distributed homogeneously in both sexes. "New" ecological part of information, that is in the male sex already, but not in the female sex yet, is transmitted through the male line only (father → son). This way it is delayed in the male subsystem for some time. Another, "old" part of information, which was already lost by males, but still present in females, transmitted through the female line only (mother → daughter). The lifetime of "new" and "old" information equals to sexual dichronism—ΔT (ΔT—significantly more than a life time of one generation). What can be the approximate size of dichronism? Since Bulgar women have "Lithuanian" dermatoglyphics while Bulgar men—"Turkish", and Turkish invasion was going for about 5 centuries, therefore for approximately 25 generations Turkish genes not yet passed to women.

A. Evteev (2008) conducted craniological study of sexual dimorphism of local groups of Karels (North-Eastern Europe) and Barabinskih Tatars (Western Siberia). All of the groups within each nation are close to each other morphologically and have a high degree of homogeneity. However there are significant differences in the sexual dimorphism levels between those groups. The author notes that "the uniqueness of men and women, apparently, can last for many generations "and therefore "it is possible that the mechanisms of predominant realization of the genome obtained from the parent of the same sex exist."

Depending on how the mixing of ethnoses was happened, different results will be observed.
Symmetric hybridization. In symmetric *hybridization* the contribution of man and woman's genotypes from both ethnoses into hybrid posterity is identical. Each ethnos provides all three genetic parts—common, man's and woman's. As only the general part of the information quickly mixes up in hybrid ethnos, there should be two types of men and women. Since distinctions between initial ethnoses frequently have divergent character man's types will be more precisely allocated.

Asymmetric hybridization. If mixing of ethnoses occurs *asymmetrically*, so that two sexes from one ethnos participate in hybridization, and only one sex—from another ethnos, then two different scenarios are possible. In the first case there can be 2 types of men and 1 type of women in the hybrid posterity (the conquerors men in the defeated country), because the "old" part of the information (explicitly female) from the ethnos–conqueror is missing. In the second case, on the contrary, there will be 1 type of men and 2 types of women (the captured women brought into the country of the conqueror), because the missing part will be "new" (man's of the defeated ethnos). Thus in an area of hybridization the geographical change of sexual dimorphism can be observed which can be linked with a historical direction of migrations. The female forms of phenotypes indicate the original ethnos, the male forms—the number of sources and vectors of gene flows occurred during dichronous evolution.

For example, the above facts indicate Ugro-Finnish origination, different on culture and language, for Udmurt and Bashkir ethnoses. The tetramodality of male Bashkir crania can be explained as a trace of three different conquests from south and east. The vector of gene flows in these populations is S.-E. → N.-W., and in Bulgarian population S. → N.

It is interesting to note, that the island population (Japanese) in full conformity with the theory appears monomodal for both sexes. Hence, population sexual dimorphism can serve as another genetic criterion for verification of historical and ethnographic concepts. Thus it can be not only morphological (for example, on dermatoglifics, epicanthus etc.), but also a physiological (for example by a spectrum of groups of blood, enzymes, antibodies), ethological or psychological.

In addition, theory allows to link population impact into hybridization with its *assimilation potential*. The more male genotypes population provides for hybridization, the higher its assimilation potential.

Sexualization in Human Culture

Evolutionary theory of sex treats dioecy as not the best way of reproduction, but as a cost-effective way of evolution. Reproduction, being at the beginning as a solely reproductive program, becomes recombination one. Then it becomes the evolution, and in advanced forms—cultural program (REP → REC → EV → CULT). For example, information about sexual behavior, which the lower animals include in the gene flow (in the zygote), is included in the flow of cultural information at higher forms. It is known that after the growth in isolation puppies and kittens fully retain their ability to reproduce, but young chimpanzees lose it: the males become impotent, the females—frigid. Even after the artificial insemination of such females the lack the maternal instinct leads to the fact that she bites off fingers, hands or head of her young. Therefore, sexual behavior in primates and humans is already the field of culture, where you need to see and learn. Incidentally, this may explain the dominant role of the visual receptor in the sexual behavior of men, role of eroticism and pornography in the culture and much more. The question even arose, not whether it is caused by a large percentage of impotence and frigidity in conditions of urbanization, as compared to rural areas, where children have the opportunity to observe the animals. There have even been successful attempts to treat these anomalies with pornography.

It is known that for the full development of the child the presence of both parents is needed. For example, it was noted that a large percentage of frigid women grew up without fathers.

In the evolution of humans, sexuality becomes further alienated from reproduction and transforms into an independent phenomenon associated with culture. This is evidenced by the wide spread of such "antireproductive" phenomena, as contraception, abortion, sterilization, masturbation, homosexuality, prostitution and pornography. Indeed, sexuality takes much more space in the human life than it's necessary for reproduction. It is difficult to explain the enormous redundancy of sex acts per one conception, the continuing need for them after menopause, during pregnancy and breast feeding, huge role of sexuality in life, culture and human creativity. And the difference of our ancestors from other animals was not in the hard work (which might be expected, according to the concept of F. Engels, that the hand created a man), but rather a year-round (all-season) sexuality. What benefits provide this unique wasteful habit? Indeed, the evolution is economical and does nothing in vain. One possible explanation is related to the fact that the continuation of coitus after conception binds man to woman and forces him to take care of the offspring (**Ch. 6**).

Yet another explanation can be offered. Embryos in the womb need to know about the environment. In animals giving multiple birth (rats, gerbils, mice), embryos are positioned randomly in bicornuate uterus like peas in a pod. It was shown that development of many sex-related traits of adults (not only the primary sex-dependent characters) depends on hormonal environment in mother's womb. It turned out that these traits are affected by sex hormones of the neighboring embryos: females positioned between two brothers (♂♀♂), are exposed to higher doses of androgens, and lower—estrogens than the female, between two sisters (♀♀♀). After birth the first had more masculine anatomy, later age of puberty, shorter life span and reproductive period, fewer litters. They were more aggressive towards other females and sexually less attractive to males. These important regulatory functions are lost with the transition towards single birth. According to evolutionary theory of sex—males is an ecological sex, bringing information about the environment. It is hypothesized that this role was assumed by androgens in the sperm of sexual partners of the mother. After all, in polygamous species, their concentration in *utero* is directly related to the tertiary sex ratio.

Acceleration of Selection

"...as well as man can change their pet birds choosing the most beautiful same as ...the females' preference for more attractive males would almost certainly bring to their change "
C. Darwin (1871)

Scientists attempt to explain the role of culture in humans under the terms of natural selection, as an improved survival due to the transfer of technical knowledge and useful to society traditions. However, it's difficult to understand the richness of culture, many of which manifestations can be viewed rather as an expensive waste of time than a struggle for survival.

According to Miller (1999), a substantially larger role of men in the production of cultural objects can be explained in terms of sexual selection. The cultural achievements can serve as a demonstration of personal qualities such as talent, creativity and taste, and can help in the competition for mating partners and their retention.

If we compare the natural, sexual and artificial selection, then its intensity in this series is growing. And the rate of evolution increases also. The abundance of males and deficit of females (from gametes to populations) in polygamous and panmiktic species leads to the substitution of slow natural selection by the sexual one (competition for the female and the female choice). But the male is able to impregnate all available females, while female had to choose only one. This gives sexual selection the features of an artificial one, whereas the female serves as the "breeder". Because males, according to evolutionary theory of sex, are experimenting in different directions, the vector of evolution is often determined by the choice of the female. For Victorian biologists at the time of C. Darwin it was hard to accept such a crucial evolutionary role of females, so the female choice was almost universally subject to ridicule and rejection (Cronin, 1991).

Can sexual dimorphism determine the tastes and preferences of females? By selecting a male that has the highest sexual dimorphism, females accelerate the evolution of their offspring. Such facts are known. Scientists investigated three species of stalk-eyed flies (*Diopsidae*) living in Southeast Asia. In these insects the distance between the eyes, placed on long stalks, often exceeds the body length. DNA studies have shown that in one species, males and females have equal eye stalks, and this species is the ancestral form for the other two, whose males have stalks twice longer than females. According to the theory of sex, long stems is a progressive character. It was found that the sexual attractiveness of males to females is directly proportional to the length of their eye stalks, that is, females prefer "wide-eyed" males to "narrow-eyed" (Wilkinson et al., 1998). In some insects the advantage of rare variants of males in reproduction, which was associated with the choice of the females, was described (Speiss, Bowbal, 1987). In humans, selection of a taller partner is also a norm (Gillis et al., 1980).

Since both at the level of genes and behavior males create sexual dimorphism, and females eliminate it, we can assume that by choosing a male that has the highest sexual dimorphism, females accelerate the evolution of their. In the acceleration of evolution, especially cultural evolution, the more important role belongs to behavioral features and learning.

Let's leave roller skates and bicycles in the woods, and hang rings over the sea surface. We'll have to wait very long for bears to learn to ride and dolphins—to jump through the rings. Contrary, in the circus they learn much faster because the tamer teaches them with a whip and cookie. May be surprisingly rapid human progress is obliged to the "trainer", too. The hypothesis that give females the role of a "trainer" and the sex—the role of a "big stick", feedback, shaping the useful for posterity behavior and accelerating learning, has an interesting confirmation. The experiments were conducted with two geographical subspecies of cowbird (*Molothrus*), with a different repertoire of male songs. It turns out that a mute female influences the repertoire of the male, who grew up in her cage (King, 1983).

Chapter 15

Sexual Dimorphism
in Pathology

*"- The medicine of an individual is opposite to the interests of the population. ...
when we save the current generation, we spoil
the health of the future (generations)."*

V. Geodakian, from interview

The theory can be applied to pathology, if we will consider an organism's response to harmful environmental factors as an attribute. Any initially **pernicious** for the organism factor of environment, as a result of selection and evolution, gradually becomes only **harmful** to it, then **indifferent**, over time it becomes **useful**, and ultimately **necessary**. Such a situation occurs with prolonged exposure of antibiotics to bacteria, insecticides to insects, and other similar situations.

According to the theory, an ecological sex (male), being evolutionary vanguard of a population, should precede a female one then going on the sequence of adaptation steps: *pernicious → harmful → indifferent → useful → necessary*. That is, it is possible that a factor that has became useful, or necessary for the male, is still harmful, or fatale for the female. Therefore *epidemiological sex ratio (ESR)*—a ratio of ill men and women, in the given array changes accordingly. It is at its maximum at the beginning of the adaptation array (in a pernicious stage of adaptation), and falls to a minimum at the end (at the stage of necessity).

If any illness strikes males more frequently (*ESR » 1*) ("man's" diseases: gout, cancer of the larynx, tongue, esophagus, lungs, rectum etc.), it means, that the new pernicious factor of environment has appeared and male sex conducts "vanguard fights" in the search of new ways of evolutionary development. Applying chess terminology, it is a debut phase of a population game with the environment, played (and endows) mainly by males. At this stage males are mostly the victims. (For example, on a cancer of the larynx, or lungs in some countries *ESR = 25–30!*).

If men and women suffer in equal proportions (*ESR ≈ 1*), this means that for a population the factor of environment became indifferent or neutral. The population as a whole is adapted. The adaptation of a males is genotypic, because after the appropriate selection they acquired a new distribution of genotypes. The adaptation of females is phenotypic, modificational. It is achieved by means of wider reaction norm compare to males. So, the certain "distance" between males and females is established on the adaptation scale of "pernicious—necessary" (for example the environmental factor can be already useful for males, but still remain harmful for females). This corresponds to the middle game of chess played by both sexes. The only difference is that males battle with environment for the new information and pass it to females without selection, by means of bloodless (genetic) way.

If women get sick more often than men (*ESR « 1*) (*"feminine"* (but not gynecological!) illnesses: cholecystitis, obesity with a diabetes, thyroid tumors, etc.), this means that environmental factors became already useful or necessary for men, but still remain pernicious or harmful for women. For example, the man's liver, "have learned" safely utilize alcohol calories, because it is already acquired the appropriate enzymes. Contrary, women's liver doesn't have them yet and as a result gets poisoned, and have cirrhosis 10 times more frequently. This corresponds to a chess endgame, which is played mostly by females: "rearguard" fights when old evolutionary positions are abandoned.

Such an interpretation can easily explain many epidemiological phenomena and predict new regularities. For example, the relationship between the type of disease, it's severity and an *epidemiological sex ratio*. For "masculine" illnesses (a stomach or skin cancer) the ESR is decreasing with increasing the rate of disease, and for "feminine" ones (endemic and diffuse toxic goiter, cholecystitis) on the contrary, it grows.

From this pattern we can make some, very interesting, epidemiological conclusions. For example, theory allows connecting different ESR values on AIDS in different populations with duration of "exposure" of the given population with this infection. If for Asia *ESR ≈ 15*, with the minimal number of patients, for America and Europe *ESR ≈ 10*, and for Africa *ESR ≈ 1*, with a maximum infected, the theory predicts that African AIDS "is more old" then American and European, and Asian AIDS is the "youngest".

When analyzing the data, of course, it's needed to make adjustment for ways to spread the disease and different probabilities to get infected. For example, while AIDS was not spread beyond risk groups (male homosexuals), the infection of men would be much higher, with the transmission of the virus through blood (drug addicts) both sexes can be infected, by, more probable transmission of the virus through blood (addicts) can be infected by both sexes, and with heterosexual contacts the different rate of infection transmission is possible from woman to man and vise versa.

The higher frequency of disease in men, the more severe form of illness is observed in women. Besides high ESR values tell that the disease is new, the cause of disease can be found in the changes of environment, and illness should be treated as "vanguard fights". Higher frequency of disease in women (*ESR « 1*), tells that the cause can be found in sexual dimorphism (male sex already has a trait, but female sex still doesn't). The illness should be considered as a "rearguard" fights. And in this case, accordingly, heavier form of illness is observed in man. In other words, the maximal rate of disease takes place at an equal sex ratio (*ESR = 1*), and both sexes have the minimal severity of illness. Minimal disease incidence rate is observed at sex ratio shifts in any direction (for *ESR » 1* and *ESR « 1*). In both cases the disease proceeds more easily for the sex which has higher incidence rate, so that for each sex is much worse to suffer from the opposite sex's disease.

Out of this analysis can be concluded that "new" masculine diseases of today will strike women in the future. For example, nuclear form of schizophrenia in the past occurred in men only. Now, as psychiatrists notice, this form is more and more frequently appears in women. Contrary, some forms of dementia (ex. oligophrenia), typical for the female sex, should gradually disappear.

If there is no sexual dimorphism on a specific quantitative character, its distribution curves for males, females, and for the whole population (without taking sex into account) will match. On both sides of the norm (in the medical meaning) in these distributions there are two areas of pathology. In the presence of sexual dimorphism the distribution curves for males and females, while remaining within the overall curve, are shifted on the magnitude of sexual dimorphism. Therefore, one "general population" area of pathology is enriched by males, and the opposite—by females. These arguments allow us to predict the existence of two opposite types of diseases and even characterize some of their features.

Epidemiological Rule of Sex Ratio

It's known that *epidemiological sex ratio* depends on age. From *"Ontogenetic rule of sexual dimorphism"* one can expect, that among males there should be more illnesses, that manifest themselves during mature stages of ontogeny, while among females, on the contrary, there should be more illnesses of the juvenile age (*"Epidemiological Rule of Sex Ratio"*).

EPIDEMIOLOGICAL RULE OF SEX RATIO

WOMEN ARE MORE FREQUENTLY SUFFER FROM CHILDREN'S ILLNESSES, MEN—MORE OFTEN FROM ILLNESSES OF ADVANCED AGE.

Such prediction of the theory also is traced: many "children's" illnesses are also "female's" diseases. For example, rheumatism, whooping cough and pyelonephritis. Contrary, adult's illnesses: tumors, atherosclerosis etc., as a rule, have men as a target. The same law can be traced inside the illnesses of one system. For example, dental caries—illness of the juvenal stages of ontogeny affects women more frequently, but parodontosis—illness of teeth more characteristic for the final stages strikes men more often.

Immune Diseases

From the theory's representations of females that implement the evolutionary trends in conservation, and the male's, realizing trends of change, we can predict the existence of sexual dimorphism on autoimmune and immune deficiency diseases. The first should occur more often in women, the second—in men. This theoretical prediction is also confirmed (Glücksmann, 1981).

Alcoholism

Analysis of alcoholism problem was conducted by G. Skorobogatov (Balluzek et al., 2009). Combining together facts of increased alcohol consumption by men over time and applying to them *"Phylogenetic rule of sexual dimorphism"* he concluded that the evolution process of adaptation to a new food product takes place.

On **Figure 15.1** distribution of men and women depending on alcohol consumption (d), based on the data on the developed countries during the period since 1966-1968 is resulted (cit. by Balluzek et al., 2009).

The figure shows that the greatest amount of alcohol is consumed by men who drink an average of 100-120 g of ethanol daily (half-bottle of vodka or two bottles of wine), whereas among women, the greatest amount of alcohol is consumed by those who drink on average 40–50 g of ethanol (half a glass of vodka or half a bottle of wine). It is visible that the vector of alcohol consumption evolution is directed from women to men and in the direction of increasing consumption.

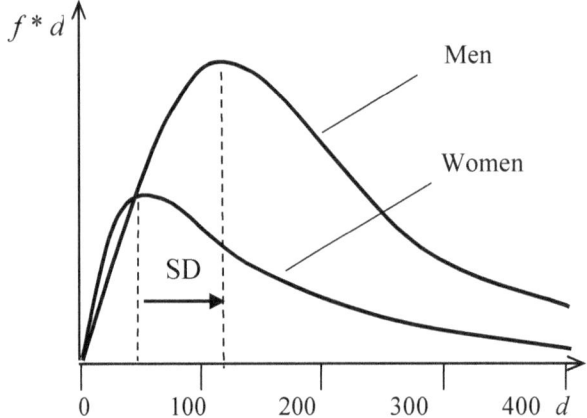

Figure 15.1. Distribution of men and women depending on the amount of consumed alcohol d (g. of C_2H_5OH/day) (Balluzek et al., 2009).

The existence of a clear paternal effect on alcoholism was shown by Ehrman and Parsons (1981). It was noted that children of an alcoholic father become alcoholics 10 times more frequently (26%) compare to children of an alcoholic mother (2%). Similarly brother have more children alcoholics than sister (21 and 0.9% accordingly).

Energy Calculations

What benefits can new food product give? It's known that ethanol does not have vitamins, microelements, or essential amino acids and fiber.

G. Skorobogatov believes that such benefits are related to energy. Alcohol oxidation provides heat and working energy. Speed of alcohol oxidation is nearly constant and does not depend on organism energy needs. Each person oxidizes about 20 g of 40% alcohol per hour. If the subject consumes alcohol with such speed, the alcohol does not accumulate in an organism. However, people who consume higher doses, produce too many calories, and have a tendency to refuse other food products.

For replenishment of daily energy consumption man requires approximately 400 g of ethanol (two bottles of vodka). Non-drinking person will need about 2 hectares of land to grow food, while alcoholic will need only 2 acres of forest which can be used to produce ethanol.

Why People Drink Alcohol?

An ordinary man alcoholic, of course, does not know about evolution and his progressive role in it. Then why he uses a product to which he is not quite adapted? In many societies drinking is traditionally associated with certain events: birth, death, weddings, holidays, successful or unsuccessful deals and events. People drink because they are happy, or sad, for "courage" that is, in all stressful situations. It is known that alcohol relieves stress, and stress can lead to various disorders and diseases.

Men who abuse alcohol, are 4.4 times less susceptible to the endocrine system diseases, 3.6 times less to malignant tumors, 1.5 times—to heart attacks and other cardio-vascular system diseases, and 1.3 times—to the diseases of the nervous system and sensory organs. All of these diseases are diseases of civilization, which are caused predominantly by stress (Lisitsyn, Kopit, 1978).

[It should be noted that alcoholics die 3 times more often from traumas and poisonings, and 1.5 times more often from infectious and skin diseases].

Some of the data confirm abovementioned assumptions. Despite the fact that among alcoholics there are 3–4 times more men, the incidence of and mortality from alcoholic cirrhosis is almost 3 times higher among women. It was also suggested that alcohol and its associated mortality from cardiovascular diseases and external causes act as a tool for removal of intra-family and social aggression through increased mortality of male part of the population (Akopian, 2009).

Evolutionary role of Cancer

Evolutionary aspects of cancer were discussed in several articles (Eaton et al., 1994; Simpson, Camargo, 1998; Greaves, 2000).

The wider the population variation, the safer it is, because it can determine changes of environment earlier. At the same time, the selection is higher and the population has to pay high cost for its evolution. On the contrary, if the variation is narrow, the evolution is less costly, but the risk is higher. So, in every environment a certain optimum of variation exists that minimizes both the risk and the cost of evolution.

Entropy forces of environment (outbreeding, mutations, errors of translation, radiation and heat) increase variation. Natural selection narrows variation and therefore regulates its optimum. But it works too coarse. For example, the first to die from hunger will be children, the most valuable, but weakest age group. To prevent this, more precise mechanism was developed. This mechanism is cancer.

Cancer is considered as one of the mechanisms of natural selection, eliminating individuals of a population according to their biological value—sex, age and information role. Certainly, natural selection uses all illnesses, but cancer is a tool specially created for it. It precisely discharges children with birth defects, as well as mature individuals not serving the tasks of their age group.

The concept can explain many features of cancer: its antiquity, generality, and close relation with sex (Male > Female) and age (increases). It can explain why children have high frequency of cancer of relatively young connective tissue (blood and bones in particular) while adults—cancer of more ancient epithelial tissue which lines both the outside (skin) and the inside cavities. It can also explain why the same type of cancer can be caused by different reasons and, on the contrary, that one factor can cause different forms of cancer.

Cancer and Age

Cancer may affect people at all ages, even fetuses, but risk for the more common varieties tends to increase with age (Cancer Research UK, January, 2007). There exists a specific regulation of the age groups having different biological value: children > reproductive age > an active old age > a passive old age.

CANCER AT CHILDREN

Spectrum of children cancer considerably differs from that of adults. Adults have about 90 % of epithelial tissue cancers, while children—only 0.8–2 %. The cancer of children is often combined with various developmental anomalies, therefore it is possible to consider, that it prevents anomalies of growth and development from spreading. Thus, Wilms' tumor (kidneys) happens at developmental anomalies of urinary system, lymphomas—at hereditary agammaglobulinemia, a cancer of bones at anomalies of osteogenesis, leucosis—at Down syndrome. Unless the testes descent before the age of about five years the thermal damage to spermatogenesis is irreversible and individuals with undescended testes often develop testicular cancer.

REPRODUCTIVE AGE

The main mission of reproductive age is to leave offspring. Therefore the cancer of reproductive age should eliminate those who do not fulfill its purpose. Based on this assumption one can expect that long periods of abstention, lack or small amount of pregnancies, and refusal of breastfeeding should raise risk of cancer.

In 17th century B. Ramazzini wrote "…cancerous tumors are very often generated in women's breasts, and tumors of this sort are found in nuns more than in any other women. Now these are not caused by suppression of the menses but rather, in my opinion, by their celibate life." Other similar observations that breast cancer was common in convent nuns were made in Italy and France. Rigoni-Stern conducted survey in Verona from 1760 to 1839 and found that nuns were five times more likely to die from breast cancer than married women (cit. by Greaves, 2000, p. 144-145).

A large fraction of women in the USA and Europe now delay their first child-birth until their late twenties or early thirties and some have no children at all. M. Greaves (2000) notes that in less developed societies the gap between post-pubertal fertility and first pregnancy is about 2.5 years (or 30 cycles of ovulation) compare to 17.5 years (or ~200 cycles of ovulation) in the USA. Greaves argues that early and more frequent pregnancies reduce proliferative stress and risk to ovarian tissue and the breast. Interestingly, that mimicking the protective effect of early pregnancy by administering gonadotropin or high doses of estradiol and progesterone prevents breast cancer in rats (Guzman et al., 1999).

In addition, breast-feeding for two to three years was the norm for ancient societies. Today, in rural African communities breast-feeding of more than 1.5 year is common and breast cancer rate is small. In the USA and Europe, most women breast-feed less than 3-6 months. With unilateral breast-feeding (boat people in southern China and Hong Kong, Canadian Inuit women), the unsuckled breast has significantly greater risk of cancer (Greaves, 2000, p. 149).

According to new general concept of sex hormones the surplus of the hormones has "carcinogenic" action, while hormones of an opposite sex act as "anticarcinogens". This view provides an explanation why testosterone (poison for spermatozoids) is present in sperm. The concentration of testosterone in man's blood after sex drops approximately 20 times. It can be assumed that without condoms his partner will get the hormone.

Usage of condoms prevents pregnancy and sexually transmitted diseases. At the same time the risk of developing breast cancer increases tenfold. Medical statistics of 19th century confirms this. In Japan and Asian republics of Russia rate of breast cancer was 10 times less compare to US and developed European countries.

POSTREPRODUCTIVE AGE

Overall, around 80 % of cancers occur at post reproductive age. The main mission of post reproductive age is to transfer the cultural information to the descendants. Regulation in this age group can also be carried out through a cancer, instead of other illness. It's well known that active lifestyle prevents many diseases, including cancer. In special studies, rats infected by an experimental cancer were divided into two groups. One of them twisted the squirrel wheel and has remained alive, while the control group, without physical activity, perished.

Sexual Dimorphism and Cancer

The analysis made by P. M. Rajewski and A. L. Sherman (1976) for the malignant tumors, has shown, that phylogenetically younger organs or systems of organs have higher values of sexual dimorphism (higher incidence of tumors in males) compare to old ones. For example, such phylogenetically young formations as lungs, larynx, tong and esophagus had the greatest sexual dimorphism values. On the contrary, small sexual dimorphism values were characteristic for the reproductive system and thyroid glands (**Table 5.3 Appendix C**). Organs and tissues that are in contact with environment have higher values of sexual dimorphism. Based on *phylogenetic rule of sexual dimorphism* authors have predicted, that lung cancer incidents will grow, while stomach cancer will decrease, because the sexual dimorphism is increasing in the first case, and decreasing in the second.

Opposite (Mirrored) Types of Cancer

It was noted that an overdose of estrogen treatment of a prostate cancer in men leads to its disappearance but at the same time breast cancer develops (effect of mirror symmetry of Gaussian curve). This tells us that deviation of hormonal ratio from norm can lead to cancer. It's been known for over a hundred years that removal of the ovaries (source of estrogen) reduces the risk of breast cancer. At the same time, continual administration of estrogen to mice causes mammary cancer. Similar picture exists in breast and endometrial cancer statistics—those two forms are discordant over time (Data of National Cancer Institute for the years 1937, 1947 and 1969) (Epidemiology of cancer in the USSR and the United States, 1979).

Another observation. In the USA 87 million women were treated by estrogens from osteoporosis. Analysis of the results showed that the probability of such patients to develop breast cancer was 1.44 times higher than average. In 2002 group research has been stopped because it was discovered that women taking estrogen-progestin pills had high frequency of breast cancer and heart diseases. Contrary, administration of antiestrogen (tamoxifen) provides a 50 % reduction in breast cancer for North American women (Jordan, Morrow, 1999).

In 2003 in the USA frequency of breast cancer has reduced by 7 %. In that year millions of women have stopped taking hormones for prevention of menopause effects after it has been shown that these medicines raise risk of tumors (number of prescriptions in 2001 and 2003—22.8 and 15.2 million accordingly). The most considerable reduction in frequency was observed for estrogen-dependent tumors.

Hormonal Treatment of Cancer

Such understanding combined with the concept of "fractional" sex leads to an important conclusion—breast and prostate cancers are not always possible to treat by hormones based on a passport sex (**Figure 15.2**). The fact is that sperm androgens act as a "carcinogen" for men and a preventive therapy of breast cancer for women.

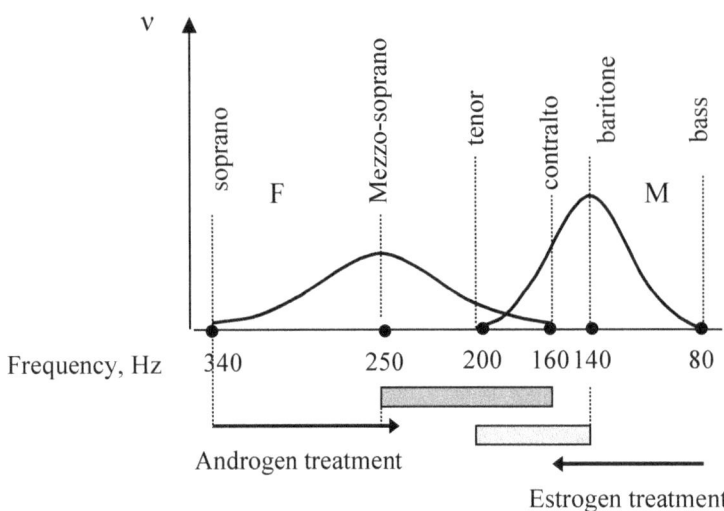

Figure 15.2. Breast and prostate cancer treatment.
Abscissa: Frequencies of women and men voices (Hz).
Ordinate: Frequencies of phenotypes in population (v).
Zones of wrong breast cancer treatment in females ▓
and prostate treatment in males ░.

If the goal of cancer hormone therapy is to bring the extreme groups in each sex to their norm (middle of the distribution) then it is necessary to administer different hormones (one side of the curve should get androgens and another—estrogens).

Doctors explain the ineffectiveness of cancer treatment by the existence of drug-resistant types of cancer. But it's not the cancer that is "not suitable" for treatment but the patient. Different side effects demonstrate different reactions of patients to the most anticancer drugs. The same substance can cause completely opposite effects in different patients: diarrhea or constipation, loss or increase of appetite, drowsiness or insomnia.

In 2006 about 600 thousand women and 300 thousand men died from breast and prostate cancer. They were treated by the hormones of opposite sex. Some of them could have been saved by using more selective therapy based on the person's hormonal status.

Congenital Malformations of the Heart

Congenital malformations of the heart and major blood vessels occupy an important place among other congenital anomalies of human development (Jonas, 1960; Vishnevsky, Galankin, 1962; Keith e. a., 1959; Kjillberg et. al., 1959). According to the made assumption, female congenital diseases of the heart and main vessels should carry the features preserved from the last embryonic stages of intra-uterine development, or have some attributes peculiar to a species, from low steps of evolutionary ladder (nearest past). The anatomic attributes determining man's congenital defects should not have precedents at phylogenetic predecessors of the humans or in the in the embryo. According to the theory they are unsuccessful tests of the evolution process. Defects equally frequently occuring in men and women we shall name neutral.

Two sources of data were used for the analysis—data collected in several large cardiological hospitals of Russian Federation and also the published data (31814 patients total) (**Table 15.1 Appendix C**). The majority of the diagnoses were confirmed at the time of operation or pathoanatomical research, and in some cases as a result of intracardial diagnostic procedures.

The most well defined feminine congenital defects are patent ductus arteriosus (sex ratio 0.37), Lutembaher's syndrome (0.47), ostium secundum (0.54), ventricular septal defect and an open arterial channel (0.66), and a Fallot's triad, which is a combination of atrial septal defect, stenosis of lung artery and a hypertrophy of a right ventricle (0.69).

As is known, the arterial channel makes an integral part of blood circulation of late stages of development of a fetus and normally gets closed during the first year after birth. The oval window, which with some clauses can be identified as atrial septal defect of secondary type, is the second channel connecting the big and small circles of blood circulation of a fetus. If closing of an arterial channel and an oval window does not occur, the appropriate formations are considered as defects. These formations as necessary attributes of a structure of adult normal cardiovascular system can be found at representatives of the lowest (down to reptiles inclusive) classes of vertebrata (Zhedenov, 1954; Dzhagaryan, 1961). Thus, these developmental anomalies can be considered as a return to a near in ontogenetic and phylogenetic sense to the past, and female prevalence among ill persons is in agreement with a new hypothesis.

Lutembaher's syndrome involves two components: atrial septal defect (common atrium) (almost always of secondary type) and a mitral valve stenosis. First of them, as it was shown, is a congenital heart disease of a female type. A mitral valve stenosis related to Lutembaher's syndrome is usually an acquired defect which as is known, in women happens much more often, than in man (Jonas, 1960; Burakovsky, Kolesnikova, 1967). Hence, here there is a combination of two female components; therefore Lutembaher's syndrome is a typical female heart disease.

For a combination of ventricular septal defect (neutral defect) and a patent ductus arteriosus (female defect), presence of a neutral component brings the ratio to 0.66, while for the patent ductus arteriosus only it is equal to 0.37. The similar picture is observed in case of Fallot's triad, which represents a combination of female defect (common atrium) with neutral defect (a stenosis of lung artery). The third component—hypertrophy of right ventricle is not an independent anatomic formation, but a consequence of the former two.

So, all examined female group disorders include an atavistic component and from the formal point of view represent a return to near onto- and phylogenetic past that confirms the initial assumption.

Most well defined man's congenital defects are: congenital aortal stenosis (2.66), coarctation of the aorta (2.14), transpositions of the great arteries (1.9), an abnormal inclusion of all lung veins (1.39), coarctation of aorta and an open arterial channel (1.37).

For simple man's defects (aortic stenosis and coarctation of the aorta) the sexual dimorphism is higher than for the complex defects (transposition of the great arteries, an abnormal inclusion of all lung veins and combination of coarctation of aorta with an open arterial channel). The explanation can be that at transpositions of the great vessels there is always either an atrial or a ventricular septal defect or an open arterial channel, or a combination of these anomalies. Two of these defects are feminine and one neutral. The same can be applied also to a combination of coarctation of aorta with an open arterial channel (feminine defect).

It is possible to say, that no one of man's components of congenital heart diseases have a corresponding similar formation at normal embryo or at phylogenetic predecessors of the person (Zhedenov, 1954; Dzhagaryan, 1961), that is also in accordance with the hypothesis presented.

Other congenital heart diseases are of neutral type. Among them it is also possible to allocate simple (aortopulmonary window, ostium primum, and a stenosis of lung artery) and complex (partial and full atrioventricular canal, Ebstein's anomaly and tricuspid atresia) defects.

Simple defects of this group, as well as female defects, can be considered atavistic. The difference between them is that these defects contrary to female ones represent a return to the past far in ontogenetic and phylogenetic sense. They can be considered as consequence of a block in heart development at early stages of embryogenesis (the first 2–3 months of embryo's life during which the anatomic formation of the heart occurs), and on earlier in comparison to female's defects stages of phylogeny. This assumption is not obvious, probably, only for a stenosis of a lung artery valve. For complex defects of neutral group the sex ratio depends on which of their components prevail—female or male.

The hypothesis presented appears to be more general, than known concepts of Rokitansky, Spitzer (Rokitansky, 1875; Spitzer 1923) and Krimsky (1963) as it explains the genesis not only feminine and neutral, but also masculine defects. The offered division of congenital heart diseases and large vessels into *masculine, feminine and neutral* represents not only theoretical, but also the certain practical interest. It allows considering sex of the patient as a diagnostic symptom.

The measures of a diagnostic information value were used to obtain a quantitative estimate of a symptom "sex" (Sherman, 1970). It appeared, that male and female type defects have rather big value of a diagnostic coefficient. For example, it equals 1.32 for an open arterial channel (that is, taking into account the patient's sex increases probability to have this type of defect, approximately by 1.32). Average value of this coefficient for a symptom "sex" as a whole for all congenital heart diseases is small— 1.10. However, it is stable and there is no mistake in the determination of a patient's sex. There is no time or equipment required to reveal it and there is no harm to the patient. These properties allow recommending a symptom "sex" for the diagnostics of congenital heart diseases.

*   *   *

Discussing the problems of the philosophy of medicine, V. Razumov (2011) notes the importance of treatment of medical problems from the view point of the evolution theory, and in particular the interpretation of "entities of developmental abnormalities, cancer, age-sensitive nosology and causation of sex differences in their frequencies based on the Evolutionary theory of sex, and ontogenetic and phylogenetic rules of sexual dimorphism ... ".

Chapter 16

Mechanisms of Regulation

For the regulation of population parameters such as density, mutation rate, sex ratio, variation and sexual dimorphism the ecological information is needed. Individuals of a population need some kind of "sensors" reacting to the changes of ecological factors of environment. The received information then will be transformed on the organism and chromosomal levels.

Receiving Information from the Environment

Since there are many different environmental factors (temperature, pressure, humidity, amount of food, numbers of enemies and pests etc.) it is natural to think that evolution could not combine each environmental factor directly with the population parameters by means of independent mechanism. Evolution must have produced some "generalizing lever" by means of which any environmental factor could affect the population parameters.

A definite range of values exists for each factor which corresponds to comfort conditions. Discomfort and elimination zones adjoin the comfort one on both sides. Similar zones can be distinguished in the population areal where in a stable environment the comfort conditions are more often in the areal center, the discomfort ones—in the periphery; and the elimination zones correspond to the territories outside of the areal.

The control mechanisms of population parameters are switched on as a result of ecological information received by the organisms in the discomfort zones. They get the information because of their contact with the front of the environment factor. Particular nature of the environmental factor which causes the discomfort of the organism seems to have no significance for starting up these mechanisms. The cause of the discomfort (frost, dry periods, famine or enemies) makes no difference. It means that "generalized" ecological information is as though a "one-dimensional" one (only "good" or "bad"), and their cause is unimportant.

Stress in plants

Hermaphrodite plants have a tendency to increase ratio of quantity of pollen to seed production in such stressful conditions as drought (Galen, 2000), deficiency of nutrients (Helsop-Harrison, 1957), damage by planteating animals (Cobb et al., 2002) or pathogen infection (Lokesha, Vasudeva, 1993).

Pollen amount as a transmitter of ecological information in plants

Plants are linked to their habitat and can not move. What is the plant's source of ecological information? How the plants "know" about ecological situation? Of course, plants can get information about temperature or humidity directly. But how, for example they can get information about sex ratio changes?

When analyzing the areal of a plant species it is clear that in the average more optimal comfort conditions are in the center of the areal (in its depth), while more the extreme, the discomfort ones exist in its periphery (on its borders). The existence of the areal boundary itself indicates that in its natural expansion, the species came upon a "wall" of the ecological niche at this place because of one environmental factor or another. Such factors may be either low or high temperature, or various concentrations of moisture, nutritive substances, enemies, parasites, etc. It is evident that in the center of the areal the density of the population is maximal, and at the periphery, minimal.

In realizing the informational connection between plants, the principal role may belong to the amount of pollen. In dioecious or cross-pollinated species in otherwise equal circumstances, the falling of a large amount of pollen on a female plant means that there are many male plants in the area. Conversely, a small amount of pollen means that there are few male plants in the vicinity. In the center the amount of pollen getting on female flower is always on the average greater than that on the periphery due to different population density. (The pollen of the same species is meant, as the amount of other species' pollen seems to be higher in the areal periphery). Male plants are first who die from the extreme environmental conditions. This also decreases pollen count.

The pollen amount gives information to the female flower about population density, tertiary sex ratio around it, about its location in the areal center or periphery, and about optimal or extreme environmental conditions. Reception of the great amount of pollen always carries on the information about favorable environmental conditions and requires increased production of females with small phenotypical variation. On the contrary, reception of the small amount of pollen carries on the information about unfavorable conditions. This can occur either at the periphery where the density of the population drops sharply, or at the center, but when extreme conditions occur at the center, they eliminate the male individuals first. Both require higher production of male offspring with higher phenotypic variation to speed up the search of evolutionary pathways.

So, in any optimal environmental conditions female flower receives more pollen than in extreme conditions. In other words, small amount of pollen always mean "bad", and big amount of pollen always mean "good".

Consequently, the statement of classical genetics should be reconsidered, that pollen carries only genetic information and the pollen amount which gets on the female flower is of no importance, since one pollen grain is sufficient for fertilization. According to new view, pollen transmits not only genetic information, but also the ecological one, which controls the ratio between stabilizing and leading selection. The information is coded by the pollen amount. The sequence of events is illustrated in **Table 16.1**.

We have presented the theoretically expected picture. Now let us see to what extent this corresponds to the picture observed in reality. Does pollen carry regulating information in addition to genetic information? Does dependence really exist between sex ratios, diversity of offspring, and amount of pollen falling on the female stigma of the flower?

Table 16.1 Characteristics of the Areal (Geodakyan, 1970, 1971).

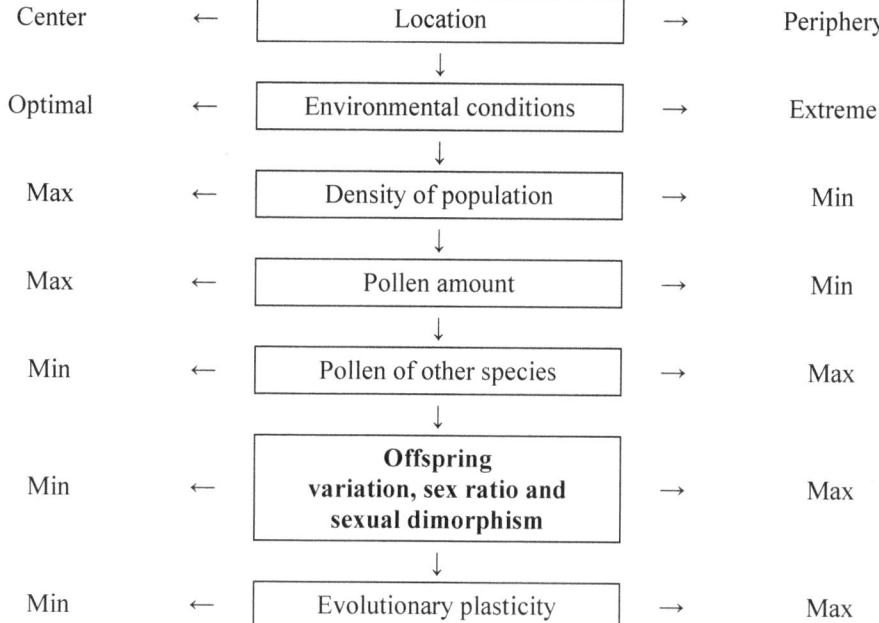

Center	←	Location	→	Periphery
Optimal	←	Environmental conditions	→	Extreme
Max	←	Density of population	→	Min
Max	←	Pollen amount	→	Min
Min	←	Pollen of other species	→	Max
Min	←	**Offspring variation, sex ratio and sexual dimorphism**	→	Max
Min	←	Evolutionary plasticity	→	Max

Dependence of phenotypic variation on pollen amount was discovered by Ter-Avanesian. Fertilization of cotton plant (*Gossypium*), black-eyed pea (*Vigna unguiculata*), and wheat with a small amount of pollen, increased the offspring diversity (**Table 16.2 Appendix C**). Ter-Avanesian writes that as a result of limited pollination "instead of gomogenous sorts we get populations" (Ter-Avanesian, 1949, 1978).

Use of limited pollination on interspecies hybrid *Lycopersicon esculentum x Solanium pennellii* resulted in increase of variation on some quantitative attributes of sprouts (number of leaves, hipocotyl length and weight of sprouts), and on a hybrid *lut1C x racemigerum*—tendency to increase of a recombinant form output on marker genes (Zhushenko, Korol, 1985, p. 318).

There is no evolutionary explanation of the observed phenomena in these and other articles. Thus, dependence of a sex ratio and variation of posterity from the amount of pollen really exists. It's very unlikely that somebody investigated specially the influence of pollen amount on sexual dimorphism. This prediction of the theory is probably easy to check not in special experiments, but by making observations in the center and periphery of a plant species areal. For example, it is known, that a poplar leaves of female trees have more oblong form, and male's—more rounded. Means, on borders of an area of a poplar these distinctions should be more abruptly expressed, than in the middle of an areal.

Taking these dependencies into account would increase the effectiveness of selection. Selection conditions should vary depending upon the task that need to be accomplished. In order to preserve existing forms, let us say in a genotype collection, then, it is necessary to pollinate with a large amount of pollen and to maintain plants in optimal conditions. It is probably done that way.

If there is an opposite—selection problem, it is necessary to create conditions of a "periphery" of an areal, which means to use small amount of pollen and to expose plants, whenever possible, to extreme conditions (do not water, do not fertilize etc). A few selectioners will act this way.

Contrary, in order to obtain viable offspring, the selectioner evidently never spares the pollen in hybridization and always uses a large amount for pollination. He also strives to provide the plants with optimal conditions. This leads to the production of a minimal number of male offspring and decreases phenotypic variety, that is, it stabilizes the population. The result is that the selector decreases his chances of finding new forms. He works as if under conditions of the center of the area while he should be working under conditions of the "periphery". How can this be done?

First of all it is necessary to investigate the dependence of variety of offspring on pollen quantity. With an increase in the number of fertilizing pollen grains, the phenotypic variation of offspring must at first increase to a certain maximum and with a further increase in amount of pollen; it must fall to a minimum corresponding to the condition of the center of the areal (**Figure 16.1**). Maximum diversity corresponds to the optimal value of amount of pollen for the periphery of the areal of the given species.

Figure 16.1

Theoretical relationship between variation of offspring and the amount of pollen.
A', A"—optimum for areal periphery (most suitable for selection).
B', B"—minimum for areal center (most suitable for genotype preservation).

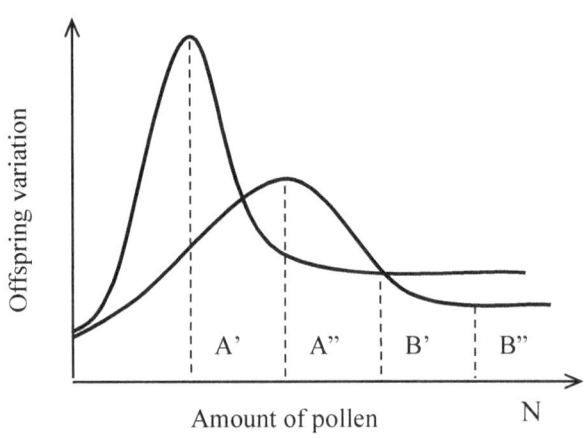

It is obvious that the value of the maximum will vary for various species, depending on their biology and ecology (entomophilous or anemophilous species). For example, according to Ter-Avanesian's data (Ter-Avanesian, 1949), this value for cotton plant is about hundred pollen grains, for black-eyed pea—40–50, and for wheat—10–20. After this maximum is determined it is then necessary to fertilize female plants with just this minimal, but adequate amount of pollen and to provide the plant, as far as possible, with the extreme conditions of the peripheral area, This will increase the effectiveness of selection. It is known, of course, that natural form formation proceeds intensively especially at the borders of the areas. Consequently in selection which has artificial form formation as its goal, it is necessary to duplicate as much as possible the conditions of the boundaries of the areas (extreme conditions, small amount of pollen of the species and much pollen of other species, etc.). From these positions it is necessary to think over also the effect of foreign pollen, which has been described a number of times, the understanding and explanation of which is inadequate in existing representations. Here we have a clear example of how the theory, being incapable to explain the observed facts, simply ignored or rejected them.

People used selection of plants for centuries and it is more art than a science. So, we can assume that some outstanding selectioners may intuitively come to the same conclusions. It can be illustrated by this very interesting statement of great Russian selectioner I. V. Michurin. When speaking about conditions of success in obtaining of new sorts by hybridization, he writes "... Never give seedlings fat soil and furthermore it is necessary to avoid application of any fertilizers. ... Necessity of such mode of education of hybrids was so sharply expressed in business that has compelled me to sell in 1900 former under nursery a black earth site of the ground and to find for the nursery the other site with the leanest sandy ground. Otherwise I never would achieve a success in creating a new grades of fruit plants and in introduction new species of plants into a culture" (Michurin, 1950).

The evolutionary regularities revealed are not only of theoretical but of practical value as well; their good knowledge may among other things increase the efficiency of artificial selection.

Stress as a transmitter of ecological information between the animals

Such a "generalizing lever", a non-specific factor which transmits ecological information from the environment to population in animals is the mechanism realized through the *stress*. Animals actively "move and communicate" with each other (fight, look after etc.) and thus receive the ecological information. For example, if a population of animals luck females, males more often need to fight for the female, or experience a sex starvation longer.

All the unfavorable environmental factors regardless of their specific nature bring about stress. Stress, initiated as a result of discomfort conditions, transforms ecological information into the physiological one, which is coded by concentrations of various hormones in the organism. Further control in the organism is performed by the hormones.

Consequently, frequent stresses in animals and man must enlarge phenotypic (and, probably genetic) variation of the offspring, increase the quota of males and produce more pronounced sexual dimorphism in the progeny. One can also think, that in such situations greater variability of sex chromosomes, and first of all Y-chromosome will be observed. Y-chromosome serves as a link between environment (cytoplasm) and autosomes and in this sense can be named an "ecological" chromosome.

Stress and variation. Chronic stress can create steady changes of hormonal balance which may lead to sharp raise of the form modification process. This can be due to modifying influence of hormones on regulatory processes in ontogeny, and it is also possible, by virtue of a mutagen role of hormones (Salganik, 1968; Kerkis, 1975). Stressed mice had high frequency of lethal dominant mutations, and males also had high frequency of chromosomal aberrations in spermatocites (Borodin, Gorlov, 1984). The chronic stress during mice pregnancy promotes more distinct display genotypic variability of endocrine gland weight (Schuler et. al., 1976). Under stress conditions the value range of arterial pressure in rats increases (Markel, 1984).

Stress and sex ratio. The shift toward males in the offspring sex ratio was observed in colonies of organisms that were subjected to various stress factors (temperature shock, dehydration, starvation, mechanical damage) (Hughes, 2003). Usually animals under stress produce more sons (Krackow, Hoeck, 1989; Pratt, Lisk, 1989; Pratt et al., 1989; Perret, 1990). Demographic data suggest that during the big natural or social shifts (sharp changes of a climate, a drought, war, famine, and resettlements) the secondary sex ratio in humans increases (Sex-ratio, 1960).

Stress and sexual dimorphism. Concerning the dependence of sexual dimorphism from environmental conditions, we shall refer to research at *Mustelidae*. It was shown, that the less favorable conditions are, the usually more abruptly expressed is sexual dimorphism on body weight. Optimum conditions cause the difference in body weight between males and females to decrease (Schubin, Schubin, 1975). "Peak" of boys birth rate (increase in secondary sex ratio) can be observed in certain years in places of strong and long earthquakes (Tashkent, 1966, California, and other).

Sex hormones

Sex hormones are not [just] "sex" [ones], but general system hormones
V. Geodakyan, 2006

It is well-known that male and female sex hormones *androgens* (**An**) and *estrogens* (**Es**), which are chemical antagonists, are developed and are present in both genders, but in different proportions. Androgen concentration in man is approximately hundred times higher, and estrogen—hundred times higher in women. This means that the sexual dimorphism on **An** / **Es** ratio is about 10^4 !

Since the evolutionary theory of sex considers sexual dimorphism as ecological vector (**F** → **M**) and sex determination in ontogeny is controlled by sex hormones, this means that androgens are the ecological, centrifugal hormones which bring the system closer to the environment, and estrogens, on the contrary, are centripetal hormones, removing the system away from the environment. This generalized "ecological" interpretation of sex hormones means that they are not only sex hormones, but also universal. They determine the "distance" between the conservative (female) and operative (male) subsystems.

Sex hormones form phenotypic realization of sex; therefore they should determine the relation of an organism and environment in ontogeny. Androgene /estrogene ratio in the body regulates its information contact with the environment. This leads to a new interpretation of androgens as environmental hormones "bringing the system closer" to the environment, while their antagonists, estrogens, insulate or protect the system from the environment.

By analogy to a sex ratio (**M/F**), ratio **An/Es** is a regulator of a "distance" from the environment, and *evolutionary plasticity* as well. The evolutionary plasticity should **decrease** in **optimum** environment, and **increase** in **extreme** one. Thus, estrogens should increase the norm of reaction, and androgens, on the contrary, should narrow it.

Previously described phenomenon of the increased male mortality represents that total price which the male sex has to pay to the environment for the new information. This price develops of many components: at animals brighter, appreciable coloring and plumage, more risky behavior, in humans—a choice of dangerous professions, the higher susceptibility to "new" illnesses (the illnesses of a "century" or "civilization") and to social defects (smoking, alcoholism, drug addiction, crime and gambling).

Let's consider the action of estrogens and androgens on the characters of males and females (**Figure 16.2**). It can be the same on both sexes. For example, with an increase of the estrogen concentration in both sexes, the reaction norm increases. The reaction norm increases plasticity in ontogeny, and the ability of organisms to leave the areas of discomfort and selection (i.e. move far from the environment). This action increases the average duration of life and slows down evolution (Group 1, women's characters). And an increase in androgens in both sexes narrows reaction norm, thus rising social status, physical strength, aggression, lability, and the rate of evolution (Group 2, male characters).

There can be different action of hormones on males and females. Reproductive rank determining sexual attraction to the opposite sex, fertility, duration of reproductive period, early onset of puberty in females grow with the growth of estrogens, but in men, they grow with an increase of androgens (Group 3, female characters). And the intelligence (and a group of it's derivatives: altruism, humanism, tolerance), on the contrary, in women increases with increase of androgens, and in men—with increase of estrogens (Group 4, male characters). Since the common effect of hormones on both sexes more substantial than different action, it confirms more generalized interpretation of sex hormones (Geodakyan, 2006).

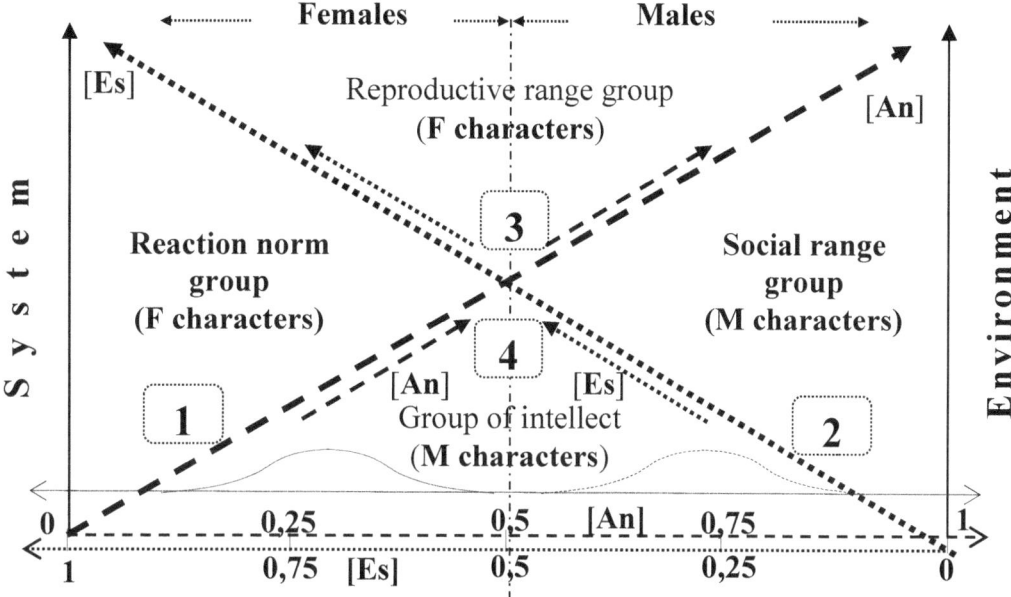

Figure 16.2. Four types of hormonal action (Geodakyan, 2006).
Equally on both sexes: (**1**) estrogens (**Es**) increase female (**F**) characters;
(**2**) androgens (**An**) – increase male (**M**) characters.
Differently on each sex: (**3**) characters increase from "own" sex hormones;
(**4**) characters increase from hormones of opposite sex (paradoxical effect).

According to V. Geodakyan hormones or their precursors can be present in asexual forms, because they also need preservation and change. Some of them, more exposed to the environment, become first subjected to selection die more often. But if they were able to reproduce, they transmit new characters to the next generation. Other organisms, more distant from the environment, preserve themselves for posterity. The peculiar analogue of "men" and "women."

Regulation of Variation in Unitary Systems

Three main characteristics of dioecious population: sex ratio, variation, and sexual dimorphism (SD), are considered by the theory not as constants, specific for given species, as it was believed earlier, but as variables closely related to the environment and determining evolutionary flexibility of a species. The species should have more plasticity in changing (extreme) environment compare to stable (optimum) one. Accordingly all main characteristics should increase in changing environment compare to a stable one. Regulatory mechanisms are similar to those in a refrigerator or thermostat. In all cases there are two mechanisms: a) the mechanism establishing the optimum of an adjustable parameter for the given evolutionary situation and b) the mechanism of a *negative feedback*, watching the deviations from an optimum and keeping up with this optimum (Geodakian, 1987).

As compared to sexual dimorphism, variation is a more general phenomenon: it is important not only for dioecious forms but also for asexual and hermaphrodites. Therefore more general control mechanisms should exist. All reasons about conservative role of heterozygotes and operative one of homozygotes are true also for autosomes, i.e. for all diploid forms. Regulation of variance (σ) at the level of mono-modal, unitary system can be done by changing the ratio of homo-heterozygote. Earlier, the conservative role of heterozygotes and operative role of homozygotes was shown for all diploid forms (Geodakian, 1987; 2005a). In the simplest case of monohybrid crossing:

$$2Aa \leftrightarrow AA + aa$$

growth of variation means an increase of the proportion of homozygotes, while its decrease, vice versa, is related to increases of heterozygotes proportion (small circles **Figure 16.3**).

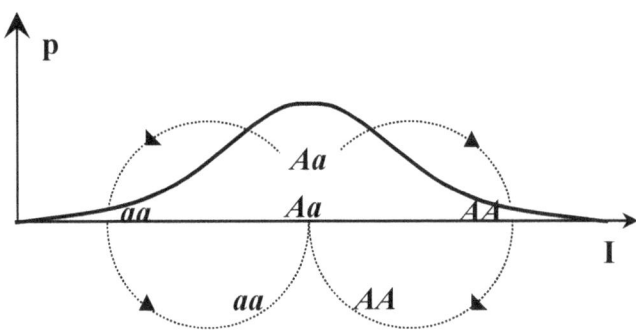

Figure 16.3. A generalized scheme of the variance (σ) regulation (small circles) in unitary systems. (Geodakyan, 2005a).
I - genetic information of the trait, p - frequency of occurrence of the trait.
Three classes of traits: the modal (heterozygotes, Aa).
and deviations from it (homozygotes: aa, AA).

This mechanism is able to regulate the variation and automatically maintain its optimum not only for dioecious forms, but also for asexual and hermaphrodite ones. It was shown (Chetverikov, 1926; Fisher, 1930; Mather, 1953) that the change of equilibrium to the left (hybridization, outbreeding) increases heterozygocity, raises potential variability and narrows variation. On the other hand, change of equilibrium to the right (inbreeding) decreases heterozygocity, raises free variability and variation.

Regulation of Variation of the Sexes and Sexual Dimorphism

Variation of the sexes and sexual dimorphism in dioecious forms is regulated by sex chromosomes. For the realization of the predicted feedback controlling the value of sexual dimorphism, it is necessary, that the posterity of parents with optimum sexual dimorphism had also optimum sexual dimorphism. Posterity of the parents with maximal sexual dimorphism ($♀_{min}$ x $♂_{max}$), should have minimal sexual dimorphism ($♀♀_{max}$, $♂♂_{min}$), and vise versa (**Figure 16.4**). For this to happen, sons should inherit quantitative traits more from the mothers and daughters—from their fathers. The paternal X chromosome cross-inheritance can provide such mechanism, thus maintaining optimal (opt) values of genotypic sexual dimorphism. And Y-chromosome inheritance from fathers to sons can provide its change. For this purpose sons should inherit quantitative attributes more from mothers, and daughters—from fathers. The well-known genetic mechanism of sex-linking (X-coupling) can provide such connection.

Hence, we come to a conclusion that X-chromosomes have greater regulatory modification role in determination of reaction norm and quantitative attributes compare to autosomes. Higher correlations "mother—son" and "father—daughter" on quantitative attributes in comparison with "mother—daughter" and "father—son" were repeatedly described (Beilharz, 1961,1963; Brumby, 1960; Morley, Smith, 1954; Pahnish et al., 1961).

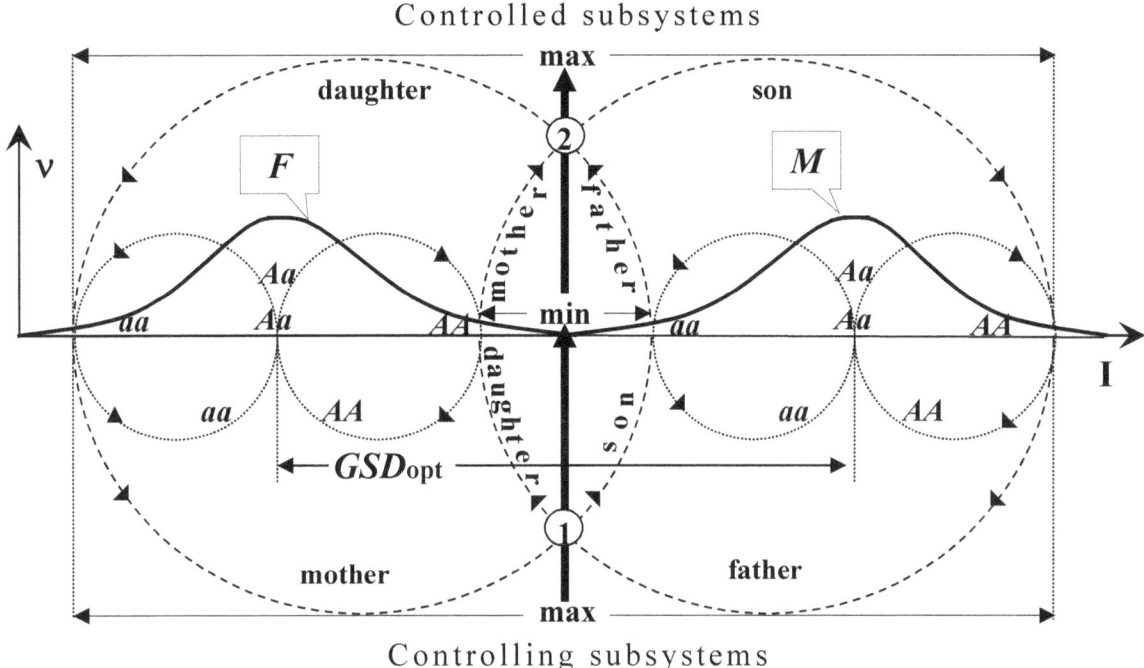

Figure 16.4. Regulation of variation (σ) and dimorphism in binary systems
(large circles) of humans (Geodakyan, 2005).
Cross- inheritance of paternal X chromosome keeps optimal
value of genotypic sexual dimorphism (GSD).
The role of Y- chromosome is to change GSD value.
Max dimorphism of the parents produces *min* dimorphism of offspring (1)
and vice versa (2). I —genetic informationof the trait, n —frequency.
F – females, M – males. Three classes of characters are marked: modal
(heterozygotes Aa), *min* and *max* deviations from it (homozygotes: aa, AA).

Small circles show the regulation of variation at the level of monomodal, unitary systems (sub-population of the same sex). Larger circles indicate the regulation of variation and dimorphism for the whole system.

Regulation of Sex Ratio

Two negative feedback mechanisms are possible (Geodakyan, Geodakyan, 1985):

1. The initially genetically determined probability to have offspring of given sex is equal for all males and females in population, but environmental conditions may bring about change in this probability. This mechanism may be called **organismic**, or **physiological**. It may regulate sex ratio only in polygamous or panmictic population.

2. Different organisms have genetically determined different probabilities to produce offspring of a given sex. With this mechanism the regulation may be on population level, through greater or lesser participation in reproducing individuals, giving birth to a greater number of males or females. This mechanism may be called **populational**. Unlike the organismic mechanism, the populational mechanism, may regulate the sex ratio not only in polygamous or panmictic, but also in fully monogamous populations. Consider a monogamous population consisting of couples and singles of excessive sex. Then, primarily the animals of high reproductive rank will get marriage partners of the deficit sex. Only the negative feedback will be able to minimize the number of bachelors, while a positive feedback will maximize their number and cause the death of the population. For the negative feedback to exist, α-males should produce more daughters, and ω-males—more sons. It should be noted that in males their social and reproductive ranks match, while in females they are usually opposite (social α-females have reproductive ω- rank). This happens because social rank in both sexes is determined by androgens, but reproductive rank is determined by androgens in males and by estrogens in females.

To confirm the existence of negative feedback mechanisms it seemed possible to discuss both the direct and indirect experiments. The direct experiments have shown the dependence of tertiary sex ratio on secondary sex ratio, the indirect ones—the influence of different factors, affecting from tertiary sex ratio, on secondary sex ratio.

Organismic mechanisms of sex ratio regulation

For the organismic type of regulation these factors are:

1. the amount of pollen caught on the female flower: the amount is directly proportional to the number of male flowers surrounding the female flower, and consequently, tertiary sex ratio;

2. aging and different elimination of gametes:

   2.1. aging of pollen and delayed pollination in plants, because the more male plants are around female plant, the less time in average is required to pollinate it and vise versa;

   2.2. aging and elimination of gametes in the body of males or females (delayed fertilization), because the aging probability in the organism of abundant sex is always greater than in the organism of deficient sex, so there exists a relationship between gamete aging and tertiary sex ratio.

3. sexual activity for polygamous animals, which is directly related to the number of the same sex, and reversely proportional to the number of the opposite sex, thus also dependent on tertiary sex ratio.

Table 16.3 (Appendix C) shows the results of direct experiments on eight animal and plant species. It has been shown that an increase in the tertiary sex ratio leads to a decrease in the secondary (Geodakyan V. A., Geodakyan S. V., 1985).

At lizards the results were inconsistent: in two studies shifts of a secondary sex ratio in a direction opposite to tertiary sex ratio have been found (Olsson, Shine 2001; Robert et al. 2003), in two other studies no significant effects were found (Le Galliard et al. 2005; Allsop et al. 2006), and in the third one the production of an abundant sex, at least in the first brood (Warner, Shine, 2007) has been noted. Parkes observed the diminishing of secondary sex ratio by increasing the tertiary sex ratio in mice (Parkes, 1925; 1926). For natural rat populations Snyder (1976) cited White's observations (White, 1914) for the time of the plague epidemic in India, when almost completely adult female elimination was observed. White writes: "…for compensation of almost completely adult female disappearance only females are born".

Humans are not strongly monogamous, therefore different secondary sex ratio deviations from 1 : 1 may arise. For Nigerians Thomas (1913) has reported secondary sex ratio values in relation to wives' number (**Table 16.3** and **Figure 16.2**). It may be concluded that there exists reverse relation between secondary and tertiary sex ratio. Extensive study (14420 births) of the influence of operational sex ratio on the sex ratio of newborns in preindustrial (1775–1850) Finland showed that more sons were produced when males were rarer than females (Lummaa et al., 1998).

Furthermore, a negative feedback exists, an increase of the secondary sex ratio can be observed in harems. **Table 16.3** and **Figure 16.2** show data on three harems. These data are not easy to account for by purely stochastic sex determination. The probability of random deviation from 1 : 1 under such boy excess is 10^{-15}. The number of mothers and progeny makes the effect statistically significant; however the number of fathers is very small.

Figure 16.2

Secondary sex ratio in harems.

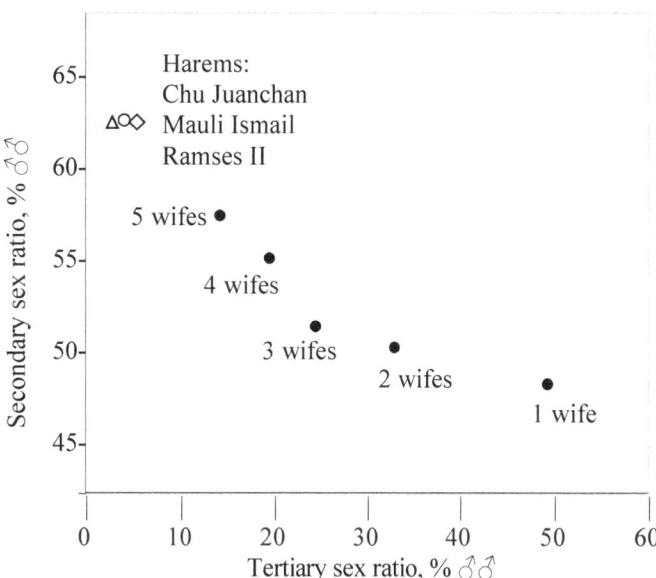

Male deficiency is observed during war-time and after (Geodakian, 1965; 1972; James, 1976). During this period an increase of male birth on 1–2% can be observed (see **Figure 3.1 Chapter 3**). These statistically significant data are known as "phenomenon of war years", because in peace time human sex ratio is approximately stable.

Theory can also predict excessive birth-rate of boys in the "female collectives" (textile towns and others), and girls - in the "male" collectives (expeditions, ship crews, geological parties, seaports and others).

Different forms of organisms may have different mechanisms of negative feedback regulation.

NEGATIVE FEEDBACK IN PLANTS

The amount of pollen and secondary sex ratio. The amount of pollen caught on the female flower in cross pollinating plants can regulate other population parameters, in particular sex ratio (Geodakian, 1977; Geodakian et al., 1967). **Table 16.4 Appendix C** shows the dependence of the secondary sex ratio on the amount of pollen for six dioecious plant species from four families. Predicted rule is confirmed for all studied species of plants: then pollen amount raises, the number of male plants in the progeny is diminished and vice versa. The amount of pollen depends on the density of planting, therefore in dense planting the share of female plants must increase.

Aging of pollen. Ciesielski (1911) studied sex ratio dependence upon pollen aging. In his experiments the seeds received from flowers pollinated with fresh pollen, give up to 90% of male plants. Old pollen pollination (12 hours and more) gives 90–100% of female offspring. However, Lilienfeld (1921) and Bessey (1918; 1933) did not confirm these data.

Delayed pollination. Female plants can estimate density of male plants around them by the time of pollination delay. Miglia and Freeman (1995) showed that spinach (*Spinaceae oleracea L.*) flowers pollinated on the day of anthesis produced a slightly female-biased sex ratio. Delayed pollination for 2 weeks produced a significantly male-biased sex ratio (**Table 16.6 Appendix C**). Similar results were obtained later by Freeman (Freeman et al., 2007) on Atriplex (*Atriplex povellii*). Pollination was accomplished by using fans to blow pollen collected from at least 3 males. Maximal effect was observed on plants that were deliberately not pollinated. Authors think that such plants were still pollinated by very small amount of pollen and that increased secondary sex ratio is a result of combined action of two effects—sparse pollination and it's delay.

NEGATIVE FEEDBACK IN INSECTS

Bees and other hymenoptera, ticks have an unusual negative feedback mechanism. From fertilized eggs only females (or females and males) are born, while from unfertilized ones—only males are developed (Flanders, 1946). Generally, the fewer males there are in the initial population, the fewer eggs are fertilized, and more males appear in offspring.

SEX RATIO AND SEXUAL ACTIVITY

Sexual activity (SA) may be the chain link between the tertiary sex ratio and the secondary sex ratio in animals (Geodakian, 1965). On the one hand, it depends on the tertiary sex ratio: for each sex the activity diminishes with the increase of the number of individuals of one's own sex, and increases, when the opposite sex number rises. On the other hand, SA is related with organismic physiological parameters. Consequently, if negative feedback relation exists, and is realized through SA, then high male birth-rate on high SA of males will be observed. Low male SA will result in high female birth-rate. For females the picture is reversed: high female SA results in increased females' birth-rate probability, low— males. So, high SA increases the probability of the birth of a child of the same sex, low SA—the birth of a child of opposite sex.

The lower SA, the older are the gametes participating in fertilization. **Table 16.5 Appendix C** shows the influence of male SA and sperm aging on secondary sex ratio. The table contains data concerning man and seven animal species (six families). We did not refer to articles dealing with: a) sperm aging achieved by ligation of male sex glands; b) sperm aging taking place in vitro, or in female ducts since these conditions differ from the natural conditions in which female deficiency can be observed. In all species listed in **Table 16.5**, a decrease of male birth-rate is shown, when male SA decreases, or old sperm is used. The results given in **Table 16.5** confirm the hypothesis on SA participation in secondary sex ratio regulation suggesting the possibility of these species having negative feedback relation.

James has analyzed in detail and on great statistical material the relation of the secondary sex ratio and SA in humans (James, 1971a,b,c;1975a,b,c; 1976). He showed, that on high men's SA an increase of boy birth-rate in their progeny is observed.

Aging of sperm. In humans and different animals' species such regularity may be due to faster death or inactivation of Y-sperm, in comparison with X-sperm (Geodakyan et al., 1967). In humans, this suggestion was experimentally confirmed. It was observed that after prolonged periods of abstention, the content of Y-chromatin in sperm decreases markedly. When abstention was 2 days, the percent of Y-chromatin was 43.5, when abstention was 14 days or more—37.2%. This effect is not due to the decrease of Y-chromatin coloring, because its structure in cell nucleus is stable. The decrease of Y-sperm concentration with time may explain its value, which is lower than the theoretically expected value (50%). Different Y-chromatin values, as well as their big scattering, have been observed by various investigators (Schwinger, 1976).

Aging of eggs. The different affinity of fresh and old eggs to X- and Y-sperm may be another possible mechanism of negative feedback relation, active in female organism linking their SA with offspring sex ratio. For negative feedback realization it is necessary that on eggs aging (low SA) their relative affinity to X-sperm should be lowered, and to Y-sperm—raised. In case of male deficiency and low female SA the offspring will be born, as a rule, from older eggs, which have more probability to be fertilized with Y- than with X-sperm. Thus the male birth-rate increases compensating, in this way, males' deficiency in population.

Table 16.6 Appendix C shows data on delayed fertilization. Predicted rule is also confirmed in all studied species, both with male and female heterogamy (15 species from 11 families). It should be noted that higher animals do not have such big secondary sex ratio changes, as lower animals possess.

We do not analyze papers containing the results of experiments, conducted with artificial insemination (in which sperm is not viewed in natural conditions); neither do we refer to works describing males and females who are not kept in natural conditions. Many of the articles concerned with sperm aging as well as with egg aging, are treated in Lanman's reviews (1968a,b); the results of studies presented by Lanman seem to be contradictory. Moreover, many of these articles do not give any description of the conditions, in which males are kept before the experiments. The result of the experiment depends on two effects: on eggs aging (promoting male birth-rate) and on sperm aging (promoting female birth-rate). To exclude the second effect males should be kept with females not involved in the experiment and should not be kept in isolation before the experiment.

Populational mechanisms of sex ratio regulation

Two things are required for the realization of the populational mechanism. First it is necessary that in a population there was a genetically caused polymorphism on probability to leave posterity with the certain sex ratio. And second, that this probability should be in reverse correlation with a *reproductive rank* of a person—the higher the reproductive rank, the more offspring of an opposite sex the person should produce. *Reproductive rank* of a male (access to marriage partners) correlates, as a rule, with their *social-hierarchical rank* (access to resources in general). Females may have a reverse correlation, because their hierarchical rank, as well as for males, is defined by their strength and aggressiveness, but reproductive one is determined more by their appeal and compliance.

For years scientists have been concerned with the question "is the birth of a child of one sex or the other a purely stochastic phenomenon?" (for review see Cluttonbrock, Iason, 1986). A vast statistical material studied (5 mill of births in Saxony in 1876-1885 yea) was established, that families in which one sex prevails, appear more frequently while families with equal sex ratio more rarely, contrary to theoretical expectation (Stern, 1960) (see also **Chapter 3**). Bar-Anon and Robertson (1975) have analyzed 150,000 births sired by 107 bulls and found significant differences in the sex of their progeny. In the progeny of some bulls they observed prevalence of calves (1.5 %) and correlation (coefficient 0.5) between the sex ratio in the progeny of the producer and his father.

Trivers and Willard (1973) have studied the relation between parents' *reproductive range* ("success") and that of their offspring sex ratio. They analyzed data on such species as deer, pig, sheep, dog, seal and human. Females of these species, which have more "reproductive success", give birth to more sons. For grey seal (*Halichoerus grypus*) sex ratio in female offspring was 58.2% ♂♂ in the early period of mating season, and 42.5% ♂♂ at the end of the mating season (Coulson, Hicling, 1961).

The shift of the sex ratio in favor of sons was observed in the offspring of subdominant *Water vole* males (Evsikov et al., 1995). Contrary, dominant males of zebra finches (*Poephila guttata)* (Burley, 1986) and yellow-headed blackbirds (*Xanthocephalus xanthocephalus)* produced more sons (Patterson, Emlen, 1980). In domestic chickens (*Gallus gallus domesticus)* no effect was observed (Leonard, Weatherhead, 1996).

It seems that the results obtained by C. Kanazawa (see **Chapter 3**) does not support the existence of populational mechanism in humans since males with high reproductive success produce more boys, and females with high reproductive success produce more girls.

The contradictory results received by different authors, on the same object at different schemes of experiment, may be accounted for by their ignoring the role of population mechanisms (for example, it is possible to ignore the social range of individuals in Drosophila).

In natural conditions disturbances of tertiary sex ratio from optimum always makes one sex "deficient" and the other—"abundant" with opposite changes of their states. Deficient sex, in average, has more SA, frequently have sex saturation, in fertilization participate more freshly gametes etc. And abundant sex, reversely, has less SA, frequently has sex starvation; in fertilization, older gametes participate etc., compared with the same characteristics of optimal sex ratio population. Deviations from tertiary sex ratio optimum firstly reflect animals with low reproductive range. In monogamous populations they remain without mating partners, and in panmictic populations—their SA decreases. Such disturbances always lead to the increase of deficit sex birth-rate independently of organismic or population mechanism occurs and where it acts—inside male or female organism (**Table 16.7 Appendix C**).

Artificial situations are also possible, when the state of both sexes is exposed to similar changes. For example, males and females have low SA and old gametes when kept in isolation, since both sexes are in abundant state. The result of sex ratio regulation depends on which of the parent organism has a negative feedback mechanism. Analysis of natural secondary sex ratio deviations may show the type of negative feedback mechanism: organismic or populational.

Predictions

The his chapter lists predictions that can be made using Evolutionary Theory of Sex. Some of them were already discussed in the book, in such case the reference is given to that chapter.

1. Three main parameters of the dioecious population—sex ratio, dispersion and sexual dimorphism—should decrease in stable environment and increase when conditions change and in any natural or social cataclysms (earthquakes, war, starvation, migrations) (Chapter 8).

2. In males the share of "genetic component" must be larger, and the "environmental" one—smaller than in females. If we compare variance between pair of identical (monozygotic) twins and variance between different pairs of twins, then intra-pair variance should be greater in female twins, while between pairs it should be higher for male twins (Chapter 7).

3. Phenotypic variance in the pure lines should be relatively larger in females and in the polymorphic (wild) populations, larger in males (Chapter 7).

4. Along with the sexual dimorphism, in the phylogeny there must exist also a "sexual dichronism", expressed in an earlier appearance of characters in males. In paleontology, it will manifest itself in the deeper layers of bedding, in which characters appear in males (Chapter 9).

5. Large vertebrates whose evolution was accompanied by increased sizes as a rule should have males larger than females (Chapter 10).

6. During the evolution many insects and *Arachnids* got smaller, therefore these species should have larger females (Chapter 10).
   Confirmed on a large group (173 species) of lower *Crustaceaean*.

7. Sexual dimorphism should be more pronounced in polygamous species than in monogamous ones. Polyandrous species should show dimorphism in the "opposite" direction of polygyny (Chapter 9).

8. The selection features in domesticated animals and plants must be more dominant in males.
   Confirmations:
   The examples are numerous: with the meat producers such as pigs, sheep, cows, birds, the males grow faster, gain weight and provide better meat than females, the stallions are better than mares in sporting and physical labor features, the fine-fleeced rams provide 1.5–2 times more fleece what the sheep produce, among the fur producers males have better fur than females, the male silkworm produces 20% more silk (Chapter 10).

9. The selection of dogs. The theory predicts larger males in large breeds of dogs and larger females in small ones.

10. In males of basset hound and spaniel (selection for the length of the ears) dogs, the relative length of the ears should be bigger compare to females.

11. Evolution of the horses went in the direction of increasing size, and horses tend to have bigger mares. However, some horses were selected towards smaller sizes (e.g. horses of *Falabella* breed have height from 40 to 75 cm and weight 20–60 kg.). Theory predicts that males of such breeds should be smaller than females.

12. When parasitic forms undergo the reduction of traits, the males should outpace females, so on disappearing characters males should be more primitive (Chapter 10).

13. Direction of reciprocal effects on the evolving traits. In reciprocal hybrids on divergent characters of parents, the paternal form should dominate over maternal one. On converging characters—maternal form dominates over paternal one (Chapter 11).

14. In multidimensional niches (tropics) paternal effect should be observed more often than maternal one. Contrary, in the small-size niches dominated by one environmental factor (frost of the Arctic Circle, or heat in the desert), the maternal effects must prevail (Chapter 10, 11).

15. The contribution of the father in heterosis exceeds the contribution of mother. The heterosis effect should manifest itself more clearly at sons than at daughters (Chapter 10).

16. The effect of heterosis should manifest itself more clearly in sons compare to daughters (Chapter 10).

17. At paternal half-siblings (descendants of one father and different mothers) viability, which is closely related to the reaction norm should be higher for daughters, while at maternal half-siblings—vice versa, for sons.

18. The higher the reproductive rank, the more offspring of an opposite sex the person should produce (Chapter 14).

19. Predictions of sex of the progeny in relation to sexual activity (Chapter 14, Table 14.7).

20. With repeated solution of the problem, including both the search and training, at first, until success is determined by searching for and finding new solutions, better results should be in males, but in the end, when the solution is already known and it's perfection becomes the determining factor, better results should be in females (Chapter 12).

21. Female preferences are determined by the direction of sexual dimorphism (Chapter 5, 12).

Pathology

22. All "new" diseases should strike predominantly males.

23. Diseases, which often strike males, are diseases of an old age. And female's diseases are often resemble diseases specific to young age (*epidemiological rule of sex ratio*) (Chapter 13).

24. In post-evolutionary phase (variation of a character is greater in females) the theory predicts the existence of "relics" of sexual dimorphism and sex variance in pathology. "Relic" of variation appears as increased frequency of congenital anomalies in females. "Relic" of sexual dimorphism is manifested by different direction of deviations (atavistic defects for females and futuristic defects for males) (*teratological rule of sexual dimorphism*) (Chapter 13).

25. Male's diseases of today are often female's diseases of tomorrow (Chapter 13).

26. Some diseases, more typical for females, must gradually disappear.

Humans

27. Abduction angles of the thumb from the palm and twisting around the axis must be greater in men than in women.

28. With age, the visual (and other) ipsi-links should be strengthened, which should lead to improved spatial-visual abilities(Chapter 12).

29. Since during the evolution the eyes gradually moved from the sides of the head to the front of the face, the eyes of men should be closer to the nose.

30. Women should have better developed peripheral vision.

31. In humans atrophy of the olfactory nerve develops with age. Consequently, its atrophy must proceed more intensively in men than women. In phylogeny olfaction gets poorer (Chapter 12).

32. Hearing in women at high frequencies should be better than men. With age hearing should deteriorate (especially hearing at high frequencies) and in men—more than in women (Chapter 12).

33. The voice frequency should be higher during childhood and decrease with age, in man more than women (Chapter 12).

34. The relative dimensions of *corpus callosum* in human ontogeny markedly increases. It means that it must be larger in men and grows in phylogeny (Chapter 12).

35. The ratio between the lengths of right and left temporal planes is higher in women. Hence, both in phylogeny and ontogeny it must decrease (Chapter 12).
Confirmation:
In infants this ratio is 0.65 and in adults— 0.55.

36. On verbal tests, which take into account the perfection of execution, women should outperform men. However, on verbal tests, based on a search (search of verbal associations or solving crossword puzzles), men should surpass women (Chapter 12).

37. On spatial-visual tests that require search, the men should outperform women, but on the same tests that require training, the advantage should be for women (Chapter 12).

38. Gender differences on the last evolutionary acquisitions should be maximal. These characters include abstract thinking, creative skills, spatial imagination, and humor. They should be more pronounced among men.
Confirmations:
Famous scientists, composers, artists, writers, comics and clowns are mostly men. There are many women among performers, and actors.

39. There should be more girls born in the "male" teams (geological party, the ships or the men's prison), and more boys—in "female" ones (textile city) (Chapter 14).

Pathology in humans

40. All "new" diseases, diseases of civilization and urbanization (atherosclerosis, cancer, heart attack, hypertension, myopia, and schizophrenia) should strike predominantly men.

41. Diseases, which often strike men, are diseases of an old age. Women's diseases are often resemble diseases specific to children (Chapter 13).
Confirmation:

Diseases of the elderly: cancer, atherosclerosis, parodontosis, etc., tend to hit men more. Many "children's" diseases (dental caries, rheumatic, whooping cough, and pyelonephritis) are also women's diseases.

42. Men's diseases of today are often women's diseases of tomorrow (Chapter 13). Thus, in the past nuclear form of schizophrenia occur only in men, but now it is increasingly appearing in women.

43. Some types of dementia (such as mental retardation), more typical for females, must gradually disappear.

44. The incidence of lung cancer should increase (sexual dimorphism increases with time), and gastric cancer—decrease (sexual dimorphism decreases) (Chapter 13).

45. Autoimmune diseases are more likely to occur in women, and immune deficiency diseases—in men (Chapter 13).

46. Developmental defects prevalent in females are a consequence of excessive conservatism, so they must have atavistic character, whereas the developmental defects, that most often occur in males, are a consequence of excessive lability, and should have the character of the search, unsuccessful trials—futuristic defects.
Verification:
- Among infants born with one kidney, approximately 2.5 times more boys and among children with three kidneys almost two times more girls
- Excess muscles are 1.5 times more often found in men than in women.
- The number of boys born with 6th fingers is 2 times higher than the number of girls.
- Confirmed by analysis of 32,000 cases of congenital heart disease (Chapter 13).

Summary

"Geodakian's theory can be summarized in one phrase:
Males are the Nature's experimental animals."

From forum discussion

In the conclusion we shall briefly formulate main statements of the new theory:

1. The behavior of controlled systems follows certain "logic". Understanding this "logic" can help to explain and predict system's behavior. Theoretical way of solving such problems is the most effective one.

2. The differentiation of adaptive, controlling systems, evolving in changing environment, on two connected subsystems with conservative and operative specialization, raises stability of system as a whole. Therefore the structure of many evolving systems (biological, social, technical, game, etc.) will consist of *"a stable nucleus"* and *"a mobile shell"*, providing information mutual relations of system with environment *(the Principle of the conjugated subsystems)*.

3. It is possible to consider all types of differentiation of alive systems (genotype—phenotype, females—males, gametes—somatic cells, nucleus—cytoplasm, autosomes—sex chromosomes, nucleic acids—proteins) as a specialization on transfer of the information on two main streams: genetic (from generation to generation) and ecological (from environment to the system).

4. From the thermodynamic point of view the living systems are open working systems exchanging matter, energy and information with the environment. There exists a deep analogy between lifeless and alive working systems. The quantity of information is a factor of extensiveness (a working substance) in living systems, and the factor of intensity is information potential.

5. From these positions a genotype—is "an information charge", that substance, which transfer makes useful work, and a phenotype—is "an information potential", the force that actuates a charge. Logic of ontogeny is carrying out the genetic information towards the factors of environment, so phenotype is "a cane of a blind man" making the information contact with the environment.

6. In self-reproductive systems one can trace the differentiation on two conjugated subsystems on all levels of organization: population, organism, cell, nucleus, chromosome (nucleoprotein). Conservative or internal (genetic) subsystems are accordingly: a female gender, gametes, nucleus, autosomes, and nucleic acids. Operative or external (ecological) subsystems are male gender, somatic cells, cytoplasm, sex chromosomes, and proteins. There is a deep analogy between members of each group. The basis of this analogy is the character of information exchange with environment.

7. Under the new concept sex differentiation is the favorable form of information contact of a population with environment. It is specialization on two main alternative aspects of evolution: **preservation** (conservative) and **change** (operative) (**Table S.1 Appendix C** and **Figure S.1**).

8. In panmictic or a polygamous population where every male can impregnate many females, the females realize rather centripetal trends of stabilizing selection (the quantitative side of duplication, a genetic stream of the information), and males—centrifugal trends of directional selection (the qualitative side of duplication, an ecological stream of the information).

9. Advantages of asexual duplication—simplicity and quantitative efficiency, the hermaphrodite one—maximal recombination (assortment efficiency), and dioecious one—evolutionary flexibility and plasticity (qualitative efficiency). More tightly link of a female gender with a quantity of posterity, and a male sex—with its quality, allows trying various solutions of evolutionary problems without risk of fastening of unsuccessful decisions.

10. *Sex ratio, variation* and *sexual dimorphism* are the main characteristics of dioecious population. They are variables, connected with the environment and the evolutionary flexibility of a species. The higher these characteristics are, the stronger are the trends of labilizing or leading selection and the lower are the trends of stabilizing selection, i.e. the higher is the population flexibility in phylogeny. Different environmental conditions require different evolutionary plasticity of a population; therefore these characteristics should depend on conditions of environment. In stable conditions (the optimum environment) they should decrease, and in changeable conditions (the extreme environment)—grow (*"The Ecological rule of sex differentiation"*). There are two mechanisms of their regulation: a) *the centennial*, establishing optimum in the given evolutionary situation and b) the *negative feedback*, supporting this optimum.

11. Wider phenotypic variation provides tighter link of a male gender with the environment. A source of a different variation of sexes to attributes can be: a) a different level of mutations at male and female individuals; б) a different degree of additivity of inheritance of parental attributes by descendants of a different gender; c) different hereditary reaction norm at male and female individuals.

12. The hereditary reaction norm or modification variability of female individuals is wider, than that of males. Males inherit more of a genetic component of a parent trait, while females—more of an environmental one.

13. The wide reaction norm of female individuals gives them higher ontogenetic plasticity (adaptability), allows leaving zones of elimination and discomfort and to group around the population norm. So, in stable environment they reduce their phenotypic variation.

14. The raised death rate and damageability of males is a general biological phenomenon. It is favorable to a population form of information contact with environment, a payment for the new (ecological) information.

15. The higher male death rate and more intensive sexual selection among males combined with their potential opportunity to impregnate many females (the wide liaison channel cross-section with posterity), lead to that the hereditary information on distribution of genotypes in a population, transferred to the following generation by a female gender, is more representative (better reflects distribution in previous generation), and the information transferred by a male is more selective (better reflects the environment requirements).

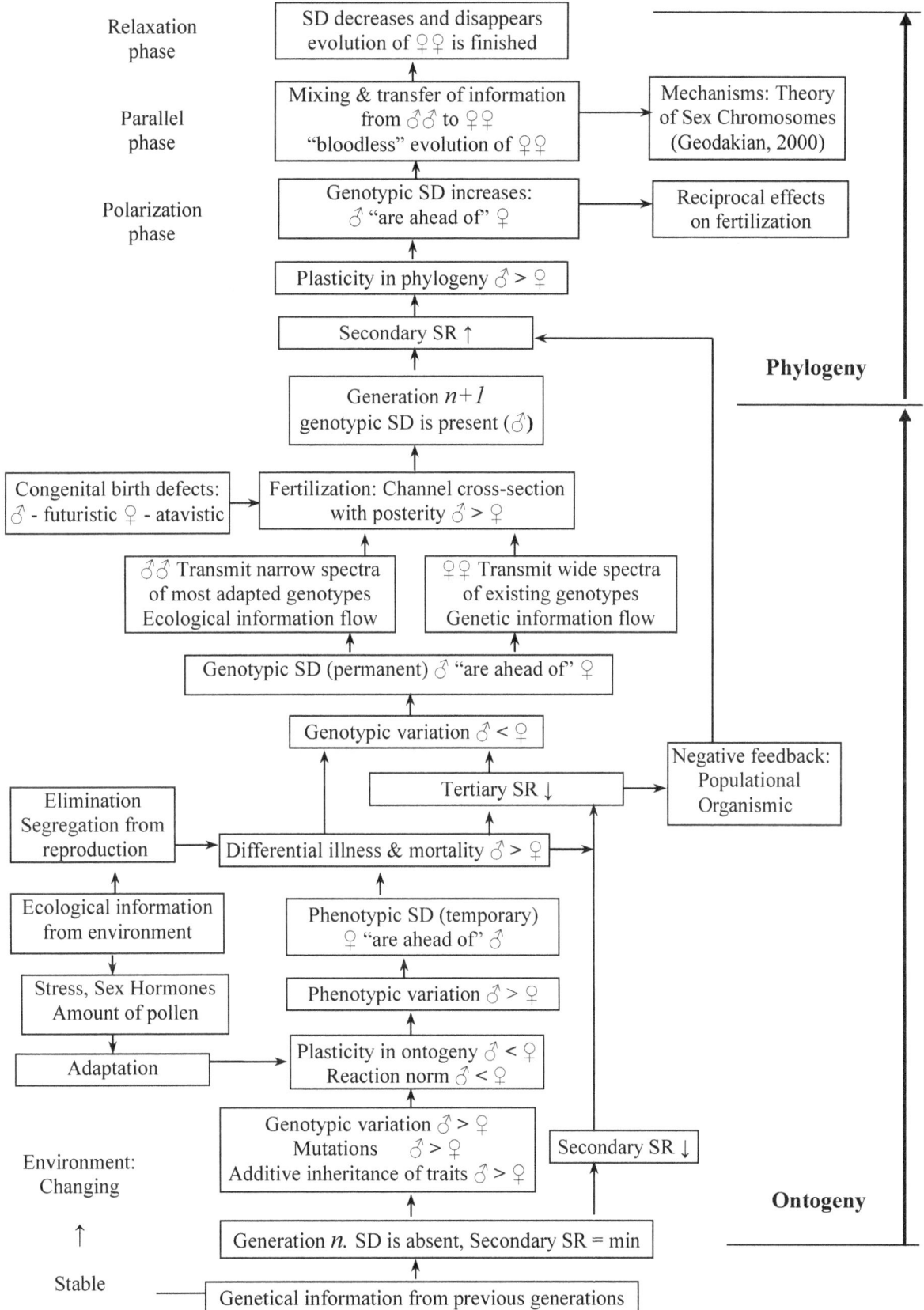

Figure S.1. Effects related to differentiation of sexes (SD – sexual dimorphism, SR – sex ratio).

16. Genotypic variation of females participating in a reproduction, is wider, than male's. Hence, in phylogeny, contrary to ontogeny, males are more plastic (changeable), therefore the evolutionary transformations of a population reflect males first. The new view on appearance of sexual dimorphism as a consequence not only sexual selection (as Darwin thought), but natural and artificial selection as well. Males can be viewed as an evolutionary vanguard of a population, and sexual dimorphism to an attribute as an evolutionary "distance" between the sexes and as a "compass" showing a direction of the attribute evolution.

17. If there exists a population sexual dimorphism on any attribute (different frequency and/or expressiveness of an attribute at males and females) that is when it is possible to speak about the male and female form of an attribute on penetrance or expressivity, then the evolution of an attribute goes from the female form to male one. In other words, the attributes that more often can be found and are stronger expressed at a female should have the "atavistic" nature, and the ones that more often can be found at a male—the "futuristic" one (search). *("Phylogenetic rule of sexual dimorphism").*

18. Variation of the trait points to the phase of evolution. If males have higher variation—the phase is divergent (beginning of evolution, appearance of the character), equal variation means parallel phase, and higher female trait variation—the convergent phase (end of evolution, disappearance of the character) (*"Phylogenetic rule of variation of the sexes"*).

19. If there is population sexual dimorphism according to a certain trait, then during ontogeny (with age), this trait changes, as a rule, from the female form to the male, i.e., the female form of a trait is more characteristic of the initial, juvenile stage, while the male form is more characteristic of the definitive stage (mature, adult). In other words, female forms of traits should, as a rule, weaken with age, while male forms should intensify *("Ontogenetic rule of sexual dimorphism").*

20. At reciprocal hybrids of divergent forms on evolving (new) attributes it should be observed *reciprocal "paternal effect"* (domination of fatherly breed, a line). On divergent attributes of parents the father's form should dominate, and on convergent attributes—mother's form. In particular, the theory successfully predicts existence of paternal effect on all economic-valuable attributes at agricultural animals and plants.

21. Considering *heterosis* as summation of the dominant attributes (adaptations) got by parental forms as a result of a divergence, and having compared to a phylogenetic rule of sexual dimorphism, it is possible to predict closer relationship of a male gender with heterosis, for parents, as well as for hybrid descendants.

22. *The phylogenetic rule of sexual dimorphism* after successful verification on the big group (173 species) the lowest Crustacean on three attributes with distinct sexual dimorphism has been used for the decision of specific problems of evolution *Chidoridae*. As a result, the new place was proposed for the *Leidigia* group in taxonomic system.

23. *The teratological rule of sexual dimorphism* easily explains a different spectrum of congenital anomalies of development of heart and the main vessels, observed at children of a different sex. Elements of defects with which girls are born more frequently, have the "atavistic" nature (ostium secundum and patent ductus arteriosus). They can be found as a norm at human phylogenetic predecessors and at human embryos at last stages of development. Elements of man's defects (stenoses, coarctations, transpositions of the main vessels) have the "futuristic" nature (search). Usage the sex of a patient as a diagnostic character increases the probability of right diagnosis on all congenital heart anomalies by 14%, and on some congenital anomalies—up to 32%. The advantage of character "sex"—stability and simplicity.

24. Proposed interpretation of sexual dimorphism as a phylogenetic "distance" between the sexes, as an evolutionary "last news" already reached males, but not yet females can be applied to all dimorphic animal and plant characters. On species specific attributes the law reveals these phenomena in the field of pathology, on populational characters—in normal state, and on sex-related characters—as a reciprocal *"paternal effect"*.

25. The mechanism of **stress** serves as a transmitter of the ecological information and a regulator of evolutionary plasticity at animals. The **quantity of pollen** getting on a pistillate flower serves the same purpose at cross-pollinating plants. A plenty of pollen corresponds to optimum conditions of environment (the center of an areal, surplus of the male's plants, favorable weather conditions), and small pollen quantity means extreme conditions (borders of an areal, deficiency of male's plants, and adverse weather conditions). The quantity of pollen getting on a pistillate flower, defines sex ratio, variation and sexual dimorphism of posterity. High pollen quantity leads to a reduction of these characteristics and stabilization of a population. Small quantity leads to their increase and destabilization of a population. Application of this conclusion of the theory in selection practice can raise the efficiency of selection.

26. Theory predicts existence of two types mechanisms regulating sex ratio, variation, and sexual dimorphism: a) the mechanism establishing the optimum of an adjustable parameter for the given evolutionary situation and b) the mechanism of a *negative feedback*, watching the deviations from an optimum and keeping up with this optimum.

27. *Sexual activity* (probably through differential ageing of gametes) acts as a link between tertiary and secondary sex ratio at animals. Thus, high sexual activity of the given sex leads to occurrence of descendants of the same sex. The quantity of pollen getting on a pistillate flower serves as a link for cross-pollinating plants: the more pollen, the less male plants are in the offspring, and vise versa. At animals two different mechanisms of realization of the negative feedback regulating a sex ratio of a population—*organismic* and *populational* are possible.

   In the first case, genotypic probability to leave posterity of certain gender is identical for all individuals of a population, but can vary depending on environmental conditions. In the second case, the genotypic probability to produce predominantly posterity of a certain sex is different and is related to hierarchical, social range of an animal. This probability is low for dominant (α-range) males (more females in the progeny), and high for subordinate (Ω-range) ones (more males in the progeny). Regulation is accomplished by changing the participation in reproduction of individuals of different range. The populational mechanism, contrary to organismic one, can regulate sex ratio in strongly monogamous populations.

28. It's proposed to consider inside male and female sexes three gradation of a "fractional sex": modal, feminine, and masculine. In humans they can be represented as average frequencies of voices (in Hz). For women it is a soprano (340 Hz), mezzo-soprano (250 Hz) and contralto (160 Hz). For men—a tenor (200 Hz), baritone (140 Hz) and bass (80 Hz).

29. Phenomenon of homosexuality is considered as adaptation to extreme environmental conditions which regulates quantity and quality of progeny. Hetero- orientation is considered as conservative subsystem, homo- —operative one. The ratio homo/hetero is closely related to environmental conditions: it is minimal in optimal environment and grows in extreme environment. Such view allows explanation of close relationship of homosexuality with gender, left-handedness, high education, and high percentage of homosexuals amongst notable culture and art workers.

30. Anthropological data derived from studies of modern populations can be interpreted in the light of historical processes of mixing of ethnic groups. Delay of new information in a man's subsystem leads to oligomodal distribution of male part of the population. It's possible to relate *geographical sexual dimorphism* in the specific area with the historical flows of male and female genotypes.

31. Considering alcoholism as result of ongoing adaptation to a new food product (ethanol) one can explain increasing alcohol consumption, existence of sexual dimorphism, and high incidence of alcoholism in females.

32. Cancer is considered as an instrument of natural selection removing individuals with developmental defects and those who do not perform the tasks of their age group. This concept explains many features of cancer, its antiquity and polyetiological nature, the close connection of cancer with gender and age, and allows proposing new ways of treatment.

Place of Evolutionary Theory of Sex compare to other theories

Dioecy carries out two functions: genetic—creation of a variety of genotypes by crossing, and ecological—maintenance of the information contact with the environment through sex differentiation. The first function provides evolutionary advantages of *sexual* forms of duplication (hermaphrodites and dioecious) before *asexual*, and second—advantages of *dioecious* forms before *hermaphrodites*. Therefore the theory can be named **"the evolutionary or genetical–ecological theory of sex"**. Taking into account specialization of the sexes by the two main evolutionary streams of the information: genetic (from generation to generation) and ecological (from the environment), conditionally it is possible to name a female sex "genetical" and male one—"ecological".

Theory explains *population* level—organisms and gametes. It analyses phenomena related to the process of *differentiation* (separation into two sexes) and explains the benefits of dioecy over hermaphroditism (**Figure S.2**). A. S. Kondrashov in his classification placed the theory under category of "Immediate benefit hypotheses", because selection amongst "cheap" males and male gametes is more efficient. The theory supplements other theories dealing with the process of *crossing* most of which describes genome, cellular or organism levels. Therefore the advantages of differentiation should not be considered as sole or "leading" factors in the evolution of dioecious reproduction (Kondrashov, 1993). The importance of the theory is that it explains the effects of sex differentiation—sex ratio, variation, and sexual dimorphism. Theories of "crossing", as a rule, cannot do that.

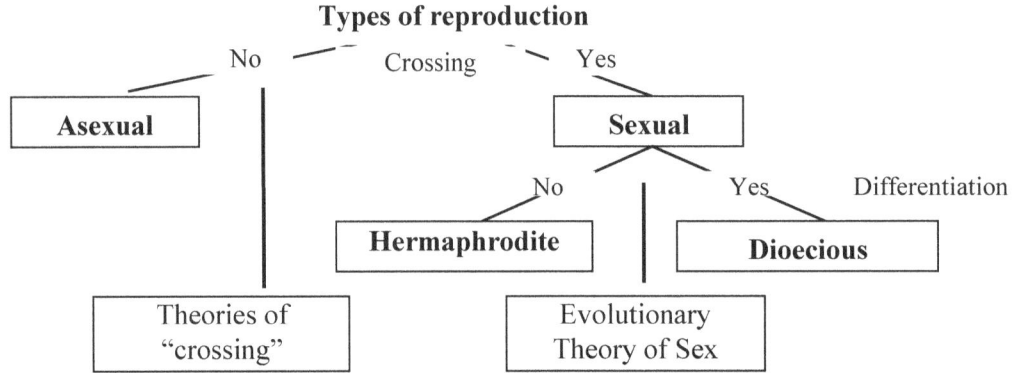

Figure S.2 Place of Evolutionary Theory of Sex compare to other theories.

*   *   *

Summing up it is possible to tell, that the new theory of sex differentiation is developed, allowing understanding and explaining from uniform position the broad spectrum of biological, medical, social, psychological and other problems related to sex. It has allowed to discover the laws unknown before and to give practical recommendations to biologists, geneticists, selectioners and other specialists. Conclusions and predictions of the theory prove to be true on the big factual material collected from the scientific literature and received experimentally that allows considering its reliability rather high.

Till now it was considered, that division into two sexes is necessary for self-reproduction, that sex is a way of replication. It appears that sex is more likely a way of asynchronous evolution.

The evolution theory became a basis of outlook of the modern human. But its problems can be considered as "black boxes" without input—it is impossible to conduct a direct experiment on them. The evolution theory received the necessary information from three sources: paleontology, embryology and comparative anatomy. Now we can add a fourth source of judgment—theory of sex, which gives the possibility to extract the valuable information about processes of individual and historical development of a human species.

Laws, with which we are acquainted with the theory of sex, are applicable for all living organisms—humans, animals and plants. But nevertheless the men, and such of its attributes, as temperament, intellectual abilities, creative abilities, humor and other psychological properties is interesting to us the most. These characters are on "an evolutionary march", and the theory describes them most precisely. It explains why the maximal distinctions between males and females are observed on these, the latest evolutionary human acquisitions, why abstract thinking, the spatial imagination, humor prevail in man.

It is considered also, that biological evolution of the person has come to an end 40-50 thousand years ago. The theory denies this point of view, enabling in the list of human attributes to distinguish evolving attributes from stable ones, at present time constant.

It is also believed that there is no difference in evolution of males and females, that they are in sync. It turns out that this is an asynchronous process, and different roles are allocated for men and women. Gender inequality can be traced to lifestyle, laws and morals throughout the history of mankind. In case of danger, women and children should be saved first, while men play the role of defenders and rescuers who are ready to sacrifice themselves. In many countries, inheritance of property and social status occurred on the paternal line, boys and girls had separate education, married women were punished more severely for adultery, women did not serve in the army, etc.

Attempts to introduce ideas of equality are often based on the idea of **similarity** and **interchangeability** of the sexes. Joint education, permission for women to serve in the military, drug testing (until recently) only on men are just a few examples, the rationality and necessity of which can be questioned. Knowledge of the theory of sex will allow us to responsibly treat the definition of social roles of men and women. The idea of social equality of sexes should be abandoned, in favor of the idea of their **difference** and **mutual complementarity**. Equal rights, cooperation and smart choice of ecological niche rather than sameness and competition should become an ideology that should be implemented at all levels of social life. At the state level, the prediction of the future must take into account the fact that deprivation of the male part of society the possibility to perform its exploratory, innovative, entrepreneurial mission dooms the society to stagnation. At the family level, a proper understanding of the differences between the sexes can remove many conflicts, ensuring the normal development of family life.

Ignorance of the evolutionary role of sex complicates understanding of many sex-related objects and phenomena: the male and female genes, hormones, chromosomes, gametes, gonads, genitalia, features, functions and organs as well as dozens of concepts that are derived from indigenous notion of sex. For example, sex was considered a scalar and sexual dimorphism had no meaning (it was known that men are on average higher than women, but was not known what that means). The evolutionary theory of sex, making sex a vector, allowed to put new meaning into all derivatives also making them vectors. Then it was found that sexual dimorphism is a "compass" of evolution and sex chromosomes are eco-evolutionary chromosomes (many species have differentiation into two sexes without the sex chromosomes) (Geodakyan, 1996,1998,2000). Sex hormones, too, are not just sex, but also system-wide hormones. Estrogens are substances that expand the reaction norm thus remove the system from the environment and slow down evolution, and androgens, on the contrary, narrow the reaction norm, bring the system closer to the environment and accelerate it's evolution (Geodakyan, 2006).

According to T. Kuhn, science goes through a coherent paradigm shift (dominant ideas, attitudes and concepts). In this case, the accumulation of a significant number of anomalies contradicting the current paradigm is the evidence about the crisis and the need for a paradigm shift. The proliferation of theories in the field of sex tells the same (Abraham, 1998).

According to evolutionary theory of sex, wider reaction norm of females allows them to leave the zones of selection. This turns monomodal population into a bimodal, direct "ecology" (contact with the environment) into mediated: **environment** $\rightarrow$ **M** $\rightarrow$ **F** (where $\rightarrow$ is the flow of information), and synchronous evolution into **dihronous**: first males, then, after many generations, females. This is the main difference between evolutionary theory of sex and all other theories, but it is so important, that allows raising the issue of changing paradigms of all biological theories (Darwin, Mendel and fundamental medicine).

Appendix A:
Main Concepts and Definitions

T he inductive way of development in biology has resulted to that basically the identical phenomena, in different areas have received different names. For example, animal species at which males and females are separate, independent organisms, are called dioecious, but similar plants—plants "of two households". Hermaphrodite in animals is an individual making both ova and spermatozoids. They can be produced simultaneously, as well as consecutive. The plant is considered a hermaphrodite if its flower has both male (pollen-producing) and female (ovule-producing) parts. The presence of separate male and female flowers on the same plant is called monoecious ("of one household"). Duplication without fertilization at animals is called *parthenogenesis*, while at plants—*apomixis*. Heterochromosomes at a male heterogamety are designated as X and Y, but if female sex is heterogametic, they are called Z and W. This situation complicates understanding, search for generalizations and creation of theories. Therefore when discussing biological theories, equally applicable for both plants and animals, we shall adhere, whenever possible, to more generalized concepts, terminology and definitions which are described below.

System—environment

The System—set of elements separated from the environment. There are two types of links: between the elements (internal) and between the system and environment (external). One can distinguish two different types of systems: organismic (endosystems) and population (ekzosystems). Population systems usually consist of a big amount of similar (and therefore interchangeable) elements with weak links between them. The same relation to the environment is what brings them together as a system. Therefore they usually compete with each other for food, shelter and other resources. In a population system the law of "selection" is valid—the failure of weak elements results in selection and evolution of the system. Organismic system consists of a relatively small amount of different (and therefore not interchangeable) elements with strong links between them. The elements cooperate with each other. Organismic systems obey the law of a "weak link" (Libich's law)—the failure of a weak link results in the elimination of the whole system.

With allocation of any material system automatically there is an appropriate environment in which the system exists. The environment is always bigger than a system; therefore the evolution of the system is dictated by evolution of environment.

Any environment can be described as a set of factors: temperature, pressure and various concentrations (chemical substances, predators, parasites, victims, individuals of ones own and other species). The system can exist only inside a certain range within each factor. Set of such ranges of existence under different factors of environment represents an *ecological niche* of the system. In such aspect the concept of an ecological niche is meaningful for all systems, and not just for live ones.

The ecological niche can be characterized by the quantity of environmental factors to which the system is sensitive—*regularity* of a niche, and by a range of existence of system on given factor—*width* of a niche. Ecological niches of living systems are multidimensional and, as a rule, have rather small width of ranges under factors of environment compare to ecological niches of lifeless, physical-chemical systems. In other words, living systems are more sensitive and demanding to environmental conditions. But the basic difference between them in mutual relation with environment not in it, but that the alive system, contrary to lifeless, dealing with environment, can change position of an "acceptable" range in a field of the factor of environment that is to adapt (in a broad sense of this word) to change the ecological niche according to changes of environment—evolve.

It's necessary to consider each system's behavior only in a field of the appropriate factor of environment to which the system is sensitive. For the description of physical-chemical phase transitions such field is temperature, for the description of ion behavior—an electric one. Only in appropriate "space" (in certain coordinates) they have a "distinctive" behavior. For living systems such coordinates are time and the generalized factor of environment. Therefore work of self-replicating systems can be presented as a "movement" in two-dimensional space: time—generalized factor of environment. One component of this movement, directed along the time axis, represents a vector of transfer of the genetic information from generation to generation. It realizes connections between generations, internal interactions within the system, or the tendency of heredity. The second component of the movement, directed along the generalized coordinate of environmental factors, represents a vector of ecological shifts. It carries out communications with environment, external interactions of system or the tendency of variability.

On borders of a range of existence the lifeless system undergoes phase (aggregate) transition (crystallization from a liquid to a solid state, boiling of a liquid with steam formation). Alive systems undergo similar transition on borders of an ecological niche. The main difference is that phase transitions in lifeless system is reversible (a solid state ↔ liquid ↔ steam), but phase transition in alive system is irreversible (life ↔ death). Hence it is possible to speak about a phase, or an aggregate condition of life and death. The other difference is that within the living system's ecological niche it is possible to allocate the central zone of comfort and two peripheral zones of discomfort, adjoining to borders of the range. In those two zones of discomfort the organisms experience difficulties, suffer from the harmful factor (freeze, starve, get sick etc.) therefore their duplication is worsened. Behind zones of discomfort there are zones of elimination (the border between zones of discomfort and elimination is actually the border of the system's ecological niche) (**Figure 7.1**, Chapter 7).

For the generalized description of mutual relations of system with environment we shall introduce the concepts of *"space of abilities"* (set of system's internal degrees of freedom, or the list of internal potential programs of the system), *"space of opportunities"* (set of external degrees the freedom—characteristic of environment) and *"space of realization"* —the set of system programs which are realized in the given conditions (**Figure A.1**) (Geodakian, 1970).

Figure A.1
"Space of abilities" *(1)*, "space of opportunities" *(2)*
and "space of realization" *(3)* (Geodakian, 1970).

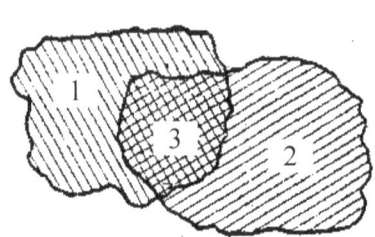

Gene—Character

Attribute (phen) is the allocated property or the characteristic of an individual. There are three types of attributes: quantitative, qualitative and alternative or Boolean ("yes"—"no"). Average value of the quantitative attributes for the population can be received by averaging character values from separate individuals. For alternative and qualitative attributes the average is represented by frequencies of occurrence of the given attribute or gradation of an attribute in a population.

Genotype—Phenotype

Genotype—set of genes of the specific organism, unlike the genome, which characterizes the species, rather than individual. Along with environmental factors genotype determines the phenotype of the organism.

Phenotype—set of characters of an organism. It is possible to define phenotype as "carrying out" of the genetic information towards factors of environment. As a first approximation it is possible to speak about two characteristics of a phenotype: a) the number of directions of carrying out characterizes number of factors of environment to which the phenotype is sensitive, we will call regularity of a phenotype; б) "distance" of carrying out characterizes a degree of sensitivity of a phenotype to the given factor of environment. Both characteristics define riches and development of a phenotype. The more dimensions phenotype has, and the more sensitive it is, the further a phenotype from a genotype, the more developed (or rich) it is. If we compare virus, bacteria, worm, frog and human the richness of a phenotype in this line will increase.

Reaction norm

Norm of reaction—the ability of a genotype to form different phenotypes in ontogenesis, depending on environmental conditions. It characterizes a share of participation of environment in realization of an attribute. The wider the reaction norm, the more influence of environment and the less influence of a genotype in ontogenesis.

The same gene in different conditions of environment can be realized in 1, 2, several or the whole spectrum of attribute values. Of course, in each particular ontogenesis, only one value from this spectrum is realized. In the same way the same genotype in different conditions of environment can be realized in the whole spectrum, potentially possible phenotypes, but in each particular ontogenesis only one is realized.

One can understand a hereditary norm of reaction as the greatest possible width of this spectrum: the wider it is, the wider the reaction norm. Phenotypic value of any quantitative attribute *(P)* is defined, on the one hand, by its genotypic value *(G)*, and on the other hand—by the influence of environment *(E)*: $P = G + E$. If we express the influence of environment as a share (χ) of phenotypic value: $E = \chi * G$, then $P = G / (1 - \chi)$. If we take extreme values of a phenotype at the maximal influence of environment, then:

$$\chi = 1 - (G / P) = 1 - H \qquad\qquad [10]$$

where *H*—heredity.

This will be the reaction norm for a given attribute. So, the reaction norm is that maximal share from phenotypic value of an attribute on which the attribute can be changed by the environment. It is the greatest possible impact of environment in the definition of an attribute.

Let's imagine, for example, that the descendant has received from the parents the genetic information on height of future adult organism G = 170 cm. It means, that if the genotype will receive an unobstructed opportunity of development, the height of an adult organism will be P = 170 cm. If the organism will develop in the extremely adverse conditions of environment (will starve, freeze, etc.) its height will be less, let's say P_{Min} = 160 cm. On the contrary, if development will occur in a maximum favorable conditions, the highest height will be P_{Max} = 180 cm. Hence, the reaction norm on height will be:

$$\chi = |(180 - 170) / 170| = |(160 - 170) / 170| = 1/17$$

It means, that the attribute "height" on 15/17 is defined by a genotype, and on 2/17 by the environment.

If instead of height, we will take blood type or color of eyes, the attribute will be completely defined by a genotype, and influence of environment will be insignificant. If on the contrary, we will examine such attributes as weight, behavioral or psychological attributes influence of environment will be more, and influence of a genotype, accordingly less, than for height. For example, at any genetic inclinations on linguistic abilities the child in the adverse environment (absence of training) will not learn how to speak (Mowgli from R. Kipling's novel).

The reaction norm on different attributes is different and can vary from 0 to 1. The more genes control an attribute, the wider its reaction norm is. The norm of reaction should be minimal for monogenic (monolocus) attributes and increase with the increasing of an attribute polygenic inheritance. As we cannot measure genotypic value of an attribute, it is possible to replace it with the value of an average phenotype (P_{Aver}), which is a phenotype of the organism grown up in normal environment.

It is possible to define the reaction norm by conducting the following mental experiment. If we take three genetically identical zygotes and place one of them in usual conditions of environment, second—in maximum adverse conditions, and third—in maximally favorable ones. Then

$$\chi = (P_{Max} - P_{Min}) / 2 P_{Aver} \qquad [11]$$

Thus the reaction norm characterizes a limit of ontogenetic plasticity, the greatest possible deviation of a phenotype from a genotype. The reaction norm is a resource of ontogenetic plasticity which decreases with the aging of an organism no matter will it be used or not.

Sex

Sex is an alternative character that distinguishes male and female individuals from each other, allows them to produce different gametes and participate in sexual reproduction. We will consider female an individual that produces large usually immobile *gametes (as eggs)*. Accordingly male is an individual that produces small usually mobile *gametes (as spermatozoids)*. Sex of an individual, as any other character, is defined by a genotype and environment. It is possible to speak about genotypic and phenotypic sex and sometimes may not match each other (a genetic male individual can have a female phenotype and vise versa). We will consider individual a *hermaphrodite* if it carries characters of both sexes on the organism level, as a plant "of one household" for example. *Heterochromosomes* will be designated as X and Y for any gamety type.

Mechanisms of Sex Determination

Sex can be determined by internal (genetic) or external to the developing individual (environmental) factors. Temperature-dependent sex determination is the most common among the external mechanisms: zygotes develop into males or females during incubation in the certain temperature ranges. In addition to temperature, sex determination may depend on other external factors—nutrition, lighting, place in the colony, etc.

Sex chromosomes in plants and animals did not occur immediately. Initially, the genes that determine sex were localized in the autosomes. During evolution, the accumulation of morphological differences within such pair of autosomes has led to the emergence of the sex chromosomes. Among many species of dioecious plants sex chromosomes have evolved in only some of them (Ming et al., 2007). There are various ways of chromosomal sex determination.

XX/XY SEX DETERMINATION

Well-known that chromosomal mechanism plays an important initial role in animal, plants and human sex determination at the time of conception (**Figure A.2**).

One sex has cells with diploid set of autosomes 2A and two equal sex chromosomes (XX). So, all gametes of this sex have one X chromosome. This is *homogametic* sex. Another sex besides of the same diploid set of autosomes 2A has two different sex chromosomes X and Y. This forms two types of gametes X and Y. This is *heterogametic* sex.

Figure A.2

Chromosomal sex determination.

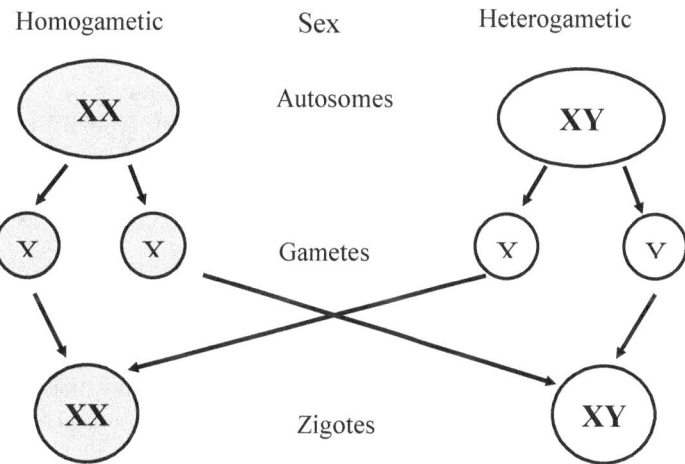

Most animal and plant species have homogametic female sex and heterogametic male sex. These are mammals, many insects and fishes, and some plants.

"... it seems that nature's basic plan is to make a female and that the addition of a Y chromosome produces a variation, the male."

XX or XY chromosomes pass the baton to the undifferentiated gonad to determine its destiny as ovary or testis. Basically, it appears that the presence or absence of a Y chromosome is critical. If no Y chromosome is present, ovaries differentiate, and female development occurs; if a Y chromosome is present, testes differentiate, and male development occurs. Apparently, the Y chromosome produces a substance which induces the differentiation of testes (Jost, 1970). "Thus it seems that nature's basic plan is to make a female and that the addition of a Y chromosome produces a variation, the male." (Hyde, 1979). If the person has XY chromosome setup and testosterone production is blocked the female develops. There are a number of very well-known sex researchers who call females the "default" sex.

The gonad passes the baton to the hormonal secretion of its cells, and the process of fetal differentiation into the anatomy of a male or female continues. By birth, the first part of the program is completed. After birth, the baton is passed to environmental variables that play a determining role in shaping the individual's sex identity—usually, but not always, in accordance with his genetic sex.

The final part of the program is focusing primarily on learning and may lead to heterosexual, bisexual, or homosexual patterns of life-styles (Money, Ehrhardt, 1972).

ZZ/ZW SEX DETERMINATION

ZZ/ZW system is similar to XX/XY system. The difference is that homogametic sex is male one, and heterogametic sex is female. Only limited number of species have such system (birds, reptiles, some amphibians, fishes, butterflies, from plants—wild strawberry). Species with male heterogamety belong to a *Drosophila* type, and with female heterogamety—to an *Abraxas* type (see **Table 7.2 Chapter 7**). XX/XY as well as ZZ/ZW system provides approximately equal number of males and females.

X:A SEX DETERMINATION

In the system of genetic balance sex is determined by the ratio of X chromosomes to the number of autosomes. This type can be found in the genus Drosophila and some plants (sorrel (*Rumex*) and hops (*Humulus*)) (Ming et al., 2007).

XX/X0 sex determination. With this mechanism one of the sexes (homogametic) has two X-chromosomes, while the other (heterogametic) has only one. This type is a variety of X : A mechanism of sex determination, because, as in Drosophila, the sex is also determined by the ratio of X chromosomes and autosomes.

HAPLO-DIPLOID MECHANISM OF SEX DETERMINATION

Gaplo-diploid mechanism of sex determination is widespread among insects (more than 200 thousand species). With this method females (or males and females) are produced from fertilized eggs, and males are developed from unfertilized eggs. species using gaplo-diploid mechanism of sex determination have a huge advantage, because this method allow identification of all harmful recessive genes which reduce species' viability and adaptive capacity. The expression of these genes leads to the death of individuals, carrying them, which allows the population to be free from the "genetic load". It gives a surprising plasticity to the genome of insects, which allowed them to populate almost all the continents.

Population characteristics

For the description of an individual object (element) it is necessary to list its properties (attributes)— weight, size, shape, and color. In order to describe a set of similar elements we will need a new characteristic—number of elements (N), and the list of attributes turns to the list of distributions (variations) on each attribute (σ).

Distribution of an attribute in a population can be described by the average value of an attribute (Δx) and its variation, scatter around this average. Hence, for the description of *asexual* or *hermaphrodite* population as a first glance, three parameters should be enough: number of elements, average value of an attributes, and their variations.

If the system has two subsystems (a population divided into two sexes) for their description 6 characteristics are necessary. However it is possible not to consider the characteristics of subsets itself, but their combinations in pairs. Then it's possible to reduce the description of a two-component system to three main characteristics: concentration of one kind of elements, ratio of average values and variations by the given attribute.

Sex ratio (the concentration of males), *variation* (the ratio of diversity within each sex), and *sexual dimorphism* (difference between average trait values between males and females), are the main characteristics of *dioecious* population. These parameters are closely related and can be mutually derived. For example, sex ratio can be presented as sexual dimorphism on quantity, and variation as sexual dimorphism on variation. Contrary, sexual dimorphism can be presented as sex ratio of organisms with the certain value of the trait, and variation as sex ratio of organisms with the equal deviation from the population norm. Therefore one can speak of the extent of sex differentiation, meaning the common impact of all three parameters.

Sex Ratio

Sex ratio (SR) is one of the main characteristics of sexual population. Generally it is determined by the number of males per 100 females, or in percents. We used the last approach:

$$SR = (N_M * 100\%) / (N_M + N_F) \qquad\qquad [12]$$

where N_M —number of males, N_F —number of females, $N_M + N_F = N$ —total number of organisms.

In relation to ontogenesis stage we distinguish *primary* (I SR), *secondary* (II SR) and *tertiary* (III SR) sex ratio. Primary is zygote sex ratio after fertilization; secondary—sex ratio at birth; and tertiary— sex ratio of mature organisms.

Operational sex ratio is defined as a ratio of a number of sexually active males to a number of sexually receptive females.

Sexual Dimorphism

Set of distinctions between males and females. It is convenient to consider sexual dimorphism *(SD)* on separate attributes which can be expressed in several different ways. For example as a difference of average values of an attribute for males *(M)* and females *(F)*, normalized to the value of this attribute at a female, as a ratio of values of an attribute (or frequency of its occurrence) for males (P_M) and females (p_F), the ratio of a difference of values of an attribute (or frequencies) at males and females to the sum of these values:

$$SD = (M - F) / F \qquad\qquad [13]$$

$$SD = M / F \qquad\qquad [14]$$

$$SD = p_M / p_F \qquad\qquad [15]$$

$$SD = (M - F) * 100\% / (M + F) \qquad\qquad [16]$$

The last approach allows uniform characteristic of sexual dimorphism on all types of attributes and gives symmetric estimations of sexual dimorphism on a scale -100% — +100% (-100% —the attribute occurs only at a female, 0—sexual dimorphism is not present, +100% —an attribute is solely man's).

One more aspect of sexual dimorphism ought to be discussed. Notions of *genotypic* and *phenotypical* sexual dimorphism should be distinguished. In case of organismic sexual dimorphism the character which lies in its basis is realized in the phenotype of only one sex, while in the other it is absent. Therefore sexual dimorphism is of discrete, alternative pattern ("yes"–"no"). Consequently phenotypical distribution of such character proceeds only within one sex, while in another it equals zero. The genotypic sexual dimorphism of such characters is of populational pattern, since they are genotypically distributed in both sexes (e.g. information concerning egg yield of the breed lies in genotypes of hen and rooster as well). So unlike phenotypical sexual dimorphism which can be organismic or populational the genotypic one seems to be always of populational pattern. It means that for all the characters, typical for one sex only (apparently, primary sex characters among them) genetic information is contained by both sexes. Therefore one can speak about egg yield in roosters, milk productivity, butter-fat yield in bulls etc (although it sounds as a paradox). It means that the existence of genotypic sexual dimorphism is principally possible. And according to such characters the roosters may turn genotypically more (or less) egg yielding than hens of the same breed.

Variation of the sexes

By analogy to sex ratio it is possible to allocate also secondary and tertiary sexual dimorphism and variations of sexes for the different stages of ontogeny. Because both variation and sexual dimorphism are related to attributes, and in a zygote the majority of attributes aren't developed yet, the initial characteristics make sense only for the sex ratio. In ontogeny not all attributes are shown from the very beginning as a sex, the majority is shown (appears) with age (growth, weight, intelligence). Therefore, speaking about primary and secondary characteristics or attributes it is necessary to understand those potencies from which the attribute in adult stage will be realized.

Appendix B:
The Principle of Conjugated Subsystems

Separation of any material system automatically involves the appearance of a corresponding environment where the given system exists. Since the environment is always larger than the system, the evolution of the system is dictated by the changes of the environment.

The idea of evolution implies two main and to some extent alternative aspects: **conservation** and **variation**. For better realization of the first aspect, the system ought to be stable, unchangeable, i.e. to be if possible "farther" (in the informational terms) from destructive factors of the environment (**Figure B.1**). But these factors simultaneously carry on useful information about changes in environment. And if the system has to change in accordance with the environmental alterations (the second aspect of evolution) it must be sensitive, labile and variable, i.e. to be as far as possible, "closer" (again in information terms) to the environment. Consequently there exists a conflicting situation, when the system should be simultaneously "farther" from and "closer" to the environment.

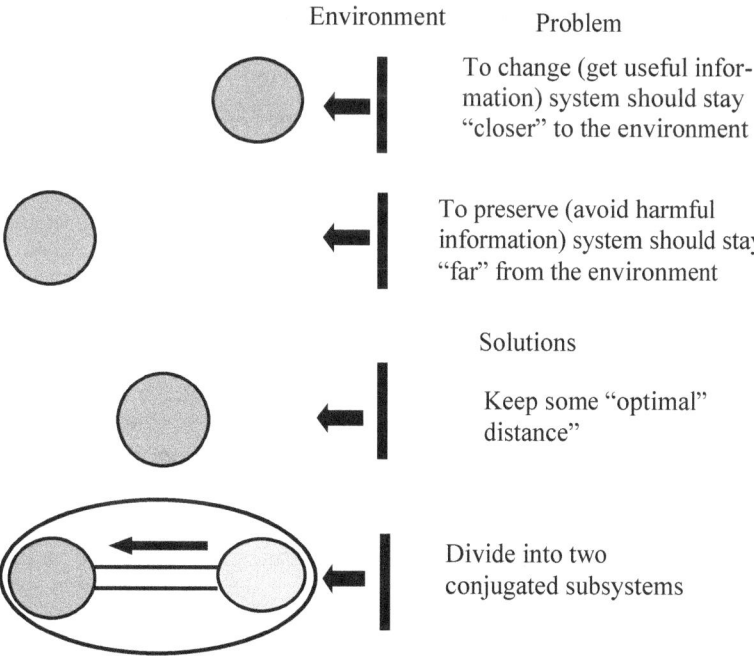

Figure B.1 System—environment relationship.

The first possible solution: the system should be at some optimal "distance" from the environment. The second one: the system should be differentiated into two conjugated subsystems, one of them to be removed "farther" from the environment for preserving the accumulated information, another one to be drawn "nearer" to the environment for receiving new information. The second solution overcomes the conflict to some extent and increases the stability of the system as a whole. Therefore it can be assumed that structures consisting of two conjugated subsystems, i.e. stable "nucleus" and labile "shell", and must be often found among evolving, adaptive systems. This conclusion lays the foundation of the new concept.

PRINCIPLE OF CONJUGATED SUBSYSTEMS

ANY ADAPTIVE CONTROLLING SYSTEM, EVOLVING IN A VARIABLE ENVIRONMENT BEING DIVIDED INTO TWO CONJUGATED SUBSYSTEMS, SPECIALIZED ONTO CONSERVATIVE AND OPERATIVE TRENDS OF EVOLUTION INCREASES THEREBY ITS EVOLUTIONARY STABILITY AS A WHOLE.

In such general form the concept is true for any evolving, adaptive systems notwithstanding their specific nature, whether they are biological, technical, game or social ones. In all the cases when the system is forced to follow the "behavior" of the environment and to shape its behavior accordingly, separation into conservative and operative "departments" increases its stability. If it is an army, it forms intelligence units and sends them to different sides against the enemy. If it is a ship, it has the keel (conservative part) and separately the wheel (operative one). The same with plane and rocket: stabilizers and control surfaces.

Initially the system was uniform and the general stream of information was: Environment → System ($E → S$). After the appearance of the connected subsystems the information flow become: environment → operative → conservative subsystem ($E → o → c$). This means that as a result of evolution the **new subsystem is always operative and arises between a conservative subsystem and environment.**

The evolutionary role of all operative subsystems in the connected differentiations (proteins, a phenotype, a male sex) is the buffer or intermediary. It is genetic information moved towards the environment (selection) for the advanced reception of the ecological information by main, more ancient conservative subsystems (DNA, a genotype, a female sex).

The Structure of Self-reproductive Living Systems

The most fundamental feature of living systems is reproductive ability. Therefore self-reproductive systems have a central place among biological systems.

Considering the structure of the basic self reproductive systems: nucleoprotein, nucleus, cell, organism and population shows, that inside each of these systems it is possible to see differentiation on two connected subsystems. In a population it is two sexes, in an organism—two kinds of cells: sexual and somatic, in a cell—nucleus and cytoplasm, in nucleus—two kinds of chromosomes: autosomes and sexual chromosomes, in nucleoprotein (a chromosome, a gene, a virus) two types of molecules: a nucleic acid and protein.

Is it accidental? Or the principle described is really a basis for these differentiations? One can find similar systems without the specified differentiation. There are populations without sexual differentiation (asexual, parthenogenetic or hermaphrodite), organisms without differentiation on a somatic cells and gametes, cells without differentiation into a nucleus and cytoplasm, and nuclei without sexual chromosomes. However, for some reason in all progressive in evolutionary sense systems, the differentiation on two conjugated subsystems is observed. Those subsystems are specialized on internal and external interactions. In all pairs it is possible to allocate one internal subsystem, specialized on the evolutionary task of **preservation** (P), and external—specialized on other main evolutionary task of **change** (C) (**Table B.1**). First, the presence of preservation and change is the main prerequisite of evolution. The absence of one of them precludes evolution: the system either disappears or remains constant. Second, the ratio between preservation and alteration (C/P) characterizes the evolutionary plasticity of the system. Third, these conditions are alternative: an increase in alteration is associated with a decrease in preservation, as their sum is unity $P + C = 1$. Therefore, without specialization of subsystems, the system must choose a compromise optimum of C/P, while, with the specialization it can maximize both aspects simultaneously.

It is necessary to understand that this division is related to information transfer and have no geometrical (or morphological) meaning. This means that information from the environment gets transmitted into internal subsystems and only after that—into external ones. In cybernetic terms one subsystem is a "constant memory" of the system (female sex, gametes, nucleus, autosomes, DNA), while the other—"operative memory" (male sex, somatic cells, cytoplasm, sex chromosomes, proteins) (**Figure B.2**).

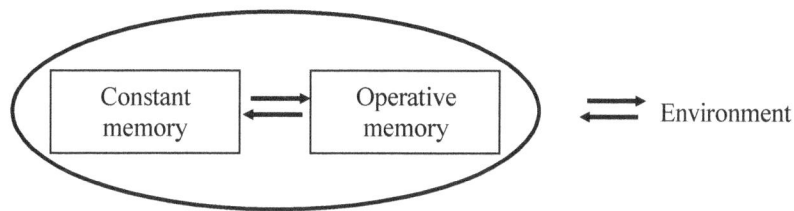

Figure B.2 Constant and operative "memory" of the system (Geodakyan, 1972).

System—Environment

Differentiation of the system into a constant and operative memory creates a structure of "stable nucleus" and "labile shell" in system—environment information relationship (**Figure B.3**). Considering the streams transmitting the genetic information from generation to generation one can tell that capacities of constant memory form an axial (genetic) line whereas capacities of operative memory make a lateral (ecological) line, "carrying out" of a part of the information towards the environment.

Figure B.3

On "system—environment" axis system is divided onto "stable nucleus" and "labile shell".

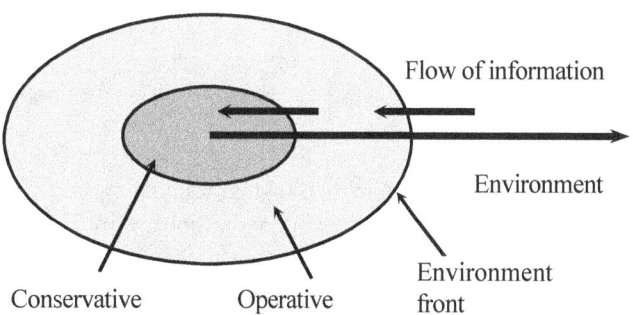

The centripetal information flow from the environment goes first into operative memory of the system, undergoes transformation and selection there. Only after that the part of the information gets into the constant memory (**Figure B.3**). Such structure of the system assumes existence of barriers of some kind between the subsystems that prevent mixing of the information. The described way of storage of the genetic information, in two volumes connected by a liaison channel of controllable section, gives to system the special properties raising its stability.

For maintenance of information contact of system with environment basically through operative memory, it is necessary, that elements of operative memory had the greater variation of attributes in comparison with elements of constant memory, that is the first should be more various than the second.

Realization of primary contact of environment with operative memory due to shift of average values of attributes, instead of a different variation of attributes will not solve a problem, because of the difficulties with alternative attributes. For example, the same individual can be resistant to both heat and cold (nonalternative character), but cannot be big and small at the same time (alternative character).

Conservative and Operative Subsystems

The necessity of processing the new information in the operative memory before it can get into a constant one makes constant memory of a system inert. Inertness of constant memory, its delay from operative in reception of the new information from environment gives its elements features of **perfection**. On the contrary, elements of operative memory get features of **progressiveness**. Consider system evolution in time aspect, we can say that operative subsystem has new features, which will move into conservative one in the future. So, operative subsystem can be considered an evolutionary "vanguard" of the system (**Figure B.4**).

Such a differentiation and specialization of subsystems on alternative tasks of preservation and change provides optimal conditions for the realization of main method of living systems evolution—method of selection, which is a trial and error method. The operative memory is a location for both trials and also errors and findings. This allows the system **to try different solutions of evolution problems without the risk of fixation on wrong ones.**

Figure B.4

On time axis operative subsystem can be considered as "vanguard" compare to conservative one.

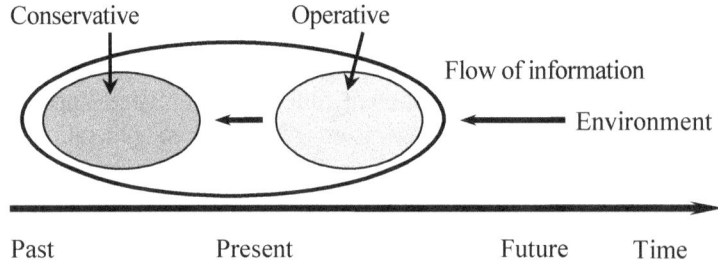

Examples of conjugated binary subsystems

DNA — proteins

Considering the nucleoprotein as a system with subsystems of DNA and protein, we can see that DNA is more stable. The temperatures of protein denaturation are much lower than that of DNA. The goal of differentiation is to surround more resistant to environmental factors DNA molecules by less stable proteins. For example, if the temperature of denaturation of DNA is around 65°C, and protein—about 45°C, then the virus, which is the DNA in a protein shell, will denaturate at 45°C. So, due to the bonded protein, the insensitive to 45°C DNA becomes sensitive. Hence, the protein is an information intermediary between DNA and the environment, able to "warn" the DNA in advance about the warming. Other proteins allow DNA to "feel" cold, to "see", "hear" and thus "learn" about the dangers of the environment.

Morphological structure of known viruses as nucleoproteins also suggests that DNA (or RNA in some viruses), in them represents the "core" and proteins—the "shell". Therefore, the proteins are the first to interact with the environment and the information flow should be **environment ↔ protein ↔ DNA**. Centrifugal flow of information, DNA → protein, has been well studied. This flow of information performs the synthesis of proteins and determines the behavior of the system. Centripetal flow of information: environment → protein → DNA, and in particular, the part protein → DNA, is more interesting. For a long time, the existence of such a flow was denied. Of course, one can imagine that in the process of evolution, when higher levels of organization (cells, organisms) were formed appropriate feedbacks were created, the feedback at lower levels (in this case, molecular) have lost its significance and disappeared. However, some facts are very difficult to explain without assuming the presence of the flow protein → DNA. These include above all the phenomena associated with the formation of adaptive enzymes and antibodies.

Autosomes — Sex chromosomes

Differentiation on autosomes and sex chromosomes do not exist in all species. In addition, there are no visible barriers between the chromosomes. However, there are numerous facts that support the view of the autosomes and sex chromosomes as a permanent memory of the nucleus. Evolutionarily sex chromosomes are much "younger" than autosomes and were derived from them, and Y-chromosome is 20–80 thousand years younger than X-chromosome. It was noted that sex chromosomes are more labile compare to autosomes. Among the nuclei with an abnormal number of chromosomes, vast majority of abnormalities are associated with sex chromosomes (in humans there is known even set XXXXY (Stern, 1960)). Sex chromosomes are the first to be destroyed by ultrasound. Radioactive substances label sex chromosomes (particularly Y-chromosome) more intensive than autosomes. Sex chromosomes (especially Y-chromosome) have higher variation compare to autosomes. There are indications on the predominantly peripheral location of sex chromosomes in the nucleus in humans (Barton et al., 1964).

Nucleus — Cytoplasm

At the cellular level of organization one can notice the differentiation into the nucleus and cytoplasm. The morphology of the cells makes the relationship between the nucleus, cytoplasm and environment rather obvious, both in terms of information flow (environment → cytoplasm → nucleus), and in terms of higher diversity of the cytoplasms (cells of different tissues) and the monotony of the nuclei. It is shown that the nuclei of any cell in the body (at least in some species) contain all the genetic information (read-only memory), while the cytoplasm of the cell determines the specialization of the specific type of cells (operative memory), determines what genetic information should be extracted from nucleus in each case.

Gametes — Somatic cells

At the organism level of organization the elements are cells. Cells of the body are divided into haploid (gametes) and diploid (somatic). First of all, there exists a wide variety (morphological and physiological) of somatic cells as compared with the monotony of gametes. Also obvious is the specialization of these subsystems on the objectives of heredity and variation. Gametes represent a conservative trend, and somatic cells, on the contrary, the tendency of change. Somatic cells receive information from the environment (organisms grow, develop, age, etc.), and only after processing in operative memory for the total life, the final information is transferred into gametes (in the form of elimination, discrimination or privileges of a given individual or in the form of transfer of any mutagenic effects).

Society (right-handed and left-handed individuals)

Differentiation of society into righties and lefties creates two behavioral modes: the conservative regime of right-handedness—the psychology of preservation (analog of females), and the operative regime of left-handedness—reformative psychology of changes, search for new solutions (analog of males). The first group is maximally adapted to the optimal (stable) environment, the second—to an extreme one.

*   *   *

Conjugated pairs of subsystems described are not along. It is possible to find this structure inside other self-reproductive systems. The same relationships probably exist between pares bacteria—bacteriofage, gene—character and others.

Differentiation of the system into conjugated subsystems takes place not only in self-replicating structures, but also in many biological systems and generally in many evolving, adaptive systems. For example, such biological systems as coferment—apoferment, antigen—antibody, right—left hemispheres and subcore—core of the brain, sympatic—parasympatic vegetative nervous system and others, have some features typical for constant and operative memory.

The described principle of the connected subsystems lays also in a basis of differentiation of some social institutes: production—science, a hospital—clinic, school—college. For example, there exists a similarity between science and manufacturing, on the one hand, and subsystems in the population (Gurevitch, 1966). Really, such features of a science as the greater variety of directions, relative lability, the right on search, with following from here hope of finds and risk of mistakes, give to a science of property of driving potential, progressiveness (operative memory), while a smaller variation of directions, relative stability and inertness, the requirement of profitability, give to manufacture of feature of volumetric factors, perfection (conservative memory). It is possible to think, that for all evolving systems closely connected to the environment, such differentiation raises stability and promotes their further evolution.

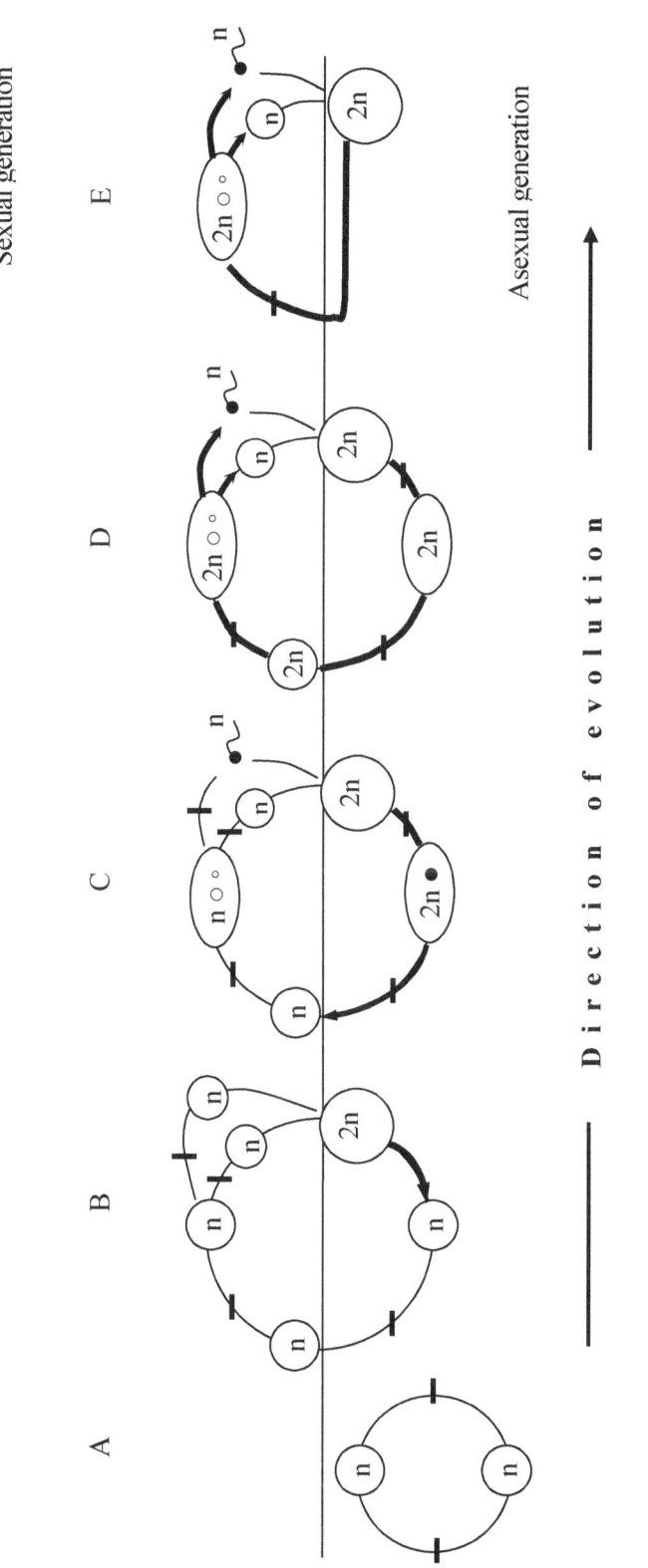

Figure 7.3

Evolution of reproduction. A—asexual reproduction only; B, C, D—alteration of asexual and sexual generations; E—sexual reproduction only.

━━ 2n—Diploid phase of the cycle, ━ ‑ ‑ n—Haploid phase of the cycle, ━━► —Meiosis, ━┤— —Mitosis, ● —Spore,

⬭ —Uni- or a multicellular organism, $\overset{2n}{\bigcirc}$ —Zygote

ⓝ — male and female gametes; — Fertilization process

n ⤳●

◯

Table 3.1 Sexual dimorphism on principal causes of death (the USA, 1967)* (Waldron, 1976).

The reason of death	Death rate per 100,000		Sex ratio, % ♂♂
	men	women	
Tumors of respiratory system	50.1	8.5	85.5
Other illnesses of respiratory system (emphysema - 71%)	24.4	5.0	83.0
Cirrhosis of a liver	18.5	9.1	67.0
Atherosclerosis, including illnesses of heart	357.0	175.6	67.0
Symptoms of an old age and illnesses	14.9	8.3	64.2
Pneumonia (newborns excluded)	32.3	19.5	62.4
Other illnesses of heart	17.9	11.1	61.7
Other illnesses of vascular system	18.2	11.1	62.1
Tumors of the digestive organs and peritoneum	53.0	36.2	59.4
All other diseases	32.4	22.4	59.1
Tumors of other and uncertain places	20.5	14.7	58.2
Patrimonial damages, postnatal asphyxia and atelectasis	11.9	8.4	58.6
Some illnesses of the early childhood	29.2	21.6	57.5
Other illnesses, especially early childhood and youth	15.3	11.7	56.7
Nonrheumatic chronic endocarditis and a myocardium damages	26.8	20.5	56.7
General atherosclerosis	17.2	14.8	53.8
The damages of vessels influencing the central nervous system	96.3	83.3	53.6
Hypertension	22.3	22.2	50.1
Genital tumors	17.9	20.1	47.1
Diabetes mellitus	14.9	16.8	47.0
Breast cancer	0.2	24.6	0.8
All cases:	**891.2**	**565.5**	**61.2**

*— All reasons of death giving the contribution of more than 1 % are included.

Table 5.3 Sex ratio for malignant tumors (summary from world statistics) (from Rajewski, Sherman, 1976 with changes).

Systems and organs	Phylogenetic "age" of an organ and system	Anatomical location of the organ within the system	Sex ratio, ♂♂ : ♀♀		Congenital anomaly
			On cancer	On anomalies**	
Nervous system	Rather ancient				
Brain			1.21*	0.54	Anencephaly
			1.28***	0.57	Craniocele
Spinal cord	New		1.27*	0.73	Spinal herniation
Peripheral nerves	Ancient		1.04*		
Respiratory system					
Pharynx	New	System entrance	0.68*		
Lungs	New	Internal part	11.66*		Agenesis of lung
			9.1***		Lung aplasia
			6.09*	0.73	Lung hypoplasia
including:			6.66***		
Central cancer			11.05*		
Peripheral cancer			7.7*		
Bronchiolar cancer			0.75*		
Digestive system	Ancient				
Tong		System entrance	1.3*	1.5	All Digestive system organs
Mouth and oropharynx	Relatively new	System entrance	3.09*		
			2.3***		
	Relatively new	System entrance	2.2****		
Esophagus			3.1*	1.3	Esophageal atresia
			2.8***	0.7	Diverticula of the esophagus

Continued

Table 5.3 continued

Systems and organs	Phylogenetic "age" of an organ and system	Anatomical location of the organ within the system	Sex ratio, ♂♂ : ♀♀		Congenital anomaly
			On cancer	On anomalies**	
Stomach	Relatively new		1.7*	0.7	
Pancreas			2.0***		
Liver			1.24****		
			2.2****		
Intestine	Ancient	Internal part	1.1*	1.03	Atresia of small intestine
				M >> F	Meckel's diverticulum
Colon	Ancient	Internal part	0.8*	1.50	Diverticula of the colon
			1.0***	M >> F	
					Congenital megacolon (Hirschprung's disease)
Rectum	Relatively new	Internal part	1.3*	1.55	Atresia
			1.5***		
Urinary system	Rather ancient		1.9*		
Kidneys	Rather ancient	Internal part	1.5*	2.60	Bilateral renal agenesis
			1.8***	2.10	Unilateral renal agenesis
Renal pelvis	New	System exit	1.75*		
Bladder	New	System exit	2.58*	2.07	Schistocystis
			3.57***		
System of internal tissue	Ancient	Internal part	1.34*		Marfan syndrome
Bones (osteosarcomas)			1.56*	0.19	Congenital hip dislocation
			1.52***	0.87	Hypoplasia of the tibia and femur

Continued

End of Table 5.3

Systems and organs	Phylogenetic "age" of an organ and system	Anatomical location of the organ within the system	Sex ratio, ♂♂ : ♀♀		Congenital anomaly
			On cancer	On anomalies**	
Cartilage (hondrosarcomas)	Relatively new		1.45***		
Blood cells (leucosis)			1.45***		
Smooth muscles (leiomyosarcomas)	Ancient		0.48*		
Skeletal muscles (rhabdomyosarcomas)	Relatively new		1.12*		
Sinovial tissue (sarcomas)	Relatively new		1.24* / 1.20***		
Fat tissue	Relatively new		1.30*		
Vessels (angiosarcomas)			.97*		
Fibrous connective tissue (fibrosarcomas)	Ancient		1.09*		
Endocrine and reproductive systems	Rather ancient		0.18*		
Hypophysis	New		1.71*		
Thyroid gland	Rather ancient		0.41* / 0.43***		Congenital cretinism
Adrenal glands core	Relatively new		1.49*		
Gonads (testes / ovaries)	Rather ancient		0.09*		Genital malformations

Authors: *—Cancer Incidence in Finland, 1971a; Cancer Incidence in Sweden, 1971b; Doll e.a., 1966, 1970; Dorn, Culter, 1955. **—Adrianovski, 1969; Conway, Wagner, 1965; Hay, 1971; Kurland e.a., 1973. ***—Ashley, 1969; **** .

Table 5.5 Change in death rate from a cancer of the various locations, corrected for age (Japan, 1950-1971) (Segi, 1973).

| Location | Average death rate per 100,000 | | Sex ratio, %♂♂ | Change in rate for (1970–1971)/(1950–1951) | | Sex ratio, %♂♂ |
	Men	Women		Men	Women	
Pharynx	1.60±0.12	0.39±0.10	4.1	0.88	0.48	1.8
Esophagus	7.1±0.039	2.2±0.1	3.2	1.18	0.88	1.3
Lungs	9.70±4.67	3.41±1.84	2.8	6.13	5.06	1.2
Mouth and pharynx	1.37±0.08	0.63±0.03	2.2	1.14	0.89	1.3
Bladder	1.95±0.43	0.93±0.16	2.1	1.96	1.75	1.1
Stomach	67.97±1.91	39.6±1.4	1.7	0.96	0.93	1.0
Liver	14.03±1.14	8.6±1.2	1.6	1.0	0.78	1.3
Pancreas	3.08±1.49	1.94±0.85	1.6	5.38	4.88	1.1
Skin	0.87±0.06	0.64±0.06	1.4	0.90	0.88	1.1
Leukemia	3.14±0.71	2.37±0.62	1.3	2.18	2.41	0.9
Rectum	4.55±0.43	3.43±0.25	1.3	1.29	1.23	1.0
Thin intestine	2.97±0.63	2.98±0.5	1.0	1.69	1.42	1.2
Thyroid	0.20±0.07	0.41±0.12	0.5	2.55	2.67	1.0
All tumors	**131.4±11.4**	**93.8±2.0**	**1.4**	**1.29**	**1.01**	**1.3**

Table 5.6 The sex ratio of patients with congenital malformations.

Congenital anomaly	Sex ratio, ♂♂ : ♀♀	Authors
Defects with female predominance		
Congenital hip dislocation	0.19	Rajewski, Sherman, 1976
	0.20	Montagu, 1968
	0.13 *	Cui, e.a. 2005
	0.27 **	Riley, Halliday, 2002
Cleft palate	0.3	Montagu, 1968
Anencephaly	0.53	Rajewski, Sherman, 1976
	0.5	WHO (reports), 1966
	0.64 **	Riley, Halliday, 2002
Craniocele	0.57	Rajewski, Sherman, 1976
Aplasia of lung	0.66	
Spinal herniation	0.71	
Diverticulum of the esophagus	0.71	
Stomach	0.71	
Neutral defects		
Hypoplasia of the tibia and femur	0.83	Rajewski, Sherman, 1976
Spina bifida	0.91 **	Riley, Halliday, 2002
Atresia of small intestine	1	Rajewski, Sherman, 1976
Microcephaly	1.19 **	Riley, Halliday, 2002
Esophageal atresia	1.3	Rajewski, Sherman, 1976
	1.5 **	Riley, Halliday, 2002
Hydrocephaly	1.32 **	Riley, Halliday, 2002

Continued

End of Table 5.6

Congenital anomaly	Sex ratio, ♂♂ : ♀♀	Authors
Defects with male predominance		
Diverticula of the colon	1.5	Rajewski, Sherman, 1976
Atresia of the rectum	1.5 2.0 **	Rajewski, Sherman, 1976 Riley, Halliday, 2002
Unilateral renal agenesis	2 2.1 **	Rajewski, Sherman, 1976 Riley, Halliday, 2002
Schistocystis	2	Rajewski, Sherman, 1976
Harelip	2 1.47 **	Montagu, 1968 Riley, Halliday, 2002
Bilateral renal agenesis	2.6	Rajewski, Sherman, 1976
Congenital anomalies of the genitourinary system	2.7 *	Cui, e.a. 2005.
Pyloric stenosis, congenital	5 5.40 *	Montagu, 1968. Cui, e.a. 2005.
Meckel's diverticulum	More common in boys	Rajewski, Sherman, 1976
Congenital megacolon (Hirschprung's disease)	More common in boys	Rajewski, Sherman, 1976
All defects	**1.29** *	Cui, e.a. 2005.

* — Data Cui e. a., (2005) obtained on opposite-sex twins. ** — data in the period 1983–1994.

Table 11.1 Structure of sexual dimorphism, characteristics of forms (Geodakyan, 2000).

Characteristic	Sexual dimorphism (SD)		
	Reproductive (RSD)	Evolutionary (ESD)	
		Modificational (MSD)	Selectional (SSD)
According to SD form	Reproductive (RSD)	Modificational (MSD)	Selectional (SSD)
According to ESD phase		Phenotypic (PSD) (common genes, different hormones)	Genotypic (GSD) (different genes and hormones)
According to basis (determine)			
Localization of gene basis (commonality)		In autosomes (common for males and females)	In sex chromosomes (different for males and females)
May be according to traits (examples)	Sexual, reproductively stable (gametes, gonads, genitals, mammary glands)	Any adaptive (thickness of fat layer, thickness of fur, learning ability, etc.)	Any evolving (size, proportion, amount of selectable traits)
Result of the change (mechanism)	Males and females in embryogenesis (differentiation)	Females in ontogeny (modifications)	Males in phylogeny (elimination, selection)
Rank (ratios). Dependence	Constitutive (primary). Independent	Intermediate (secondary). Depends on RSD	Facultative (tertiary). Depends on RSD and MSD
Time of existence	Permanent in phylogeny	Appears during each new ontogeny	The phase of trait evolution in phylogeny
Goal, function, purpose	To create two sexes	To remove females from the areas of selection	To provide for dichronous evolution of males

Table 12.1 Sexual dimorphism and ontogenetic dynamics of several human traits (Geodakyan, 1983).

Trait	Form of trait				Authors
	female	male	juvenile	definitive	
Relative length of:					
foot	Less	Greater	Less	Greater	Harrison et. al, 1964
forearm	≪	≪	≪	≪	
4th/2nd digits	≪	≪	≪	≪	
Cephalic index	≪	≪	≪	≪	Ginsburg, 1963; Roginskij, Levin, 1963
Circumference of dental arch	≪	≪	≪	≪	Roginskij, Levin, 1963
Epicanthus	Sharper	Smooth	Sharper	Smooth	
Protuberant back of nose	Rarer and weaker	More often and stronger	Rarer and weaker	More often and stronger	
Hair on body and face	Less	Greater	Less	Greater	Harrison et. al, 1964
Hair on head	Greater	Less	Greater	Less	
Erythrocyte concentration in blood	Less	Greater	Less	Greater	Pulse, 1962
Pulse rate	Greater	Less	Greater	Less	
Rate of gall-bladder evacuation	Less	Greater	Less	Greater	Gallbladder, 1978
Brain asymmetry	≪	≪	≪	≪	Searle, 1972
Reaction time	Greater	Less	Greater	Less	Boiko, 1964
Perception of bitter taste of phenylthiourea	More often and stronger	Rarer and weaker	More often and stronger	Rarer and weaker	Harrison et. al, 1964
Olfaction	Stronger	Weaker	Stronger	Weaker	

Table 13.1 Inheritance of New Characteristics by the Reciprocal Hybrids (Geodakyan, 1979,1981a).

Trait	Initial breeds		Direct cross		Back cross		Initial breeds		r	Author
	father - mother	inheritance of the trait	father - mother	inheritance of the trait	father - mother	inheritance of the trait	father - mother	inheritance of the trait		

Chickens

Trait	Initial breeds		Direct cross		Back cross		Initial breeds		r	Author
Brooding instinct, %	L – l	~0	L – c	37	C – l	88	C – c	~100	0.45	Roberts, Card, 1933
	L – l	~0	L – a	17	A – l	55	A – a	~100	0.38	Morley, Smith, 1954
	L – l	~0	L – n	37	N – l	85	N – n	~100	0.50	Saeki e.a., 1956
Early maturity of daughters, days	$L_e – l_e$	—	L – a	181	A – l	191	$A_l – a_l$	—		Morley, Smith, 1954
	$L_e – l_e$	—	L – n	189.5	N – l	231.4	$N_l – n_l$	—		Saeki e.a., 1956
	$R_e – r_e$	222.7	$R_e – r_l$	217.9	$R_l – r_e$	244.8	$R_l – r_l$	269.0	0.59	Warren, 1934
Egg laying, number	L – l	212	L – r	214	R – l	201	R – r	172	0.32	Warren, 1942
	L – l	200	L – r	230	R – l	202	**R – r**	**210**	-2.8	
	L – l	194	L – a	232	A – l	187	A – a	152	1.07	
	L – l	170	L – w	173	W – l	168	W – w	124	0.11	
	R – r	178	R – p	199	P – r	188	P – p	154	0.46	
	L – l	185	L – r	258	R – l	233	R – r	163	1.14	Dubinin, 1967
	L – l	167.6	L – m	202.1	M – l	160.1	M – m	152.1	2.71	Dobrinina, 1958
Egg weight, g	M – m	2433	M – l	2277	L – m	2085	L – l	55.5	-1.0	Dobrinina, 1958

Continued

End of Table 13.1

Trait	Initial breeds		Direct cross		Back cross		Initial breeds		r	Author
	father -mother	inheri-tance of the trait	father -mother	inheri-tance of the trait	father -mother	inheri-tance of the trait	father -mother	inheri-tance of the trait		

Pigs

Trait	father-mother	inheritance	father-mother	inheritance	father-mother	inheritance	father-mother	inheritance	r	Author
Weight at 12 mo, g	**M – m**	**2433**	M – l	2277	L – m	2085	L – l	1805	0.30	Dobrinina, 1958
Number of vertebrae	S – s	28.35	S – w	28.11	W – s	27.26	W – w	27.18	0.72	Aslanian, 1962
	S – s	28.93	S – w	28.86	W – s	27.97	W – w	27.74	0.74	Aleksandrov, 1966

Cattle

Trait	father-mother	inheritance	father-mother	inheritance	father-mother	inheritance	father-mother	inheritance	r	Author
Milk yield per year, kg	**H – h**	**6417**	**H – j**	**5808**	J – h	5588	J – j	3582	0.07	Dubinin, 1967
	H – h	**6417**	**H – k**	**6725**	K – h	6352	K – k	5481	0.39	
	K – k	**5481**	**K – j**	**5659**	J – k	5223	J – j	3582	0.23	
% of fat	**J – j**	**5.55**	J – h	4.61	**H – j**	**4.88**	H – h	3.50	-0.13	Dubinin, 1967
	J – j	**5.55**	J – k	4.78	**K – j**	**5.08**	K – k	3.94	-0.19	
	K – k	3.94	K – h	3.93	**H – k**	**3.95**	H – h	3.50	-0.05	
Amount of fat per year, kg	**H – h**	**224.6**	**H – j**	**282.6**	J – h	254.6	J – j	198.8	1.08	Dubinin, 1967
	H – h	**224.6**	**H – k**	**264.4**	K – h	249.0	K – k	216.0	1.79	
	K – k	**216.0**	**K – j**	**271.3**	J – k	265.4	J – j	198.8	0.34	

Note. Breeds of chickens: L—Leghorn; C—Cornish; A—Australorp; N—Nagoya; M—Moskovskaya; R—Rhode Island; W—New Hampshire; P—Plimutrock; L_e, R_e—early maturing; A_l, N_l, R_l—late maturing. Breeds of pigs: S—Swedish Landras, W— large white. Breeds of cattle: H—Holstein; J—Jersey; K—red Dutch. The father is denoted by a capital letter, the mother by a small letter. Breeds and hybrids with a more significant useful characteristic are printed in bold-faced type. A dash means no data were cited.

Table 15.1 Sex ratio of patients with congenital heart diseases and large vessels (Geodakyan, Sherman, 1970,1971).

Congenital malformation of the heart and great vessels	ICD-10	Number of patients			Sex Ratio, ♂♂ : ♀♀
		total	men	women	
Defects with female predominance					
Patent ductus arteriosus	Q25.0	8441	2268	6173	0.37
Lutembaher's syndrome		160	51	109	0.47
Ostium secundum	Q21.1	4588	1613	2975	0.54
Ventricular septal defect and patent ductus arteriosus	Q21.0 + Q25.0	113	45	68	0.66
Fallot's triad		297	121	176	0.69
Neutral defects					
Eisenmenger's complex	Q21.8	218	91	127	0.71
Partial atrioventricular canal	Q21.2	322	137	185	0.74
Ostium primum	Q21.1	304	138	166	0.83
Partial anomalous pulmonary venous connection	Q26.3	449	205	244	0.84
Ventricular septal defect	Q21.0	3324 2046	1644 1009	1680 1037	0.98 0.97 ***
Aortopulmonary window	Q21.4	139	69	70	0.99
Atrioventricular canal	Q21.2	440	219	221	0.99
Ebstein's anomaly	Q22.5	416	210	206	1.02

Continued

End of Table 15.1

Congenital malformation of the heart and great vessels	ICD-10	Number of patients			Sex Ratio, ♂♂ : ♀♀
		total	men	women	
Stenosis of lung artery	Q25.6	2337	1192	1145	1.04
Tricuspid atresia	Q22.4	371	198	173	1.16
Truncus arteriosus	Q20.0	292	160	132	1.21
Tetralogy of Fallot	Q21.3	2517	1447	1070	1.35
		259	156	103	1.51 ***
Defects with male predominance					
Coarctation of aorta and open arterial channel	Q25.1 + Q25.0	154	89	65	1.37
Total anomalous pulmonary venous connection	Q26.2	330	192	138	1.39
Transposition of the great arteries	Q20.3	1294	848	446	1.90
					4.58*
					1.17**
		367	241	126	1.91 ***
Coarctation of the aorta	Q25.1	3579	2438	1141	2.14
		444	263	181	1.45 ***
Aortic stenosis	Q25.3	1729	1257	472	2.66
Total:		**31814**	**14632**	**17182**	**0.85**

* — whites; ** — blacks (Storch, Mannick, 1992); *** — data for the period 1983–1994 (Riley, Halliday, 2002)

Table 16.2 Difference between Extreme Variants (Ter-Avanesian, 1949).

Cotton plant

Number of pollen grains	5	20	100	300	1000	Control
Yield per plant, g	**155.5**	115.5	85.0	57.4	62.8	71.5
Number of pods	22.2	**37.8**	28.5	112.3	8.4	7.3
Height of plant, cm	23.0	60.0	**63.0**	20.0	7.0	9.0
Weight, 1000 seeds, g	21.0	**34.4**	29.0	22.4	7.8	8.3

Black-eyed pea

Number of pollen grains	15	30	50	100	Control
Weight of green mass, g	508	**536**	328	134	222
Length of main runner, cm	1.45	**2.05**	1.45	0.95	0.70
Number of first order branches	**6**	4	1	1	1
Length of pod, cm	1.7	**2.5**	1.0	1.0	1.4
Number of seeds in pod	**1.0**	0.2	0.5	**1.0**	0.7
Weight of 100 seeds, mg	**3.60**	1.87	0.76	0.69	1.08

Wheat

Number of pollen grains	1	20	100	Control
Yield per plant, g	4.6	**4.9**	1.9	1.6
Height of straw, cm	21.0	**23.0**	7.0	9.0

Table 16.3 Plants, animals and humans secondary (II SR) in relation to tertiary (III SR) sex ratio (Geodakyan, Geodakyan, 1985).

Species	Sex ratio, % ♂♂		Authors
	tertiary	secondary	
Plants	92	44.0±2.5	Mulcahy, 1967
Melandrium album	60	48.0±1.4	
	52	48.5±1.4	
	28	46.5±1.2	
	4	55.0±4.25	
Fish (guppi)	91	32.7±1.8	Geodakian e. a., 1967
Lebistes reticulatus peters	50	50.4±1.5	Geodakian, Kosobutski, 1969
	9.1	60.7±1.4	
	91	51±1,8	Brown, 1982
	50	49±1.5	
	9.1	42±2.7	
Ticks: *Macrocheles glaber*	II SR = -0.0141 * III SR		Filipponi e.a., 1972;
M. scutatus	II SR = -0.0236 * III SR		Filipponi, Petrelli, 1967,1975
M. preglaber	II SR = -0.0362 * III SR		
Bedbugs (Southern Green Stinkbug)	Increase[*]	Decrease[*]	Mclain, Marsh, 1990
Drosophila	80	Increase of II SR[*]	Terman, Birk, 1965
Drosophila melanogaster	20		(cit.by Coyne, 1971)
	50	47.50 48.22 48.76	Lutsnikova, Petrova, 1972
	20	50.89 50.82 51.67	
Lizards	68	46	Olsson, Shine, 2001
Niveoscincus macrolepidotus	60	54	

Continued

End of Table 16.3

Species	Sex ratio, % ♂♂ tertiary	Sex ratio, % ♂♂ tertiary	Authors
Lizards			
Niveoscincus macrolepidotus	54 Decrease*	61 Increase*	Olsson, Shine, 2001
Eulamprus tympanum			Robert e.a., 2003
Mousses	Increase*	Decrease*	Parkes, 1925,1926
Rats	Increase*	Decrease*	White, 1914
	91	47.6±2.31	Geodakyan, Geodakyan, 1985
	9.1	56.5±1.72	
Marmota monax	67	31±4.1	Snyder, 1976
	50	50±2.5	
Humans	50 (1)**	49	Thomas, 1913
	33.3 (2)	51	
	25 (3)	52	
	20 (4)	55	
	16.7 (5)	57	
	Decrease*	Increase*	Lummaa e.a., 1998
Harems:			
Chu Juanchan	—*	62,0±7.5	U Han, 1980
Ramses II	1.35 (74)**	62.0±3.6	Ebers, 1965
Mauli Ismail	—*	61.8±1.6	Aisha and Africa Today, 1970
Suleiman the Magnificent	—*	59.5	

*—The numerical values not cited. **—Wives number is in parenthesis. By the sign criterion negative feedback effect is valuable (P=0.01).

Table 16.4 Dioecious plants secondary sex ratio (II SR) dependence on the amount of pollen (Geodakyan, Geodakyan, 1985).

Species (Family)	Character of pollination	Pollination conditions	Secondary SR, % ♂♂	Authors
Begonia gracilis— (Begoniaceae)	Natural	A lot of pollen	Decrease	López, Domínguez, 2003
Rumex acetosa (Polygonaceae)	Artificial	A lot of pollen Not much pollen Very small amount	8.92 30.87 42.1	Correns, 1922
	Artificial	Abundant Poor	18 45	Correns, 1922
	Natural	1920 grains/4 sm^2/24 days 72 grains/4 sm^2/24 days 72 grains/4 sm^2/24 days	43±5.5* 35±5.8* 43±5.8*	Rychlewski, Kazimierez, 1975
Rumex nivalis (Polygonaceae)	Natural	A lot of pollen	Decrease	Stehlik, e.a., 2008
Melandrium album (Cariophyllaceae)	Artificial	A lot of pollen Not much pollen	31.65 43.78	Correns, 1928
	Artificial	III SR, % ♂♂ Abundant 90 60 Average 52 Limited 28 4	44.0±2.5 48.0±1.4 48.5±1.4 46.5±1.2 55.0±4.2	Mulcahy, 1967
Cannabis sativa (Cannabinaceae)	—	Poor	Increase**	Riede, 1925
Humulus japonicus (Cannabinaceae)	Artificial Natural	A lot of pollen Normal amount	30.2±5.8 44.9±2.3	Kihara, Hirayoshi, 1932
Atriplex povellii (Chenopodiaceae)	Natural	A lot of pollen Sparse + delayed pollination	39 50	Freeman e.a., 2007

*—Cariological method of determination. **— The numerical values not cited. By the sign criterion the negative feedback effect is valuable (P = 0.01).

Table 16.5 Secondary sex ratio (SR) dependence from the sexual activity (SA) of males (Geodakyan, Geodakyan, 1985).

Species (Family)	Low SA		High SA		Authors
	Experimental conditions	Secondary SR, % ♂♂	Secondary SR, % ♂♂	Experimental conditions	
Hen (Phasianidae)	Cocks were without hens for: 7 d / 25 d / 30 d / 60 d	47.8±4.7 / 48.5±5.0 / 42.1±3.3 / 38.2±5.1	56.1±5.0 / 56.1±5.0 / 53.7±3.5 / 47.7±6.2	Cocks were with hens before the experiment	Mamzina, 1955
	Cocks were used every other day	42.4±2.9 / 45.7±4.9	55.5±2.9 / 56.7±4.6	Cocks were with hens constantly	Kurbatov, 1965
Mouse (Muridae)	Low SA	Decrease*	Increase*	Low SA	Parkes, 1925
Rabbit (Leporidae)	Coitus every 5 d	34.6±4.2	49.8±2.9	Coitus more often then 5 d	Kurbatov, 1965
	2-3 matings/day for 3 d, 3 d rest, etc.	47.1±1.2	51.2±1.6 / 54.3±2.0 / 58.3±2.6 / 53.7±1.1	1-5 d / 6-10 d / 11-15 d / 1-15 d 4-5 matings/d for 15 d	Antonian, 1974
Pig (Suidae)	< 10 matings	48.6±1.5	51.6 / 52.2±0.3	10-15 matings / 15-19 matings	Kamalian, 1962
Horse (Equidae)	Number of matings: 20-34 / 35-39 / 40-44	49.22±0.20 / 49.13±0.16 / 49.09±0.13	49,19±0.13 / 48.47±0.13 / 50.19±0.13 / 50.29±0.13	Number of matings: 45-49 / 50-54 / 55-59 / 60	Düsing, 1884
Sheep (Caprinae)	Normal SA	51.7±1.1	56.9±2.3	Increased SA	Markarian, 1965
	2 matings/d for 3 d, 3d rest	50.6±2.8 / 50.1±1.3	48.8±1.8 / 57.1±4.0 / 55.8±3.2	2 matings/d for 6 d, 1d rest / 4 matings/d for 65 d / 10 matings/d for 45 d	Antonian, 1974

Continued

End of   Table 16.5

Species (Family)	Low SA		Secondary SR, % ♂♂	High SA		Authors
	Experimental conditions			High SA	Experimental conditions	
Cattle (*Bovidae*)	80–100 matings/season		42.6 54.2 58.3		100–120 matings More than 120 matings	Miller, 1909
Human (*Hominidae*)	Low SA		Decrease* Increase*		High SA	James, 1971a,b,c; 1975a,b,c; 1976

*— The numerical values not cited. By the sign criterion the negative feedback effect is valuable (P = 0.01).

Table 16.6 Secondary SR (II SR) dependence from the delayed fertilization of eggs and pollination in plants (Geodakyan, Geodakyan, 1985).

Species (Family)	Delay value, or time of fertilization	Secondary SR, % ♂♂		Authors
		after delay	before delay	
Plants				
Spinaceae oleracea L. —spinach (*Chenopodiaceae*)	2 weeks	Increase*	>≈50	Miglia, Freeman, 1995
Atriplex povellii —atriplex (*Chenopodiaceae*)	2 weeks pollination delay & sparse pollen	42 50	39	Freeman, 2007
Animals				
Taleporia tubulosa (*Psychidae*)	4 d	**Old eggs** 59	**Control** 42.5	Seiler, 1920
Rana esculenta (*Ranidae*)	18 h 42 h 56 h 64 h 89 h 94 h	37.3 50 58 94 86 100 100	48.6	Hertwig, 1912
			53±4.7	Kuschakevich, 1910
Rana temporaria	80–100 h	77.9±4.3	—	Witschi, 1914
Bombix mori (*Bombicidae*)	— —	Increase* Increase*	— —	Lombardi, 1923 Golanski, 1959
Salmo iridens (*Salmonidae*)	4–7 d 21 d	Slightly decrease* 62.5	50	Mrsic, 1923,1930

Continued

Table 16.6 continued

Species (Family)	Delay value, or time of fertilization	Secondary SR, % ♂♂		Authors
		after delay (old eggs)	before delay (control)	
Salmo trutta	21 d	Increase*	—	Huxley, 1923
Tisbe dobzhanskii	5 d	81.39	58.88	Volkman-Rocco, 1972
Tisbe clodiensis (Venice)	4–5 d	60.92	52.71	
Tisbe clodiensis (Ponza)	4 d	69.99	59.13	
Tisbe holothuriae	4 d	62.57	54.86	
Pseudococcus	6 weeks	64.3	50.5	James, 1937
	8 weeks	76.5		
	10 weeks	90.8		
Drosophila melanogaster — (Drosophilidae)	7 d	52.23	51.23	Hannach, 1955
	14 d	52.83		
	21 d	53.23		
Mousses (Muridae)	3.25-4.25 d	47.3±4.8	—	Vickers, 1969
	9.5-11.5 d	44.4±5.2	51.5±3.5**	
	17.5 d	54.8±3.5	44.4±3.6	
Rats (Muridae)	First and last 3 h of oestrous cycle	48,43±3,38**	51.84±3.37**	Crew, 1927
	10 h before ovul. norm	—	47	Hammond, 1934
	5-3 h before ovul.	50.7		
	2-1 h before ovul.***	57.4		
	0-1 h after ovul.	41.2		

Continued

End of Table 16.6

Species (Family)	Delay value, or time of fertilization	Secondary SR, % ♂♂		Authors
		after delay (old eggs)	before delay (control)	
Rats (*Muridae*)	1-3 h after ovul.	60.0±2.3 62.7±5.9	50.0	Hart, Moody, 1949
	3-5 h after ovul. 4-6 h after ovul.	51.2±7.8 71.8±4.4		
	5-7 h after ovul.****	69.7±14.5		
Hamsters (*Cricetidae*)	End of estrus	Increase*		Pratt e.a., 1987
Rabbits (*Leporidae*)	10-12 h after ovul.***	71±9	49±4.1	Ivanova, 1953
	2.5-16 h after ovul.***	58.0±2.6 Increase*	50	Szemere, 1958 Hammond, 1934
Cattle (*Bovidae*)	Beginning of estrus Middle of estrus	49.6±3.2 —	53.6±4.4	Russell, 1891
	End of estrus	60.7±4.7		
Sheep (*Caprinae*)	End of estrus	Increase*	—	Rorie, 1999
Deer (*Cervidae*)	End of estrus	Increase*	—	Rorie, 1999; Verme, Ozoga, 1981
Human (*Hominidae*)	2-3 d*****	54.8	48.7	Guerrero, 1970,1974
	2 d	Increase*	—	Harlap, 1979
	2 d	Increase*	—	Shettles, 1978

* —The numerical values are not cited. ** —Differences are not significant. *** —Ovulation was caused by mating with vasectomyed male. **** —Groups of 6 males and 15 females, ovulation was determined by the beginning of the estrus and time of the day. ***** —Ovulation was determined by the temperature test. By the sign criterion the negative feedback effect is valuable (P = 0.01).

Table 16.7 Theoretically expected sex of offspring in the relation to sexual activity (SA) for organismic or populational types of feedback, acting inside male and female organism, for natural and artificial disturbances of SA optimum (Geodakyan, Geodakyan, 1985).

Disturbance of SA optimum	SA of		Negative feedback			
			organismic		populational	
			In the organism of:			
	father	mother	father	mother	father	mother
Natural	High	Low	♂	♂	♀	♀
	Low	High	♀	♀	♂	♂
Artificial	High	High	♂	♀	♀	♂
	Low	Low	♀	♂	♂	♀

Table S.1. Characteristics of conservative and operative subsystems and sex differentiation.

Characteristics	Conservative subsystem Females	Operative subsystem Males	Apply to systems
Thermodynamically more	equilibrium	nonequilibrium	All
Factor of	extensiveness	intensity	
Plays role of generalized	"charge"	"potential"	
On the time axis	old ("rearguard")	new ("vanguard")	
On the coordinate "system – environment"	internal ("nucleus")	external ("shell")	
Provides information	genetical	ecological	Living
Intra- and interpopulation variance	less	more	
Implements	selection and consolidation	search and trial	
More	perfect	progressive	
More	universal	specialized	
Reaction norm	wider	narrower	
Plasticity in ontogeny	more	less	
Plasticity in phylogeny	less	more	
Nature of the anomalies	"atavistic"	"futuristic"	
Perceive frequency of environment changes	low	high	
Mutation, abnormalities, cancer, stroke	less frequently	more frequently	

Table B.1 Some informational binary conjugated subsystems.

System	Subsystems	
	Conservative	Operative
Nucleoprotein	DNA (RNA)	Protein
Gene (in the organism)	Dominant (A)	Recessive (a)
Gene (in population)	Heterozygote (Aa)	Homozygote (AA, aa)
Genome	Autosomes	Sex chromosomes
Cell	Nucleus	Cytoplasm
Brain (down—up)	Subcore	Core
Brain (back—front)	Occipital lobe	Frontal lobe
Brain (right—left)	Right hemisphere	Left hemisphere
Organism (morphology)	Left side	Right side
Organism (genetics)	Gametes	Somatic cells
Organism (physiology)	Estrogens	Androgens
Organism	Genotype	Phenotype
Population	**Females**	**Males**
Society	Right-handed individuals	Left-handed individuals

Abbreviations

Detailed definitions see in the Glossary.

I SR	Primary sex ratio (zygote's after fertilization)	**M**	Males
II SR	Secondary sex ratio (at birth)	**Max**	Maximal
III SR	Tertiary sex ratio (adults in the population)	**MhS**	Maternal half-sibs
ADH	Alcohol dehydrogenase	**Min**	Minimal
ALDH	Aldehyde dehydrogenase	**MSD**	Modificational sexual dimorphism
AER	Androgens to estrogens ratio[An]/[Es]	**MUT**	Program of mutation
AIDS	Acquired immune deficiency syndrome	**Opt**	Optimal
An	Androgens	**PhS**	Paternal half-sibs
AR	Asexual reproduction	**PSD**	Phenotypic sexual dimorphism
BCD	Binary conjugated differentiations	**REP**	Program of reproduction
DIF	Program of differentiation	**RN**	Reaction norm
DNA	Deoxyribonucleic acid	**RSD**	Reproductive sexual dimorphism
ES	Estrogens	**SA**	Sexual activity
ESD	Evolutionary sexual dimorphism	**SC**	Sex chromosomes
ESR	Epidemiological sex ratio	**SD**	Sexual dimorphism
EV	Program of evolution	**SDC**	Sexual dichronism
F	Females	**SH**	Sex hormones
FS	Full sibs	**SR**	Sexual reproduction
GSD	Genotypic sexual dimorphism	**SSD**	Selectional sexual dimorphism
HR	Hermaphrodite reproduction	**SV**	Variation of the sexes (σ)
IQ	Intelligence quotient		

Glossary

Adaptation	Any change in the structure or function of an organism that allows it to survive and reproduce more effectively in its environment.
AIDS	Acquired immune deficiency syndrome; a disease caused by the human immunodeficiency virus (HIV).
Alteration of generations	Interchange of asexual and sexual stages of reproduction.
Androgens	Male *sex hormones*.
Anisogamy	Different size of *gametes* (small, mobile male and big, motionless and with a lot of resources—female).
Anthropogenesis	(from Greek anthropos—human and genesis—origination)—the process of human origination and formation.

Apomixis	*Parthenogenesis* in plants.
Arrhenotoky	Parthenogenesis in which only males are produced. Unfertilized eggs of a queen bee produce only males by arrhenotoky
Artificial selection	Mechanism by which man selects a favorable trait in a population and preserves it through controlled breeding.
Asexual reproduction	Reproduction from a single parent; there is no fusion of nuclei.
Autosome	Any chromosome other than those determining sex. Autosomes are of the same number and kind in both males and females of a species.
Binary fission	*Asexual reproduction* in which a one-celled organism divides into two equal parts (by *mitosis* when a nucleus is present).
Biogenic law	See *Recapitulation theory.*
Breeds of animals	New forms of animals with economically valuable characters introduced as a result of a selection.
Budding	Form of asexual reproduction in which a new individual is produced as an outgrowth of an older one; cell division involving unequal division of the cytoplasm.
Cardiovascular	Pertaining to the heart and blood vessels.
Central nervous system	The brain and spinal cord.
Characteristic	Any well-marked phenotypic feature that helps to distinguish one species from another. In genetics—any readily defined feature that is transmitted from the parent to the offspring.
Chromosomes	Chainlike structures within cell nucleus that contain genes.
Coarctation of the aorta	A narrowing of the aorta, typically found after the vessels are given off to the left arm.
Coefficient of cephalisation	Relation of brain mass to the body mass.
Concordance	The presence of a given trait in both members of a pair of twins.
Congenital	Present at birth. The term is commonly used in the context of congenital diseases, which are not necessarily genetic in origin.
Conjugation	Type of *sexual reproduction* found in some one-celled organisms, in which similar *gametes* from two individuals fuse.
Crossing over	Breaking of linkage groups due to mutual exchange of parts between chromatids of homologous chromosomes during meiosis.
Cross-pollination	In gymnosperms (cone-bearing plants) and angiosperms (flowering plants), the transfer of pollen from the male reproductive organs to the receptive part of the female reproductive organ on a different plant of the same species.
Depere's laws	1. phylogenetical growth; 2. phylogenetical specialization. Many Vertebrata evolve from small to large forms and from nonspecializing to specializing.
Dioecious	Male and female reproductive structures appear in different individuals.
Dioecious plants	Pistillate flowers are on one plant and staminate flowers—on another. Also called "of two households" plants.
Diploid	Double set of chromosomes in each cell.
Disruptive selection	A form of *natural selection* that increases the number of individuals displaying two extremes of a trait, but acts against individuals showing intermediate forms. This gives rise to two distinct groups in the population, showing different phenotypes, and may lead to speciation if selection acts on such traits as breeding season.
Dyslexia	Disturbed ability to read.
Divergence	Deviation of the trait.
Dizygotic (fraternal) twins	Twins that develop from two separate eggs. May be of the same or opposite sex.
DNA	Deoxyribonucleic acid, principal component of genes.
Dollo's law—irreversibility of evolution	Evolution is not reversible; i.e., structures or functions discarded during the course of evolution do not reappear in a given line of organisms.
Dominant gene	A *gene* that expresses its phenotype even in the presence of a *recessive gene.*
Ebstein's deformity	The primary abnormality in Ebstein's Anomaly is of the tricuspid valve, the valve which lies between the right atrium and right ventricle.

Ecological niche	Place that species has in the system of ecological relationships with other organisms and environmental factors.
Ecology	The science, which deals with the interrelationships between organisms and their physical environment.
Eisenmenger's Complex	Complication of untreated ventricular septal defect. First described by German physician in 1897.
Embryo	Organism in an early developmental stage.
Embryology	The study of the embryo's development.
Environment	The sum total of all the conditions and elements, which make up the surroundings and influence the development and actions of an individual.
Estradiol	Estrogen; stimulates development and maintenance of female sex traits.
Estrogens	Female hormones produced by the ovaries.
Evolution	(from Latin word evolutio—unfolding, development) in biology—irreversible historical development of living nature.
Expressivity	The degree to which a particular genotype is expressed in the phenotype. The degree of expression of a genetically controlled trait.
Fallot's tetrad	The four abnormalities shown on the right characterize this fairly common condition: 1. There is a ventricular septal defect. 2. There is narrowing of the valve leading to the pulmonary arteries (pulmonic stenosis) 3. The aorta "overrides" the ventricular septal defect. 4. There is thickening (hypertrophy) of the right ventricle.
Female	An individual that produces large usually immobile *gametes (as eggs)*. Bears young. A pistillate plant. Carries Venus sign ♀.
Fertilization	The fusion of dissimilar *gametes* to form a *zygote* in *sexual reproduction*.
Gamete	(from Greek words gamete—wife, gametes—husband)—cell that unites with another cell in sexual reproduction; a sex cell.
Gametogenesis	Process by which *gametes* are produced.
Gametophyte	The phase in the life cycle of a plant or algae that reproduces sexually, producing *gametes*. It is usually haploid.
Gene	The basic unit of inheritance that controls a characteristic of an organism. Ultramicroscopic area of DNA responsible for transmission of hereditary traits.
Genetics	Science of heredity.
Genitalia	Organs of reproduction, especially the external organs.
Genome	Information contained in the haploid set of chromosomes.
Genotype	Genetic characteristics inherited by an individual.
Gonads	The sex glands that produce sex cells.
Haeckel-Müller's law)	See *Recapitulation theory*.
Haploid (monoploid)	Single set of unpaired *chromosomes* of a given species. Examples are the gametes produced in sexual reproduction and the gametophyte stage of many plants.
Haplodiploidy	A system of sex determination whereby females develop from fertilized eggs and are diploid, while males develop from unfertilized cells and are haploid, as in honeybees.
Heredity	Genetic transmission of characteristics from parents to their children.
Hermaphrodite	Organism that possesses the sexual organs of both sexes.
Heterochromosome	1. Chromosome, consisting from heterochromatin. 2. Sex chromosome.
Heterogametic sex	The sex that have double set of *autosomes* and two different sex chromosomes (XY).
Heterogamy	*Gametes* of different size.
Heterosis	(from Greek geteroiosis—change, transformation). Acceleration of growth, increasing in size, vitality and productivity of first generation *hybrids* compare to their parents.
Heterozygous	Possessing two different alleles of a particular gene on a pair of homologous chromosomes.
Homogametic sex	Sex which has diploid set of autosomes and two X chromosomes.
Homosexuality	Sexual preference for member of one's sex.
Homozygous	Possessing identical alleles of a particular gene on a pair of homologous chromosomes.

Hormones	Chemicals produced by the endocrine glands that regulate growth, development and function of organisms.
Hybrid	Offspring from crossing two different breeds.
Hybridization	Mating or crossing two different breeds (races, forms) of animals or plants.
Inbreeding	Mating between closely related individuals (brother—sister, cousins etc.). A population of inbreeding individuals generally shows less genetic variation than an outbreeding population, with many alleles present in the homozygous state.
Intelligence quotient (IQ)	Measurement of "intelligence" expressed as a number or position on a scale. Comparable to term intellectual level.
Internal fertilization	Union of egg and sperm inside the female's body.
Isogamy	Same size of male and female *gametes*. Both gametes can move. Can be found in many Protozoa.
Male	An individual that produces small usually mobile *gametes (as spermatosoids), which* fertilize the eggs. A staminate plant. Carries Mars sign ♂.
Maternal inheritance	A mechanism of inheritance in which certain characteristics of the offspring are determined by the cytoplasm of the egg.
Meiosis	Cell division during which the *chromosome* number is reduced to monoploid and *gametes* are produced.
Mitosis	Form of nuclear division characterized by *chromosome* replication and specific chromosome movements, which maintains the diploid chromosome number in the daughter cells.
Modification	Nonhereditary changes of the organism's phenotype under environmental conditions.
Monoecious	1. Flowering plants that have separate female (pistillate) and male (staminate) flowers on the same plant. Also called "of the same household" plants. 2. Other plants and algae that produce male and female *gametes* on the same organism. 3. Animals that have both reproductive organs on the same individual (see *hermaphrodite*).
Monogamy	Single marriage; marriage with but one person, husband or wife, at the same time; opposed to *polygamy*. Also, one marriage only during life; opposed to *deuterogamy*.
Monozygotic twins	Identical twins developed from one fertilized egg. Have the same sex.
Multiple-factor hypothesis	A hypothesis that explain quantitative variation by assuming the interaction of a large number of genes (*polygenes*) each with a small additive effect on the character.
Mutagen	Any agent that initiates or increases the rate of gene *mutation* in a population.
Mutation	Change in the composition of a gene, usually causing harmful or abnormal characteristics to appear in the offspring.
Natural selection	Survival of a well-adapted organisms and modification or elimination of less-adapted. The main force of *evolution*.
Normal distribution	Tendency for most members of a population to cluster around a central point or average with respect to a given trait, with the rest spreading out to the two extremes.
Ontogeny	The development of an individual from egg to adult.
Oogamy	See *anisogamy*.
Oogenesis	Process by which egg cells are produced from a female primary sex cell.
Operational sex ratio	Ratio of reproductive males and females in the population
Osborn's law	New forms develop in different directions depending on the environmental conditions surrounding them after migration.
Ostium Primum	Atrial septal defect. Congenital opening in septum near AV valves.
Ostium Secundum	Atrial septal defect. Congenital defect at the fossa ovalis.
Outbreeding	Mating between unrelated or distantly related individuals. A population of outbreeding individuals shows more genetic *variation* than an *inbreeding* population.
Ovulation	Process during which an ovary discharges a mature egg.
Ovum (plural, ova)	Female gamete or germ cell.

Paleontology	The science, which studies the fossils of organisms.
Panmicsy	Random mating.
Parthenogenesis	A form of *asexual reproduction* in which an unfertilized egg develops into a new individual. It is common in lower animals, especially insects
Patent Ductus Arteriosus	Failure after birth of obliteration ductus arteriosus. Leaves communication between aorta and pulmonary artery.
Pathology	Abnormal physical or mental condition.
Penetrance	The proportion of individuals with a specific genotype who manifest that genotype at the phenotype level.
Phen	Character
Phenotype	The observable and measurable characteristics of an organism, a result of the interaction of *genotype* and environment.
Phylogeny	(from Greek phylon—tribe, clan and genesis—origin)—the evolutionary history or genealogy of a species.
Plasticity	Adaptability to environmental change.
Ploidy	The number of sets of *chromosomes* in a cell. See *haploid, diploid* and *poliploid*.
Pollen	The small granules, produced in the anther of a flower, in which the male gametes of flowering plants develop.
Pollination	Transfer of pollen from anther to stigma. See *cross-pollination* and *self-pollination*.
Polyandry	One female is mating with several males.
Polygamy	Organism of one sex has several mating partners. See *polyandry* and *polygyny*.
Polygene	See *multiple-factor hypothesis*.
Polygenic inheritance	Quantitative inheritance. The mechanism of genetic control of traits showing continuous variation.
Polygyny	One male is mating with several females.
Polyploid	An individual with three (triploid), four (tetraploid) or more sets of *chromosomes*.
Population	(from Latin populus—people) All members of a species inhabiting a given location and therefore are able potentially to mate with each other.
Prenatal	Before birth.
Primary sexual characters	Internal and external organs of reproduction (ovaries and testes).
Race	A *population* of organisms within a *species* that is geographically, ecologically, physiologically or chromosomally distinct from other members of a species.
Reaction norm	The ability of a *genotype* to form different *phenotypes* in *ontogeny*, depending from environment conditions. Characterizes the influence of an environment in the realization of a trait.
Recapitulation theory	The theory developed by the German biologist E. Haeckel (1834-1919) that each organism in its individual embryonic development passes main stages of its predecessors' development—that is, *ontogeny* tends to recapitulate *phylogeny*.
Recessive gene	*Gene*, which is effective only when paired with an identical gene.
Reciprocal hybrids	Direct hybrid is formed from mating a male of breed A with female of breed B. Reciprocal—from male of breed B with female of breed A.
Recombination	Swapping of DNA between paired chromosomes when eggs and sperm are made.
Replication	The process of duplicating or reproducing, as the replication of an exact copy of a polynucleotide strand of DNA or RNA.
Reproduction	The process by which living organisms give rise to new organisms of the same species.
Reproductive system	Set of organs responsible for *sexual reproduction*.
Secondary sexual characters	An external feature of an organism, not including the reproductive organs themselves, that is indicative of its sex. These characteristics include facial hair in man and breasts in women, combs in cockerels, plumage in many male birds and manes in male lions.
Selection	(from Latin selectio—selection, choice)—any process that increases or decreases the probability of reproduction. Creates new forms of plants, animals and microorganisms.
Self-pollination	In plants the transfer of pollen from male reproductive organ (anther) to the female reproductive organ (stigma) on the same flower.

Sex	Alternative character that distinguish male and female individuals from each other, allows them to produce different gametes and participate in sexual reproduction.
Sex chromosomes	Pair of chromosomes inherited by an individual, which determines sex and certain other characteristics. See *Heterogametic sex, Homogametic sex and Sex determination.*
Sex determination	The process by which the sex of an organism is determined. See Appendix A.
Sex linkage	The tendency for some characteristics to occur in one sex.
Sex ratio	Ratio of males to females.
Sexual dimorphism	Differences between male and female sexes.
Sexual reproduction	Method of *reproduction* that includes the fusion of the gametes.
Sexual selection	A process similar to natural selection but relating exclusively to success in finding a mate for *sexual reproduction* and producing offspring.
Siblings	Offspring of the same parents.
Somatic cells	All cells of the body other than *gametes* (sex cells).
Species	A category of organisms that consist of groups of similar individuals that can interbreed among themselves and produce fertile offspring.
Sperm	Male *gamete.*
Spermatogenesis	Meiotic process that produces sperm.
Spore	Asexual reproductive sell that can produce another individual.
Stress	The internal responses caused by application of a *stressor.*
Stressor	Any adjustive demand that requires coping behavior on the part of individual or group.
Synantropic	Species ecologically related to humans (rats, roaches).
Syngamy	Fusion of gametes.
System	An assemblage of interdependent parts, living or nonliving.
Testes	Male reproductive glands or gonads.
Testosterone	Male *sex hormone.*
Trait	Characteristic of individual, which can be observed or measured.
Transposition of the Great Vessels	Aorta and pulmonary arteries are transposed.
Truncus arteriosus	The aorta and pulmonary arteries both arise from a common "trunk"
Variance (σ)	The value of deviation (dispersion) from average value of trait.
Variation	A difference between individuals of the same species.
Vegetative reproduction	is a form of asexual reproduction in plants by which new individuals arise without production of seeds or spores.
Vivipary in animals	development of the embryo inside the body of the mother, leading to live birth
X chromosome	Sex-determining *chromosome*: all female *gametes* contain X chromosomes, and if fertilized ovum has also received an X chromosome from its father it will be female.
Y chromosome	Sex-determining *chromosome* found in male *gametes*. Y chromosome uniting with X chromosome always provided by female produces a male offspring.
Zygote	Fertilized egg cell formed by union of male and female *gametes.*

REFERENCES

If there are references to more than one publication by the same author in the same year, they are distinguished here by the addition of "a", "b", etc.

ORIGINAL ARTICLES

1965 – 1970

1. Geodakyan V. A. Role of the Sexes in the Transmission and Transformation of Genetic Information. *Problems of Information Transmission,* 1965, v. 1, N 1, January–March, p. 78–83. Translated from *Problemy Peredachi Informatsii*, 1965, Vol. 1, No. 1, p. 105-112.
2. Geodakyan V. A. Mal'shik ili devoska (A boy or a girl. Sex ratio—value regulated by Nature?). Nauka i Zhizn (Science and Life)., 1965, № 1, p. 55–58 [russ].
3. Geodakyan V. A. On the existence of the feedback that regulates the sex ratio. In: Problems of Cybernetics. Moscow, Fizmatgiz, 1965b, v. 13, p. 187-194 [russ].
4. Geodakyan V. A. Differentiation on constant and operative memory in genetic systems. Proc. of the conference "Structural levels of biological systems." M., 1967 [russ].
5. Geodakyan V. A. Kosobutsky V. I. The mechanism of the feedback regulating sex ratio. DAN USSR, 1967, v. 173, № 4, p. 938-941 [russ].
6. Geodakyan V. A., Kosobutsky V. I., Bileva D. S. Regulation of sex ratio by negative feedback. *Genetics*, 1967, № 9, p. 153-163 [russ].
7. Geodakian V. A., Smirnov N. N. Sexual Dimorphism and Evolution Lower Crustacea. *"Problemy Evolutsiĭ"* (*Problems of Evolution*), Novosibirsk, "Nauka", 1968, v. 1, p. 30–36.
8. Geodakyan V. A., Kosobutskii V. I. Nature of feedback mechanism of sex regulation. *Genetika*, 1969, v. 5, p. 119–126.
9. Geodakyan V. A. Organization of living and nonliving systems. In: Systems Research, Nauka, 1970, p. 49-62 [russ].
10. Geodakyan V. A., Sherman A. L. *Eksperimental'naja hirurgija i anesteziologija (Experimental surgery and anesthesiology)*, 1970, v. 32, N 2, p. 18–23.

1971 – 1980

11. Geodakyan V. A. Systems theory and the special sciences. In: Materials on the history and prospects of development of the systems approach and general systems theory. Moscow, Nauka, 1971a, p. 17 [russ].
12. Geodakyan V. A. Cybernetics and development. *Ontogenez*, 1971b, v. 2, № 6, p. 653-654 [russ].
13. Geodakyan V. A. On differentiation of systems into two conjugated subsystems. In: Problems of Biocybernetics. Management and information processes in living nature. M., Nauka, 1971c, p. 26 [russ].
14. Geodakyan V. A., Sherman A. L. Svjaz' vrozdennych anomalij razvitija s polom (Relation of birth defects with sex). *Zh. Obsh. Biol.*, 1971, v. 32, N 4, p. 417–424.
15. Geodakyan V. A. On the structure of self-replicating systems. In: *Developing the concept of structural levels in biology*. Moscow, Nauka, 1972a, p. 371-379 [russ].

16. Geodakyan V. A. On the structure of evolving systems. In: Problems of Cybernetics. , M., Nauka, 1972b, v. 25, p. 81-91[russ].
17. Geodakjan V. A. Adam und Eva - kybernetisch betrachtet. In: I. Jefremov, ed, *17 Weltratsel die grossen fragen der Forschung*, Stuttgart, 1972c, p. 136–149.
18. Geodakyan V. A. Differential Sex Mortality and Reaction Norm. *Biol. Zh. Arm.*, 1973, v. 26, N 6, p. 3–12 [russ].
19. Geodakian V. A. Differential Mortality and Reaction Norm of Males and Females. Ontogenetic and Phylogenetic Plasticity. *Zh. Obshch. Biol.*, 1974, v. 35, N 3, 376–385 [russ].
20. Geodakyan V. A. Concept of information, and living systems. *Zh. gen. Biology*, 1975, v. 36, № 3, p. 336-347 [russ].
21. Geodakyan V. A. Ethological sexual dimorphism. In: *Group behavior in animals*, M., Nauka, 1976, p. 64-67 [russ].
22. Geodakyan V. A. The Amount of Pollen as a Regulator of Evolutionary Plasticity of Cross-Pollinating Plants. *Doklady Biological Sciences,* 1977a, v. 234, N 1-6, p. 193–196. Translated from Doklady Akademii Nauk, Vol. 234, No. 6, pp. 1460-1463, May, 1977.
23. Geodakyan V. A. Evolutionary logic of differentiation of the sexes. In: *Mathematical Methods in Biology*, Kiev., 1977b, p. 84-106 [russ].
24. Geodakyan V. A. Evolutionary specialization of the sexes on trends of stabilizing and leading selection. 3rd All-Union Congress. N. I. Vavilov's Society of Geneticists and Breeders. Proc. L., 1977c, II (I), p. 46-47 [russ].
25. Geodakyan V. A. Number of pollen as a regulator of evolutionary plasticity of cross-pollinated plants. In: XIV Intern. Congr. of Genetics. Sections sessions. Proc., Part II, Moscow, Nauka, 1978b, p. 49 [russ].
26. Geodakyan V. A. Ethological characteristics associated with gender. II Congress of the All-Union Theriological Society. Proc., M., Nauka, 1978c, p. 215-216 [russ].
27. Geodakyan V. A. On the possibility of the existence of an adaptive selection of sperm. III All-Union Conf. on biological and medical cybernetics. Proc., M. - Sukhumi, 1978, p. 244-247 [russ].
28. Geodakyan V. A. Natural selection and sex differentiation. *Proc. Symp. Natur. Select.* Liblice, CSAV, Praha, , 1978, p. 65–77.
29. Geodakyan V. A. Natural selection of spermatozoids. *Proc. Symp. Natur. Select.* Liblice, CSAV, Praha, 1978, p. 707–713.
30. Geodakyan V. A. Existence of the "Paternal Effect" in the Inheritance of Evolving Characteristics. *Doklady Biological Sciences,* 1979, v. 248, N 1-6, p. 1084–1088. Translated from Doklady Akademii Nauk, Vol. 248, No. 1, pp. 230-234, September, 1979.
31. Geodakyan V. A. Brain asymmetry and sex. In: Antropogenetika, anthropology and sports, Vinnytsa, 1980, p. 331-332 [russ].

1981 – 1990
32. Geodakyan V. A. Sexual dimorphism and "paternal effect." Zh. gen. biology, 1981a, v. 42, № 5, p. 657-668 [russ].
33. Geodakyan V. A. Evolutionary interpretation of reciprocal effects. 4-th All-Union Congress. N. I. Vavilov's Society of Geneticists and Breeders. Proc. Kishinev, Shtiinitsa, 1981b, Part I, p. 57-58 [russ].
34. Geodakyan V. A. Sexual Dimorphism and Evolution of Duration of Ontogenesis and its Stages. *Doklady Biological Sciences,* 1982a, v. 263, N 1-6, p. 174–177. Translated from Doklady Akademii Nauk. SSSR, Genetika, Vol. 263, No. 6, pp. 1475-1480, April, 1982.
35. Geodakyan V. A. Further development of genetic-environmental theory of differentiation of the sexes. In: *Mathematical Methods in Biology*, Kiev, Naukova Dumka, 1982b, p. 46-60 [russ].
36. Geodakyan V. A. Bergman and Allen rules in the light of the new concept of sex. Mammals of the USSR. Proc. III All-Union Congress of theriological society. M., 1982c, p. 172 [russ].
37. Geodakjan V. A. Sexual Dimorphism and the Evolution of Duration of Ontogenesis and its Stages. In: *Evolution and Environment.* (Novak V. J. A., Mlikovsky J., eds.), ČSAV, Praha, 1982d, p. 229–237.
38. Geodakyan V. A. Ontogenetic Principle of Sexual Dimorphism. *Doklady Biological Sciences,* 1983, v. 269, N 1-6, p. 143–146. Translated from Doklady Akademii Nauk, Vol. 269, No. 2, pp. 477-481, March, 1983.
39. Geodakyan V. A. Evolutionary logic of differentiation of the sexes and longevity. *Priroda (Nature)*, 1983b, № 1, p. 70-80 [russ].

40. Geodakyan V. A. Sexual dimorphism in the pattern of aging and mortality in humans. In: *Problems of biology of aging*, M., Nauka, 1983c, p. 103-110[russ].
41. Geodakyan V. A. System approach and regularities in biology. In: Systems Res. M., Nauka, 1984a, p. 329-338 [russ].
42. Geodakyan V. A. Genetic-environmental interpretation of brain lateralization and sex differences. In: *Theory, Methodology and Practice of System Research* (Abstracts of Reports. All-Union Conf. Section 9), M., 1984b, p. 21-24 [russ].
43. Geodakyan V. A. Some laws and phenomena related to sex. In: *Probabilistic methods in biology*, Kiev, Institute of Mathematics, USSR Academy of Sciences, 1985a, p. 19-41 [russ].
44. Geodakyan V. A. Sexual dimorphism. In: *Evolution and morphogenesis.* (Mlikovsky J., Novak V. J. A., eds.), Academia, Praha, 1985b, p. 467–477.
45. Geodakyan V. A. Geodakyan S. V. Is there a negative feedback in sex determination? *Zurnal obschej biol.*, 1985, v. 46, N 2, p. 201–216 [russ].
46. Geodakyan V. A. On Theoretical Biology. In: *Methodological aspects of evolutionary theory.* Kiev, Naukova Dumka, 1986a, p. 73-86 [russ].
47. Geodakyan V. A. Sexual dimorphism. *Biol. Journ. Of Armenia*, 1986b, v. 39, № 10, p. 823-834 [russ].
48. Geodakyan V. A. System-evolutionary interpretation of brain asymmetry. In: *Systems Research*. M., Nauka, 1986c, p. 355-376 [russ].
49. Geodakyan V. A. Differentiation of the sexes and environmental stress. In: Mathematical Modeling in the problems of environmental management. Rostov-Don, 1986d, p. 88 [russ].
50. Geodakyan V. A. Ontogenetic and teratological rule of sexual dimorphism. V Congress VOGIS, Proc., v. I, Moscow, 1987, p. 56 [russ].
51. Geodakyan V. A. Evolutionary logic of differentiation of the sexes in the phylogeny and ontogeny. Cand. Dis. Doct. Biol. Science, M., Inst. Biol. Razvitija, 1987 [russ].
52. Geodakjan V. A. Sexual Dimorphism is a Consequence of any Selection. *Towards a New Synthesis in Evolut. Biol.* Proc. Intern. Symp. Praha. 1987. Czech. Acad. Sci. p. 168–170.
53. Geodakian V. A. Feedback Control of Sexual Dimorphism and Variation. *Towards a New Synthesis in Evolut. Biol.* Proc. Intern. Symp. Praha. 1987. Czech. Ac. Sci. p. 171–173.
54. Geodakjan V. A. Sexual Dimorphism of Brain Asymmetry and Psychology: Evolutionary Interpretation. *Towards a New Synthesis in Evolut. Biol.* Proc. Intern. Symp. Praha. 5-11 July 1987. Czech. Acad. Sci. p. 262–263.
55. Geodakyan V. A. Theory of differentiation of the sexes in the problems of man. The man in the system of sciences, Moscow, Nauka, 1989a, p. 171-189 [russ].
56. Geodakyan V. A. Panseksualizatsiya and anthropogenez. 3rd Workshop on genetics and breeding of animals. 1989b, Novosibirsk, p. 23 [russ].

1991 – 2000
57. Geodakyan V. A. The Evolutionary Theory of Sex. *Priroda (Nature)*, 1991, N 8, p. 60–69 [in Russian].
58. Geodakyan V. A. Evolutionary Logic of the Functional Asymmetry of the Brain. *Doklady Biological Sciences,* 1992, v. 324, N 1-6, p. 283–287. Translated from Doklady Akademii Nauk, Vol. 324, No. 6, pp. 1327-1331, June, 1992.
59. Geodakyan V. A. Asynchronous asymmetry. *Zh. higher. nerve. activities.* 1993, v. 43, № 3, p. 543-561 [russ].
60. Geodakyan V. A. Man and a woman. Evolutionary-biological destiny. Int. Conf. Woman and Freedom. Ways of choice in the world of tradition and change. Moscow, 1-4 June 1994, p. 8-17 [russ].
61. Geodakyan V. A. Sex Chromosomes: What Are They For? (A New Concept). *Doklady Biological Sciences.* Vol. 346, 1996. pp. 43-47. Translated from Doklady Akademii Nauk, Vol. 346, No. 4, 1996. pp. 565-569.
62. Geodakian V. A. Integrative Isomorphic Theories Of Asynchronous Evolution Of Sex Differentiations (Genes - Chromosomes - Hormones - Psychic - Organisms). XXXIII International Congress of Physiological Sciences. Abstracts. St. Petersburg, June 30 - July 5 1997.
63. Geodakyan V. A. and Geodakyan K. V. A New Concept on Lefthandedness. *Doklady Biological Sciences,* Vol 356, 1997, pp. 450-454. Translated from Doklady Academii Nauk Vol 356, No. 6, 1997, pp. 838-842.
64. Geodakian V. A. A New Approach To Ethnogeny Studies: Populational Sexual Dimorphism (PSD) As Evidence Of Gene Flows (GF) In The Past. The 14th Intern. Congr. of Anthropological and Ethnological Sciences. Abstracts July 26-August 1, 1998, p. 145.

65. Geodakian V. A. Ecological (E) Type Of Mortality Of Men And Genotypical (G) Of Women. Why Is The Average Life Span Of Men Shorter Than That Of Women? The 14th Intern. Congr. of Anthropological and Ethnological Sciences. Abstracts July 26-August l, 1998, p. 146.

66. Geodakyan V. A. Evolution of asymmetry, sexuality and culture (what is culture from the perspective of theoretical biology). Proc. Intern. Symp.: *The interaction of man and culture: an information-theoretic approach. Information outlook and aesthetics.* 1998, p. 116-143 [russ].

67. Geodakian V. A. Evolutionary Role of Sex Chromosomes: A New Concept. *Russian J. of Genetics,* 1998, v. 34, № 8, p. 986–998.

68. Geodakian K. V., Geodakian V. A. Sex Ratio (M/F) Adjusts Genotypical Evolutionary Plasticity (EP) Of a Population, Sinistrality/Dextrality (S/D)—Its Behavioral EP. The 14th Intern. Congr. of Anthropological and Ethnological Sciences. Abstracts July 26-August l, 1998, p. 145.

69. Geodakian V. A. The Role of Sex Chromosomes in Evolution: a new Concept. *J. of Mathematical Sci.* 1999, v. 93, № 4, p. 521–530.

70. Geodakian V. A. Evolutionary Chromosomes And Evolutionary Sex Dimorphism. *Biology Bulletin,* 2000, v. 27, № 2, p. 99–113. Translated from Izvestija Akademii Nauk, Serija Biologicheskaya, No. 2, pp. 133-148, 2000.

2001 – 2012

71. Geodakyan V. A. Homo sapiens on the path to asymmetrization (Theory of asynchronous evolution of the cerebral hemispheres and the cis-trans interpretation of left-handedness). Anthropology at the threshold of the IIIrd Millennium, Moscow, 2003, v. 1, p. 170-201 [russ].

72. Geodakyan V. A. Isomorphism: asynchronous sex - asynchronous asymmetry. Proc. of the Intern. Readings on the 100th anniversary of corresponding member. USSR Acad. AN Armenian SSR E. A. Asratian. May 30, 2003 [russ].

73. Geodakyan V. A. Convergent evolution of phenotype, asymmetry, and sexuality towards culture. Sexology and sexual pathology. 2003, № 6., p. 2-8; № 7, p. 2-6; № 8, p. 2-7 [russ].

74. Geodakyan V. A. Evolutionary biology in the synchronous "dead-end". XVIII Lyubischevskie reading. Modern problems of evolution. Ulyanovsk, 2004 [russ].

75. Geodakyan V. A. Evolutionary theories of asymmetrization organisms, brain and body. Uspekhi Physiological Sciences, 2005a Jan-Mar, v. 36, № 1, p. 24-53.

76. Geodakyan V. A. Hormonal sex. XIX Lyubischevskie reading. "Modern problems of evolution." Ulyanovsk, 5-7 April 2005b [russ].

77. Geodakyan V. A. Evolutionary role of cancer. Negentropy concept. Proc. of the Intern. Conf. of "Genetics in Russia and the world" dedicated to the 40th anniversary of the N.I. Vavilov's Institute of General Genetics. RAS, June 28 - July 2, 2006 Moscow p. 45 (242) [russ].

78. Geodakyan V. A. System roots of human evolution and society: the role of sex hormones. Int. Conf.: "Information culture of the society and individuum in the twenty-first century". Krasnodar, 20-23 September 2006, p. 75-80 [russ].

79. Geodakyan V. A. Why early and late children are different? Int. Conf. "Information and communication science in a changing Russia." Krasnodar, 2007, p. 150-153 [russ].

80. Geodakyan V. A. Binary conjugated differentiations, information and culture. Information, time, creativity, "Proc. Dokl. Int. Conf. "New methods in the study of artistic creativity and Int. Symp. "Informational approach to the study of culture and art." V. M. Petrov, A. V. Kharuto (eds), Moscow, 2007, p. 195-204 [russ].

81. Geodakyan V. A. A world of binary-conjugated systems: their nature and evolution. In: "VII School Kharitonov Readings (intern.) lecture. Russian Nuclear Center in Sarov (Arzamas-16). 01/05 March. 2007 [russ].

82. Geodakian V. A. Riddle of genomic imprinting - myth and reality. *Asymmetry*, 2008, Vol 2, № 4, p. 18-23 [russ].

83. Geodakian V. A. Dihronous evolution of living systems. *Asymmetry*, 2009, Volume 3, № 2, p. 3-31.

84. Geodakian V. A. Dominance, sex and age of the gene in the light of the dihronizm theory. Congress of N. I. Vavilov's All-Union Society of Geneticists and Breeders (VOGIS). June 20-27, 2009 [russ].

85. Geodakian V. A. Riddle of genomic imprinting. *V mire nauki (In the world of science)*, 2012, № 3, p. 74-79 [russ].

REFERENCES IN ENGLISH

1. Abraham J. N. "La Saboteuse: An Ecological Theory of Sexual Dimorphism in Animals." *Acta Biotheoretica* 1998, v. 46, p. 23-35.
2. Albert D. J., Walsh M. L., Jonik R. H. Aggression in humans: what is its biological foundation? Neurosci. Biobehav. Rev. 1993, v. 17, N 4, p. 405–425.
3. Allsop D. J., Warner D. A., Langkilde T., Du W., Shine R. Do operational sex ratios influence sex allocation in viviparous lizards with temperature-dependent sex determination? *J. Evol. Biol.,* 2006, v. 19, p. 1175–1182.
4. Amoore J. E., Venstrom D. Sensory analysis of odor qualities in terms of the stereochemical theory. *J. Food Sci.,* 1966, v. 31, p. 118–128.
5. Amos-Landgraf, J. M., Cottle, A., Plenge, R. M., Friez, M., Schwartz, C. E., Longshore, J., & Willard, H. F. (2006). X. chromosome-inactivation patterns of 1,005 phenotypically unaffected females. The American Journal of Human Genetics, 79, 493–499.
6. Andersson M. *Sexual Selection.* Princeton Univ. Press, 1994.
7. Andersson M., Wallander J. Ethology: Relative Size and Mating Behavior. Nature, 2004, v. 431, p. 139.
8. Ansari-Lari M, Saadat M. Changing sex ratio in Iran 1976–2000. *J. Epidemiol. Community Health* 2002, v. 56, p. 622–623.
9. Apicella C. L., Feinberg D. R., W. Voice pitch predicts reproductive success in male hunter-gatherers *Biol. Lett.,* 2007, v. 3, N. 6, p. 682–684.
10. Araki, T., Toda Y., Matsushita K., and Tsujino A. Age differences in sweating during muscular exercise. *Jap. J. Phys. Fitness Sports Med.* 1979, v. 28, p. 239-248.
11. Archer J. (1994). Male Violence. London: Routledge.
12. Arden R., Plomin R. Sex Differences in Variance of Intelligence Across Childhood. *Personality and Individual Differences*, 2006, v. 41, p. 39–48.
13. Ashley B. Sex differences in the incidence of tumors at various sites. *Brit. J. of Cancer*, 1969, v. 24, 1, p. 26–30.
14. Asker S. E., Jerling L. *Apomixis in plants.* CRC, Boca Raton, Fl, 1992
15. *Autoimmune diseases in women.* Society for Women's Health Research and the National Women's Health Resource Center, Inc. Report, 2002. < www.womens-health.org >

16. Badr F. M., Spickett S. G. Genetic variation in adrenal weight relative to body weight in mice. *Acta Endocrinol.,* 1965, Suppl., v. 100, p. 92–106.
17. Bagemihl B. *Biological Exuberance: Animal Homosexuality and Natural Diversity.* 1998, New York: St. Martin Press.
18. Baker H. G. Reproductive methods as factors in speciation in flowering plants. *Cold Spring Harbor Symp. Quant. Biol.,* 1959, v. 24, p. 177–191.
19. Baker H. G. Support for Baker's law as a role. *Evolution*, 1967, v. 21, p. 853–856.
20. Banta A., Brown L. A. Rate of metabolism and sex determination in Cladocera. *Proc. Soc. Exp. Biol. and Medic.,* 1924, v. 22, p. 77–79.
21. Banta A. M., Wood T. R., Brown L. A., Ingle L. Studies in the physiology, genetics and evolution of some Cladocera. Paper N 39, Dept. Of Genetics. Carnegie Inst. Of Washington, 1939, v. 39, p. 1–285.
22. Bar-Anon R., Robertson A., Variation in sex ratio between progeny groups in dairy cattle. *Theoret. Appl. Genet.,* 1975, v. 46, p. 63–65.
23. Bar-Or O. Invited review climate and the exercising child. *Int. J. Sports Med.* 1980, v. 1, p. 53–65.
24. Baron-Cohen S. (Ed.). (1999). The maladapted mind: Classic readings in evolutionary psychopathology. Cornwall: Psychology Press.
25. Barton D. E., David F.N., Merrington M. The positions of the sex chromosomes in the human cell in mitosis. *Ann. Hum. Genet.,* Lond., 1964, v. 28, p. 123–135.
26. Bateman A. J. Interasexual selection in Drosophila. *Heredity*, 1948, v. 2, p. 349–368.
27. Bawa K. S., Opler P. A. Dioecism in tropical forest trees. *Evolution*, 1975, v. 29, p. 167–179.
28. Beamer W., Bermant G., Clegg M. T. Copulatory behaviour of the ram, *Ovis aries.* II. Factors affecting copulatory satiation. *Anim. Behav.,* 1969, v. 17, p. 706–711.
29. Beatty R. A. The genetics of the mammalian gamete. *Biol. Rev.,* 1970, v. 45, p. 73–119.
30. Beilharz R. G. Research into sex-linked control of body weight in poultry and rabbits. *Proc. Aust. Soc. Anim. Prod.,* 1960, v. 3, p. 139.

31. Beilharz R. G. On the possibility that sex chromosomes have a greater effect than autosomes on inheritance. *J. Genet.*, 1963, v. 58, p. 441–449.

32. Bell G. *The Masterpiece of Nature.* Berkeley: Univ. of California Press, 1982.

33. Bell A. P., Weinberg M. S., Hammersmith S. K. *Sexual Preference: Its Development In Men and Women.* Bloomington, Indiana: Indiana Univ. Press, 1981.

34. Benbow C. P., Stanley J. C. Sex differences in mathematical ability: Fact or artefact? *Science.* Wash., 1980, v. 210, N 4475, p. 1262–1264.

35. Bermant G. Sexual behavior: Hard times with the Coolidge Effect. In *Psychological Research: The inside story.* (M. H. Siegel & H. P. Zeigler, Eds.), New York: Harper & Row, 1976, p. 76–103.

36. Bernstein M. E. Action of genes affecting the sex ratio in men. *Science*, 1951, v. 114, N 2955, p. 181–182.

37. Bernstein M. E. Studies in the human sex ratio. 5. A genetic explanation of the wartime increase in the secondary sex ratio. *Am. J. Human Genet.*, 1958, v. 10, p. 68–70.

38. Bessey E. A. Effect of the age pollen upon the sex of hemp. *Amer. J. Bot.*, 1918, N 5, p. 234–238.

39. Bessey E. A., 1933. Sex problem in hemp. *Quart. J. Biol.*, 1933, v. 3, p. 185–190.

40. Bishop P. O. Neural mechanisms for binocular depth discrimination. In: *Advances in Physiological Sciences. Sensory Functions* (Eds. Grastian E., Molnar P.), 1981, v. 16, p. 441–449.

41. Bisioli C. Sex ratio of births conceived during wartime. *Hum. Reprod.* 2004, v. 19, p. 218–219.

42. Black D. W. (1999). Bad boys, bad men: Confronting antisocial personality. New York: Oxford University Press.

43. Blackman R. L. Species, sex and parthenogenesis in aphids. *The Evolving Biosphere.* Cambridge, Cambridge Univ. Press, 1981, p. 75–85.

44. Blest A. D. A study of the biology of Saturniid moths in the canal zone biological area. *Smithson. Rep.*, 1959, p. 447–464.

45. Blumstein S. Goodglass H., Tartler V. The reliability of ear advantage in dichotic listening. *Brain and Language.* 1975, v. 2, p. 226–236.

46. Bodmer W. F. The evolutionary significance of recombination in prokaryotes. *Symp. Soc. gen. Microbiol.*, 1970, v. 20, p. 279–294.

47. Bonner J. Varner, V. E. (eds.) *Plant Biochemistry.* Wiley, 1968, 624 pp.

48. Booth A., Dabbs J. M., Jr. Testosterone and men's marriages. *Soc. Forces.* 1993, v. 72, p. 463–477.

49. Brackett B. G., Baranska W., Savicki W., Korovski H. Uptake of Heterologous Genome by Mammalian Spermatozoa and its Transver to Ova through Fertilization. *Proc. Nat. Acad. Sci. USA.*, 1971, v. 68, p. 353–357.

50. Brand G., Millot J-L. Sex differences in human olfaction: Between evidence and enigma. *The Quarterly J. of Exp. Psychology*, 2001, B 54, N 3, 1 August 2001, p. 259–270.

51. Brandt L. S. E., Greenfield M. D. (2004). Condition-dependent traits and the capture of genetic variance in male advertisement song. Journal of Evolutionary Biology, 17, 821–828.

52. *Breeding Methods for Cattle, Pigs and Poultry in the U.S.* Paris: OEEG, 1957, 153 p.

53. Brown G. R. Sex-biased investment in nonhuman primates: can Trivers and Willard's theory be tested? *Anim. Behav.*, 2001, v. 61, p. 683–694.

54. Brown, L. P. Can guppies adjust the sex ratio? *Am. Nat.*, 1982, v. 120, p. 694–698.

55. Brumby P. J. The influence of the maternal environment on growth in mice. *Heredity*, 1960, v. 14, p. 1.

56. Buffery A., Grey J. Sex differences in the development of spatial and linguistic skills. In: *Gender differences, their ontogeny and significance* (Ounsted C., Taylor D., eds.), Edinburgh, 1972, p. 123–158.

57. Bull J. J. *Evolution of Sex Determining Mechanisms.* Menlo Park, California: WA Benjamin/Cummings. 1983.

58. Bunce M., Worthy T. H., Ford T., et al. *Nature,* 2003, v. 425, p. 172–175.

59. Burley N. Sex-ratio manipulation in color-banded populations of zebra finches. *Evolution*, 1986, v. 40, p. 1191–1206.

60. Buss D. M. *The murderer next door: Why the mind Is designed to kill.* 2005, New York: Penguin Press.

61. Cain A. J. *Animal Species and their Evolution.* London: Princeton University Press, 1993, 216 pp.

62. Cameron E. Z. Facultative adjustment of mammalian sex ratios in support of the Trivers-Willard hypothesis: evidence for a mechanism. *Proc. R. Soc. Lond.*, 2004, v. 271, p. 1723–1728.

63. *UK cancer incidence statistics by age*, Cancer Research UK, January, 2007.

64. Cancer incidence in Finland, Iceland, Norway and Sweden. *Acta pathol. et morfol. Scandin.* Sect. A, 1971, Suppl. N 224, Copenh., 1971a.

65. *Cancer incidence in Sweden 1959–1965*. Stockholm, 1971b.

66. Caplan P. J., Crawford M., Hyde J. S., Richardson J. T. E. (1997) *Gender differences in human cognition.* Hillsdale, NJ, Oxford, Oxford Univ. Press.

67. Carlquist S. *Island life.* New-York, Natural History Press, 1965, 250 p.

68. Carlquist S. The biota of long distance dispersal, IV. Genetic systems in the flora of oceanic islands. *Evolution*, 1966, v. 20, p. 433–455.

69. Cartwright J. Evolution and Human Behavior: Darwinian Perspectives on Human Nature. Cambridge: MIT Press, 2000.

70. Cassidy J., Ditty K. Gender differences among newborns on a transient otoacoustic emissions test for hearing. *Journal of Music Therapy*. 2001, v. 37, p. 28–35.

71. Chakrabarti S., Fombonne E. Pervasive developmental disorders in preschool children. Journal of American Medical Association, 2001, v. 285, p. 3093–3099.

72. Chan S. T. H., O, W.-S. Environmental and non-genetic mechanisms in sex determination. In: *Mechanisms of Sex differentiation in animals and man.* (C. R. Austin & R. G. Edwards, eds) Acad. Press, London, 1981, p. 55–111.

73. Chan S. T. H., Yeung, W. S. B. Sex control and sex reversal in fish under natural conditions. *Fish Phys.*, 1983, v. 9, part B, p. 171–222.

74. Chapman A. B., Cassida L. E., Cote A. *Proc. Am. Soc. Anim. Prod.*, 1938, v. 30, p. 303.

75. Charlesworth L., Charlesworth B., Strobeck C. Effects of selfing on selection for recombination. *Genetics*, 1977, v. 86, p. 213–226.

76. Charnov E. L. *The theory of sex allocation.* Princeton Univ. Press, Princeton, 1982.

77. Charnov E. L., Bull J. When is sex environmentally determined? *Nature*, Lond., 1977, v. 266, p. 828–830.

78. Check, E. (2005). Genetics: The X factor. Nature, 434, 266–267.

79. Cheng K. M., Siegel P. B. Quantitative genetics of multiple mating. *Anim. Behav.*, 1990, v. 40, p. 406–407.

80. Choquet M., Ledoux S., Hassler C. Alcohol, tabac, cannabis et autres drogues illicites parmi les élèves de collège et de lycée. ESPAD 99-France. Paris, OFDT, 2001.

81. Chovanova E., Bergman K. P., Stukovsky K. Abstracts of comunications of II Congress of European Antropological Association, Brno, 1980, p. 136.

82. Ciesielski T. Quomodo flat ut max probes masculina, mox feminina oriatur apud plantas, animalia et homines. Lwow (Lemberg), 1911, S. 1–15.

83. Cluttonbrock T. H., Iason, G. R. Sex-ratio variation in mammals. *Quart. Rev. Biol.*, 1986, v. 61, p. 339–374.

84. Cohn L. D. Sex differences in the course of personality development: a meta-analysis. *Psychol Bull.* 1991, v. 109(2): p. 252–66.

85. Coie J. D., Dodge K. A. Aggression and antisocial behavior. In W. Damon & N. Eisenberg (Eds). *Handbook of Child Psychology,* 1997, Vol. 3: *Social, emotional and personality development.*

86. Cole N. S. The ETS gender study: how females and males perform in educational setting. Princeton, NJ, Educational Testing Service, 1997.

87. Cole R. I., Kirkpatrick W. F. *Rhode Island Agr. Exptl. Bull.*, 1915, v. 162, p. 43–48.

88. Collins S. A. Male voices and women's choices *Anim. Behav.* 2000, v. 60, p. 773–780.

89. Collins S. A, Missing C. Vocal and visual attractiveness are related in women. *Anim. Behav.,* 2003, v. 65, p. 997–1004.

90. Coltheart M., Hull B., Slater D. Sex differences in imagery and reading. *Nature*, 1975, v. 253, p. 438–440.

91. Comfort A. Aging: The biology of senescence. London, Routledge and Kegan, 1964. 400 pp.

92. Conway H., Wagner K. J. Congenital anomalies of the head and neck. *Plastic and Reconstr. surgery*, 1965, v. 36, N 1, p. 71–79.

93. Corbet G. B., Southern H. N. The handbook of British mammals. 2nd ed. Oxford, 1977.

94. Corbin A. *Pestdamp en bloesemgeur.* Nijmegen, Netherlands, SUN, 1986.

95. Corna F., Camperio-Ciani A., Capiluppi C. Evidence for maternally inherited factors favouring male homosexuality and promoting female fecundity. *Proceedings: Biological Sciences* 2004, v. 271, p. 2217-2221.

96. Correns C. Geschlechtsbestimmung und Zahlenverhaltnis der Geschlechter beim Sauerampfer (Rumex acetosa). *Biol. Zbl.*, 1922, v. 42, p. 465–480.

97. Correns C. Bestimmung, Vererbung und Verteilung des Geschlechter bei den hoheren Pflanzen. *Handb. Vererbungswiss.*, 1928, v. 2, p. 1–138.

98. Corso J. F. Age and Sex Differences in Pure-Tone Thresholds. *The J. of the Acoustical Society of America*, 1959, V. 31, Issue 4, p. 498-507.

99. Coulson J. C., Hicling, G. Variation in the secondary sex-ratio of the grey seal *Halichoerus grypus* (Fab.) during the breeding season. *Nature,* 1961, v. 190, p. 281.

100. Coyne J. A., Grant, B. J. Sex ratio compensation in *Drosophila melanogaster. Heredity,* 1971, v. 62, p. 205–207.

101. Crew F. A. E. Relation of sex of offspring to time of the coitus during oestrus cycle. *Brit. Med. J.*, 1927, v. ii, p. 917–919.

102. Crews D. Animal Sexuality. *Sci. American.* Jan. 1994, p. 108–114.

103. Crichton-Brown J. On the weight of the brain and its component parts in the insane. *Brain*, 1880, v. 2, p. 42–67.

104. Crow J. F. Advantages of Sexual Reproduction. *Dev. Gen.*, 1994, v. 15, p. 205–213.

105. Crow J. F., Kimura M. Evolution in sexual and asexual populations. *Am. Natur.*, 1965, v. 99, p. 439–450.

106. Crow J. F., Kimura M. *An Introduction to population genetics theory*. New York, Harper & Row, 1970, 180 p.

107. Cui W., Ma C., Tang Y., e.a. Sex Differences in Birth Defects: A Study of Opposite-Sex Twins. *Birth Defects Research (Part A)*, 2005, v. 73, p. 876–880.

108. Cullen K. R., Kumra S., Regan J. et al. Atypical Antipsychotics for Treatment of Schizophrenia Spectrum Disorders. Psychiatric Times. 2008, v. 25(3).

109. Currie C., Hurrelmann K., Setterobulte e. a. (Eds.) Health and health behavior among young people. WHO Policy Series: Health policy for children and adolescents Issue 1. International Report. 2000, Copenhagen: World Health Organization.

110. Curry F. W. K. A comparison of left-handed and right-handed subjects on verbal and nonverbal dichotic listening tasks. *Cortex*, 1967, v. 3, p. 343–352.

111. Darwin C. *The descent of man and selection in relation to sex*. London, John Murray, 1st ed., 1871.

112. Davidson G. C., Neale J. M. (1994). Abnormal psychology (6th ed.). NY: Wiley.

113. Davidsson P. Determinants of entrepreneurial intentions, *RENT IX Workshop in Entrepreneurship Research*, Piacenza, Italy, 1995.

114. Dawley R. M., Bogart J. P. *Evolution and Ecology of Unisexual Vertebrates*. Albany: New York State Museum. 1989.

115. Deary I. J., Thorpe, G., Wilson V., Starr, J. M., Whalley, L. J. (2003). Population sex differences in IQ at the age 11: The Scottish mental survey 1932. Intelligence, 31, 533–542.

116. Deary I. J., e. a. Brother–sister differences in the g factor in intelligence: Analysis of full, opposite-sex siblings from the NLSY1979. *Intelligence*, 2007, v. 35 (5), p. 451-456.

117. Degn H. I. Systematic position, age criteria and reproduction of Danish red squirrels (Sciurus vulgaris L.). Dan. Rev. Game Biol., 1973, v. 8, N 2, p. 1–24.

118. Delph L. F. Sexual dimorphism in flower size. *Am.Nat.*, 1996, v. 148, p. 299–320.

119. Di Renzo GC, Rosati A, Sarti RD, Cruciani L, Cutuli AM. Does fetal sex affect pregnancy outcome? Gender Medicine 2007; 4:19–30.

120. Dolinova, N. A. Dermatoglyphics of Udmurts, New Studies on Ethnogenesis of Udmurts, 1989, Izhevsk: Uro Acad. Sci. SSSR, p.108–122.

121. Doll R., Payne P., Waterhouse J. Cancer incidence in five continents. v. 1,2. Berl., l966, 1970.

122. Dorn H. F., Culter Sj. *Morbidity from cancer in the United States*. Wash., 1955.

123. Doty R. L., Brugger W. E., Jurs P. C., et. al. Intranasal trigeminal stimulation from odorous volatiles: psychometric responses from anosmic and normal humans. *Physiology and Behavior*, 1978, v. 20, p. 175–185.

124. Dusing C. Z. Die Regulierung des Geschlechtsverhältnisses bei der Vermehrung der Menschen, Tiere und Pflanzen. *Naturwissenschaften*. 1884, v. 10, p. 593–640.

125. Eaton S. B., Pike M. C. Short R. V., et al. Women's reproductive cancers in evolutionary context. *Quart. Rev. Biol.*, 1994, v. 69, p. 353–367.

126. Ebers G. C., Sadovnick A. D, Dyment D. A, et al. Parent-of-origin effect in multiple sclerosis: observations in half-siblings. *Lancet*, 2004 v. 363(9423), p. 1773–1774.

127. Eckhart V. M. Sexual dimorphism in flowers and inflorescences. In: *Gender and sexual dimorphism in flowering plants* (ed. M. A. Geber, T. E. Dawson, L. F. Delph), Berlin, Springer, p.123–148.

128. Edge The science of gender and science. Pinker vs. Spelke. A debate. http://www.edge.org/3rd_culture/debate05/ debate05_index.html (2005, May 16).

129. Edwards A. W. F. An analysis of Geissler's data on human sex ratio. *Ann.Human Genet.*, 1958, v. 23, p. 6–15.

130. Eguchi S., Hirch I. J. Development of speech sounds in children. *Acta Otolaryngologica, Suppl.* 1969, p. 257.

131. Ehrenkantz J.; Bliss E., Sheard M. H. Plasma testosterone: Correlation with aggressive behavior and social dominance in man. *Psyc. Med.* 1974, v. *36*, p. 469–475.

132. Ehrhardt A. A., Meyer-Bahlburg H. F. L., Rosen L. E. a. Sexual Orientation after Prenatal Exposure to Exogenous Estrogens. *Arch. Sex Behav.* 1985, v. 14, p. 57–77.

133. Ehrman L. Nuclear Genes Depending Cytoplasmic Sterility in Dr. paulistorum. *Science*, 1964, v. 145, № 3628, p. 159.

134. Ehrman L. A., Parsons P. A. *Behavior Genetics and Evolution*. McGraw-Hill: New York, 1981.

135. Ellis L., Bonin S. War and the secondary sex ratio: are they related? *Soc. Sci. Inf.*, 2004, v. 43, p. 119–126.

136. Eshel I., Feldman M. W. On the evolutionary effect of recombination. *Theoret. Pop. Biol.*, 1970, v. 1, p. 88–100.

137. Eveleth P. B. Eruption of permanent dentition and menarche of American children living in the tropics. *Hum. Biol.*, 1960, vol. 39, p. 60.

138. Evsikov V. I., Nazarova G. G., Potapov M. A. Female odour choice, male social rank, and sex ratio in the water vole. Advances In the biosciences. V. 93: Chemical signals in vertebrates VII. Oxford: Pergamon, 1995, p. 303–307.

139. Eysenck H. J., Gudjonsson, G. (1989). The causes and cures of criminality. NY: Plenum.

140. Fairbairn D. J. Allometry for Sexual Size Dimorphism: Pattern and Process in the Coevolution of Body Size in Males and Females. *Annu. Rev. Ecol. Syst.*, 1997, v. 28, N 1, p. 659–687.

141. Falk B., Bar-Or O., Calvert R., and MacDougall J. D. Sweat gland response to exercise in the heat among premid and late-pubertal boys. *Med. Sci. Sports Exerc.* 1992, v. 24, p. 313–319.

142. Fausto-Sterling A. *Sexing the Body: Gender Politics and the Construction of Sexuality*, 2000.

143. Felsenstein J. Sex and the evolution of recombination. In: *The evolution of sex: an examination of current ideas* (Michod R. E. and Levin B. R., eds). 1988, Sunderland, Massachusetts: Sinauer, p. 74-86.

144. Fernando J, Arena P, Smith D.W. (1978) Sex liability to single structural defects. Am. J. Dis. Child **132** p. 970–972.

145. Filipponi A., Mosna, B. & Petrelli, G. L'Ottimo di temperatura di *Macrocheles* muscaedomesticae (Scopoli) (Acari: Mesostigmata). *Riv. Parassit.*, 1972, v. 32, N 3, p. 193–218.

146. Filipponi A., Mosna, B. & Petrelli, G. Ricerhe autoecologiche di laboratori su tre specie del gruppo *Glaber* (Acarina, Mesostigmata) 3-rapporto-sessi e dinamica di popolazione. *Riv. Parassit.*, 1975, v. 36, p. 295–308.

147. Filipponi A. & Petrelli, G. Autoecologia e capacita moltiplicativa di *Macrocheles muscaedomesticae* (Scopoli) (Acari: Mesostigmata). *Riv. Parassit.*, 1967, v. 28, N 2, p. 129–156.

148. Fishelson L. Protogynous sex reversal in the fish Anthias squamipinnis (Teleostei, Anthiidae) regulated by the presence or absence of a male fish. *Nature* (London), 1970, v. 227, p. 90–91.

149. Fisher R. A. *The genetical theory of natural selection*. Univ. Press, Oxford, 1930.

150. Flanders S. E. Control of sex and sex-limited polymorphism in the Hymenoptera. *Quart. Rev. Biol.*, 1946, v. 21, N 2, p. 135–143.

151. Fohrman M., McDowell R., Mathews C., Hilder R. A cross-breeding experiment with dairy cattle. Techn. Bull. U.S. Dep. Agr., 1954, N 1074, p. 200.

152. Fombonne E. The changing epidemiology of autism. Journal of Applied Research in Intellectual Disabilities, 2005, v. 8, p. 281–294.

153. Ford C. S., Beach F. A. *Patterns of Sexual Behavior*. N.Y., Harper & Row, 1951, p. 136–140.

154. Fozard J. L. Vision and hearing in aiging. J. E. Birren, K. W. Schaie (eds.). *Handbook of the psychology of aging*. N-Y., Academic Press, 1990.

155. Frank S. A. Sex allocation theory for birds and mammals. A. *Rev. Ecol. Syst.*, 1990, v. 21, p. 13–55.

156. Freeman D. C., Klikoff L. G., Harper K. T. Differential resource utilization by the sexes of dioecious plants. *Science,* l976, v. 193, p. 587–599.

157. Freeman D. C., McArthur E. D., Miglia K. J., e.a. Sex and the lonely Atriplex. *Western North American Naturalist,* 2007, v. 67, N 1, p. 137–141.

158. Freeman S. The evolution of the scrotum: a new hypothesis. *J.Theor. Biol.,* 1990, v. 145, p. 429–445.

159. Frey D. G. The taxonomic and phylogenetic significance of the head pores of the *Chydoridae (Cladocera). Internation. Revue der gesamten Hydrobiologie,* 1959, Bd. 44, Heft 1, p. 27–50.

160. Freud S. *Introduction to Psychoanalysis, Lectures,* 1920.

161. Fukuda M., Fukuda K., Shimizu T., et al. Parental periconceptional smoking and male: female ratio of newborn infants. *Lancet,* 2002, v. 359, p. 1407–1408.

162. Gage T. B., Therriault G. Variability of birth-weight distributions by sex and ethnicity: analysis using mixture models. Human Biology, 1998; v. 70 (3), p. 517–534.

163. Gefou-Madianou D. (ed.) *Alcohol, gender and culture,* Routledge, London and New York, 1992.

164. Geiser S. W. The differential death rate of the sexes among animals. Wash. Univ. Studies, 1924–1925, N 12, p. 13–15.

165. Geissler A. Beiträge zur Frage des Geschlechtsverhältnisses der Geborenen. *Zeitschr. K. Sächs. Statist. Bureaus,* 1889, v. 35, p. 1–24.

166. Ghiselin M. T. The evolution of hermaphroditism among animals. *Q. Rev. Biol.,* 1969, v. 44, p. 189–208.

167. Ghiselin M. T. *The economy of nature and the evolution of sex.* Univ. of California Press, 1974.

168. Gillis J. S., Walter E. Avis W. E. The Male-Taller Norm in Mate Selection. *Personality and Social Psychology Bulletin,* 1980, v. 6, No. 3, p. 396–401.

169. Gini C. Combinations and sequences of sexes in human families and mammal litters. *Acta Genet. et Stat. Med.,* 1951, v. 2, p. 220–244.

170. Gittelsohn A, Milham S. Statistical study of twins—methods. *Am. J. Public Health Nations Health,* 1964, v. 54, p. 286–294.

171. Glücksmann A. *Sexual Dimorphism in Human and Mammalian Biology and Pathology.* Lond., N.Y., Toronto, Sydney, San Francisco, Acad. Press, 1981, part 1–2, 356 p.

172. Goble F. C. Sex as a factor in metabolism, toxicity, and efficacy of pharmacodynamic and chemotherapeutic agents. *Adv. Pharmacol. Chemother.,* 1975, v. 13, p. 173–252.

173. Goddard M. Nature, 2005

174. Godley E. J. Monoecy and incompatibility. *Nature,* Lond., 1955, v. 176, p. 1176–1177.

175. Golanski K. Rev. Ver. Soie, 1959, v. 11, Jan.-März.

176. Goldschmidt R. Physiologische Theorie der Vererbung. B., 1927, p. 48.

177. Gould S. J. *Ontogeny and Phylogeny.* The Belknap Press, Cambridge, 1977, 501 p.

178. Graffelman J., Hoekstra R. F. A statistical analysis of the effect of warfare on the human secondary sex ratio. *Hum. Biol.,* 2000, v. 72, p. 433–445.

179. Grant S., Houben A., Vyskot B., Siroky J., Wei-Hua P., Macas J. Genetics of Sex Determination in Flowering Plants. *Dev. Genet.* 1994, v. 15, p. 214–230.

180. Grant V. J. Maternal Personality, Evolution, and the Sex Ratio: Do Mothers Control the Sex of the Infant? London: Routledge, 1998.

181. Grant V. J. The maternal dominance hypothesis: questioning Trivers and Willard. *Evol. Psychol.* 2003, v. 1, p. 96–107.

182. Gray A. P. *Mammalian Hybrids.* A Check-list with Bibliography. 1972, 2nd edition.

183. Greaves M. *Cancer: The Evolutionary Legacy.* Oxford Univ. Press, 2000, 276 p.

184. Guerrero R. Sex ratio: a statistical association with the type and time of insemination in the menstrual cycle. *Intern. J. Fertility,* 1970, v. 15, N 4, p. 221–225.

185. Guerrero R. Association of the type and time of insemination within the menstrual cycle with the human sex ratio at birth. *New Engl. J. Med.,* 1974, v. 291, p. 1056–1059.

186. Gunter C. (2005). Genome biology: She moves in mysterious ways. Nature, 434, 279–280.

187. Gunter H. *Naturwiss. Korresp.,* 1923, v. 1, p. 19.

188. Guzman R. C., Yang J., Radjkumar L., et al. Hormonal prevention of breast cancer: mimicking the protective effect of pregnancy. *Proc. Natl. Acad. Sci. USA,* 1999, v. 96, p. 2520–2525.

189. Haig D., Westoby M. Parent-specific gene expression and the triploid endosperm. *Am. Nat.*, 1989, v. 134, p. 147–155.

190. Haig D., Westoby M. Genomic imprinting in endosperm: Its effect on seed development in crosses between species, and between different ploidies of the same species, and its implications for the evolution of apomixis. *Philos. Trans. R. Soc. Lond.*, 1991, B333, p. 1–13.

191. Haig-Thomas R., Huxley L. S. Sex ratio in pheasant Species crosser. *J. Genet.*, 1927, v. 18, p. 223–246.

192. Halliday T., Arnold S. J. Multiple mating by females: a perspective from quantitative genetics. *Anim. Behav.*, 1987, v. 35, p. 939–941.

193. Halpern D. F. (1992) *Sex differences in cognitive abilities (2ⁿᵈ ed.)*. Hillsdale, NJ: Erlbaum.

194. Hamilton J. B. The role of testicular secretions as indicated by the effects of castration in man and by studies of pathological conditions and the short life span associated with maleness. *Recent Progress in Hormone Research*, v. 3, N.Y., Acad. Press, 1948, p. 257–322.

195. Hamilton J. B., Hamilton R. S., Mestler G. E. Duration of life and causes of death in domestic cats. Influence of sex, gonadectomy, and inbreeding. *J. Gerontol.*, 1969, v. 24, N 4, p. 427–437.

196. Hamilton W. D. Extraordinary sex ratios. *Science,* 1967, v. 156, p. 477–488.

197. Hammond J. J. The fertilization of rabbit ova in relation to time. *J. Exp. Biol.*, 1934, v. 11, p. 140–161.

198. Hannach A. Z. The effect of aging the maternal parent upon the sex ratio in *Drosophila melanogaster*. *Ord. Abstammungs und Vererbungslehre,* 1955, v. 86, p. 574–599.

199. Harlap S. Gender of infants conceived on different days of the menstrual cycle. *New Engl. J. Med.* 1979, v. 300, p. 1445–1448.

200. Harper E. B., Howing W. K., Dubanovsky L. Young children's yielding to false adult judgment. *Child. Development.* 1965, v. 36, p. 175–183.

201. Harris L. J. Sex differences in spatial ability. In: *Asymmetry of the function of the brain* (Kempbel L., ed.), Cambridge, 1978, p. 405–522.

202. Harrison G. A., Weiner J. S., Tanner J. M., et al. *Human biology*. Oxford: Oxford University Press, 1964.

203. Hart D., Moody, J. D. Sex ratio: experimental studies demonstrating controlled variations. *Ann. Surgery,* 1949, v. 129, p. 550–571.

204. Hart R. *Children's experience of Place A Developmental Study*. New York, 1978.

205. Harvey P. H., Partridg L., Southwood T. R. E. (eds). *The Evolution of Reproductive Strategies.* London: The Royal Society. 1991.

206. Hawkes K. The Grandmother Effect. *Nature,* 2004, v. 428, p. 128-129.

207. Hawkes K., O'Connell J. F., Blurton Jones N. G., Alvarez H., and Charnov E. L. Grandmothering, menopause, and the evolution of human life histories. *Proc. of the Nat. Acad. of Sci.*, 1998, v. 95, p. 1336-1339.

208. Hay S. Sex differences in the incidence of certain congenital malformations. *Teratology*, 1971, v. 4, N 3, p. 277–280.

209. Heath C. W. Physique, temperament and sex ratio. *Hum. Biol.*, 1954, v. 26, N 4, p. 337–342.

210. Heinz A. Cormofyternas Fylogenis. *Lund*, 1927, p. 105.

211. Hertwig R. Uber den derzeitigen Stand des Sexualitätsproblems. *Biol. Zentralblatt.*, 1912, Bd. 32.

212. Hoekstra R. F. The Evolution of Sexes. *The Evolution of Sex and its Concequences* (Ed. Stearns S. C.) Basel, Birkhauser Verlag, 1987.

213. Howe H. F. Sex ratio adjustment in the common grakle. *Science,* 1977, v. 198, p. 744–746.

214. Hutchinson J. *The families of flowering plants*. L., 1926–1934, vol.1–2, 2nd ed., 1959.

215. Huxley J. S. Late fertilization and sex ratio in trout. *Science*, 1923, v. 58, p. 291–292.

216. Huxley S. S. Sex determination and related problems. *Med. Sci. Abstr. and Rev.*, 1924, v. 10, p. 11–15.

217. Hyde J. S. Understanding Human Sexuality. New York, McGraw-Hill, 1979, 565 p.

218. Hyde J. S., Linn M. C. Gender differences in verbal ability: A meta-analysis. *Psychol. Bull.*, 1988, v. 104, p. 53–69.

219. Hyde J. S. The gender similarities hypothesis. *Am. Psychol.* 2005, v. 60, p. 581.

220. James H. C. *Proc. Roy. Entomol. Soc.* London, 1937, Ser. A, v. 12, p. 92–98 (cit.: Werren J. H. & Charnov E. L., 1978. *Nature*, v. 272, p. 349–350).

221. James W. H. Cycle day of insemination, coital rate and sex ratio. *Lancet,* 1971a, v. i, p. 112–114.

222. James W. H. Coital rate, sex ratio and parental age. *Lancet,* 1971b, v. i, p. 1294–1296.

223. James W. H. Coital rate, sex ratio and season of birth. *Lancet,* 1971c, v. ii, p. 159–162.

224. James W. H. Distributions of the combinations of the sexes in mammalian litters. *Genet. Res.,*1975a, v. 26, p. 45–53.

225. James W. H. Sex ratio and the sex composition of the existing sibs. *Ann. Hum. Genet.,* 1975b, v. 138, p. 371–378.

226. James W. H. Sex ratios in large sibships, in the pesence of twins and in Jewish sibships. *J. Biosoc. Sci.* 1975c, v. 7, p. 165–169.

227. James W. H. Timing of fertilization and sex ratio of offspring. A review. *Ann. Hum. Biol.,* 1976, v. 3, N 6, p. 549–556.

228. James W. H. The honeymoon effect on marital coitus. J. Sex Res., 1981, v. 17, p. 114–123.

229. James W. H. Decline in coital rate with spouses' age and duration of marriage. J. Biosoc. Sci., 1983, v. 15, p. 83–87.

230. James W. H. The human sex ratio. Part I: A review of the literature. Hum Biol. 1987, v. 59, p. 721–752.

231. James W. H. Sex ratios of births conceived during wartime. *Hum Reprod,* 2003, v. 18, p. 1133–1134.

232. Jensen A. R. *The g Factor: The Science of Mental Ability*; Praeger: Westport, CT, USA, 1998.

233. Jerison H. The Evolution of Biological Intelligence. In: *Handbook of Human Intelligence.* (Sternberd R. J., ed.), Cambridge, Cambridge Univ. Press, 1982.

234. Jonas V. *Particular* cardiology. Prague, 1960, 400 pp.

235. Jordan V. C., Morrow M. Tamoxifen, raloxifen and the prevention of breast cancer. *Endocrine Review,* 1999, v. 20, p. 253–278.

236. Jost A. Hormonal factors in the sex differentiation in the mammalian foetus. Philosophical transactions of the Society of London, 1970, Ser. B, v. 259, p. 119–131.

237. Kaiser Permanente: Healthwise. South California, 1995

238. Kalmus H., Smith A. B. Evolutionary origin of sexual differentiation and the sex-ratio. *Nature,* 1960, v. 186, N 4730, p. 1004–1006.

239. Kanazawa S. Scientific discoveries as cultural displays: A further test of Miller's courtship model. *Evol. And Hum. Behav.,* 2000, v. 21, p. 317–321.

240. Kanazawa S. Big and tall parents have more sons: further generalizations of the Trivers-Willard hypothesis. *J. Theor. Biol.,* 2005, v. 235, p. 583–590.

241. Kanazawa S. Violent men have more sons: further evidence for the generalized Trivers–Willard hypothesis (gTWH). *J. Theor. Biol.,* 2006, v. 239, p. 450–459.

242. Kanazawa S. Beautiful parents have more daughters: a further implication of the generalized Trivers-Willard hypothesis (gTWH). *J. Theor. Biol.,* 2007, v. 244, p. 133–140.

243. Kanazawa S. Big and tall soldiers are more likely to survive battle: a possible explanation for the 'returning soldier effect' on the secondary sex ratio. *Hum. Reprod.,* 2007, v. 22, N 11, p. 3002-3008.

244. Kanazawa S., Vandermassen G. Engineers have more sons, nurses have more daughters: an evolutionary psychological extension of Baron-Cohen's extreme male brain theory of autism and its empirical implications. *J. Theor. Biol.,* 2005, v. 233, p. 589–599.

245. Kang S., Cho W. The sex ratio at birth of the Korean population. *Eugenics quart.,* 1959, N 6, p. 187–190.

246. Karlin S., Lessard S. *Sex Ratio Evolution.* New Jersey: Princeton Univ. Press. 1986.

247. Karlsmose B. et. al, Prevalence of hearing impairment and subjective hearing problems in a rural Danish population aged 31-50 years. *British Journal of Audiology,* 1999, v. 33, p. 395-402.

248. Katsenelinboïgen A. *Evolutionary Change; Toward a Systemic Theory of Development and Maldevelopment.* Newark: Gordon & Breach Publishing Group, 1997, pp. 1-217.

249. Keith L. D., Rowe R. D., Ulad P. *Heart disease in infansy and children.* N.Y., 1959.

250. Kerkis J. Some Problems of Spontaneous and Induced Mutagenesis in Mammals and Man. *Mutation Res.* 1975, v. 29, p. 271–280.

251. Kihara H., Hirayoshi J. Die Geschlechtschromosomen von *Humulus japonicus.* Sieb. et Zuce. Proc. 8th Congr. Jap. Ass. Adv. Sci., 1932, p. 363–367 (cit.: Plant Breeding Abstr. 1934, v. 5, N 3, p. 248, ref. N 768).

252. Kimura D., Durnford M. *Hemisphere function of the human brain,* 1974, p. 25.

253. King A. P., West M. J. *Nature.* 1983, v. 305, №5936, p.704-706.

254. Kinsey A., Pomeroy W. Sexual Behavior in the Human Female, 1953.

255. Kirkpatrick R. C. The evolution of human homosexual behavior. *Current Anthropololgy* 2000, v. 39, N 1, p. 385-413.

256. Kjellberg S., Mannheimer B., Rudhe V., Johnsson D. *Diagnosis congenital heart disease*. Chicago, 1959.

257. Kon I. S. *Sociology of Personality*, Moscow, "Politizdat", 1967.

258. Kondrashov A. S. Classification of Hypotheses on the Advantage of Amphimixis. *J. of Heredity*, 1993, v. 84, p. 372-387.

259. Krackow S., Hoeck H. N. Sex ratio manipulation, maternal investment and behaviour during concurrent pregnancy and lactation in house mice. *Anim. Behav.*, 1989, v. 37, p. 177–186.

260. Kurland L. T., Kurtzke L. H., Goldberg I. D. *Epidemiology of neitrologic and sense organ disorders*. Cambridge, 1973, 1310 p.

261. Lalumiere M. L., Blanchard R., Zucker K. J. Sexual Orientation and Handedness in Men and Women: A Meta-Analysis. *Psych. Bull.*, 2000, v. 126, N 4, p. 575–592.

262. Lampl M., Jeanty P. Timing is everything: A reconsideration of fetal growth patterns identifies the importance of individual and sex differences. Am. J. Hum. Biol., 2003, v. 15, p. 667–680.

263. Landauer A. B. Rate of motor reaction in man and women. *Perception and Motor Skills*, 1981, v. 52, p. 90–97.

264. Landauer W., Landauer A. B. Chick mortality and sex ratio in the domestic fowl. *Amer. Naturalist*, 1931, v. 65, p. 492–501.

265. Lanman J. T. Delays during reproduction and their effects on the embryo and fetus. 1. Aging of sperm. *New Engl. J. Med.*, 1968a, v. 278, N 18, p. 993–999.

266. Lanman J. T. Delays during reproduction and their effects on the embryo and fetus. 1. Aging of eggs. *New Engl. J. Med.*, 19688b, v. 278, N 19, p. 1047–1054.

267. Lary J. M, Paulozzi L. J. Sex differences in the prevalence of human birth defects: a population-based study. *Teratology*, 2001, v. 64, p. 237–251.

268. Latham R. M. Differential Ability of Male and Female Game Birds to Withstand Starvation and Climatic Extremes. *J. Wildlife Manag.*, 1947, v. 11, p. 139–149.

269. Lawrence P. S. Ancestral longevity and sex ratio of the descendantes. *Human Biol.*, 1940, N 12, p. 93–101.

270. Lawrence P. S. The sex ratio, fertility and ancestral longevity. *Quart. Rev. Biol.*, 1941, N 16(1), p. 35–79.

271. Le Galliard J.-F., Fitze P. S., Cote J., Massot M., Clobert J. Female common lizards (*Lacerta vivipara*) do not adjust their sex-biased investment in relation to the adult sex ratio. *J. Evol. Biol.*, 2005, v. 18, p. 1455–1463.

272. LeBoeuf B. J. Male-male competition and reproductive success in elephant seals. *American Zoologist*, 1974, v. 14, p. 163–176.

273. LeBoeuf B. J., Peterson P. S. Social status and mating activity in elephant seals. *Science*, 1969, v. 163, p. 91–93.

274. Lehre A.-C., Laake P. and Sexton J. A. (2013) Differences in birth weight by sex using adjusted quantile distance functions. Statist. Med., 32: 2962–2970. doi:10.1002/sim.5744

275. Lehre A. C., Lehre K. P., Laake P., Danbolt N. C. Greater intrasex phenotype variability in males than in females is a fundamental aspect of the gender differences in humans. Developmental Psychobiology 2009; v. 51, p. 198–206.

276. Lehrke R. S. Sex linkage: A biological basis for greater male variability in intelligence. *Human variation: The biopsychology of age, race and sex.* (Osborne R. T. et al., eds.) N. Y., etc., 1978, p. 171–198.

277. Lem F. *Arciv. Hyg.*, 1923, v. 93, p. 27.

278. Lenneberg E. H. *Biological Foundations of Language*. N. Y., Wiley, 1967.

279. Lenz F. Die Uebersterblickeit der Knaben im Lichte der Erblickeitsiehve. *Arch. Hyg.*, 1923, v. 93, p. 126–150.

280. Leonard M. L., Weatherhead P. J. Dominance rank and offspring sex ratios in domestic fowl. *Anim. Behav.*, 1996, v. 51, p. 725–731.

281. Levine L. Studies on sexual selection in mice: I. Reproductive competition between albino and black-aqouti males. *American Naturalist*, 1958, v. 92, p. 21–26.

282. Levine L., Cormody G. R. *American Naturalist*, 1967, v. 101, p. 189–191.

283. Levy J. Lateral differences in the human brain in cognition and behavioural control. In: *Cerebral correlates of conscious experience.* (Buser P., Rongeul-Buser A. eds.) New York, 1978, No. 1.

284. Lichtman M., Vaughan J., Hames C. *Arthritis Rheum.* 1967, v. 10, p. 204.

285. Lilienfeld H. Die resultate einiger Bestäubungen mit verschiedenaltrigen Pollen bei Cannabis sativae. *Biol. Zentralblatt.*, 1921, v. 41, p. 296–303.

286. Lintonen T., Rimpelä, M., Ahlström, S., e. a. Trends in drinking habits among Finnish adolescents from 1977 to 1999. *Addiction*, 2000, v. 95, N 8, p. 1255–1263.

287. Lloyd D. G., Webb C. J. Secondary sex characteristics in plants. *Bot. Rev.*, 1977, v. 43, p. 177–216.

288. Loat C. S., Asbury K., Galsworthy M. J., et al. X Inactivation as a Source of Behavioural Differences in Monozygotic Female Twins. *Twin Research*, 2004, v. 7, N 1, p. 54–61(8).

289. Lockshin M. *Arthritis Rheum.* 1989, v. 63, p. 665.

290. Lombardi L. Sulla determinizione del sesso nel *Bombix mori*. *Asistencia publica*, 1923, v. 2, p. 1.

291. López S., Domínguez C. A. Sex choice in plants: facultative adjustment of the sex ratio in the perennial herb *Begonia gracilis*. *J. Evol. Biol.*, 2003, v. 16(6), p. 1177–85.

292. Lowe C. R., McKeown T. A note on secular changes in the human sex ratio at birth. *Br. J. Soc.*, 1951, v. 5, p. 91–96.

293. Lubinsky M. S. (1997) Classifying sex biased congenital anomalies. Am. J. Med. Genet. **69** p. 225–228.

294. Ludwig W., Boost C. *Induct. Abstamm. Verebungsl.*, 1951, v. 83, p. 383.

295. Lummaa V., Merila J., Krause A. Adaptive sex ratio variation in preindustrial human (*Homo sapiens*) populations? *Proc. R. Soc. Lond. B Biol. Sci.*, 1998, v. 7, p. 563–568.

296. Maccoby E. E. *The development of sex differences* (McCoby E. E., ed.). Stranford (Cal.), Stranford Univ.Press, 1966, p. 216–261.

297. Maccoby E., Jacklin C. *The psychology of sex differences*. Stanford, 1974, 360 p.

298. Mackintosh N. J. *IQ and Human Intelligence*; Oxford University Press: Oxford, UK, 2011.

299. MacMahon B., Pugh T. F. Sex ratio of white births in the United States during the Second World War. *Am. J. Hum. Genet.,* 1954, v. 6, p. 284–292.

300. Masters W. H., Johnson V. E. *Homosexuality in Perspective*. Toronto; New York: Bantam Books, 1979.

301. Masterton B., Heffner H., Ravizza R. The Evolution of Human Hearing. *The J. of the Acoustical Society of America*, 1969, v. 45, N 4, p. 966-985.

302. Mather K. Genetical control of stability in development. *Heredity*, 1953, v. 7, p. 297–336.

303. Maurer R. R., Foote R. H. *J. Reprod. Fert.*, 1971, v. 25, p. 329.

304. Maynard Smith J. *The evolution of sex*. Cambridge Univ. Press, Cambridge, 1978.

305. Maynard Smith J. *The evolution of sexual reproduction*. Wiley, 1981, 271.

306. Maynard Smith J. In: *The Evolution of sex: an Examination of Current Ideas*. (Michod R. E., Levin B. R., eds.), Sinauer, Sunderland, 1988, p. 106–125.

307. Mazur, A.; Halpern, R.; Udry, R. Dominant looking males copulate earlier. *Ethol. Sociobiol.* 1994, v. 15, p. 87–94.

308. McArthur J. W., Baillie W. H. T. Metabolic activity and duration of life. I, II. *J. Exptl. Zool,* 1929, v. 53, p. 221–268.

309. McArthur J. W., Baillie W. H. T. Sex differences in mortality in *Abraxas* type species. *Quart. Rev. Biol.*, 1932, v. 7, N 3, p. 20–25.

310. McClung C. E. Notes on the Accessory Chromosome. *Anat. Anz.*, 1901, v. 20, p. 220–226.

311. McGlone J. Sex difference in functional brain asymmetry. *Cortex,* 1978, v. 14, p. 122–128.

312. McGlone J. Sex differences in the human brain asymmetry. A critical survey. *The Behav. and Brain Sci.*, 1980, v. 3, No. 2, p. 215–263.

313. McGrath J., Solter D. 1984. Completion of mouse embryogenesis requires both the maternal and paternal genomes. *Cell.*, 1984, v. 37, p. 179–183.

314. Mclain D. K., Marsh N. B. Individual Sex Ratio Adjustment in Response to the Operational Sex Ratio in the Southern Green Stinkbug. *Evolution*, 1990, v. 44, No. 4, p. 1018–1025.

315. Michaels R., Rogers K. *Pediatrics,* 1971, v. 47, p. 120.

316. Michod R. E. *Eros and evolution: a natural philosophy of sex*. Addison-Wesley Pub. Co., 1995.

317. Michod R., Levin B. R., Eds. *The Evolution of Sex: A Critical Review of Current Ideas*. 1987. Sinauer Associates: Sunderland, MA.

318. Miglia K. J., Freeman D. C. The effect of delayed pollination on stigma length, sex expression, and progeny sex ratio in spinach, *Spinacia oleracea* (Chenopodiaceae). *American J. of Botany*, 1995, v. 83, p. 326–332.

319. Miller G. F. Sexual selection for cultural displays. In *The Evolution of culture* (Dunbar R., et al., eds), New Brunswick, Rutgers Univ. Press, 1999, p. 71–79.

320. Ming R., Wang J., Moore P. H., and Paterson A. H. Sex chromosomes in flowering plants. *American Journal of Botany.* 2007, v. 94, p. 141-150.
321. Mitchell H.R. Teen Alcoholism, Lucent Books, 1998, 96 p.
322. Miyata T., Hayashida H., Kuma K., et al. Male-driven Molecular Evolution: A Model and Nucleotide Sequence Analysis. In: Cold Spring Harbor Symposia on Quantitative Biology. 1987, v. LII, p. 863–867.
323. Money J., Ehrhardt A. A. *Man & woman, boy & girl: Differentiation and dimorphism of gender identity.* Baltimore, Johns Hopkins Univ. Press, 1972.
324. Money J., Schwartz M., Lewis V. G. Adult Erotosexual Status and Fetal Hormonal Masculinization and Demasculinization. *Psychoneuroendocrinology,* 1984, v. 9, p. 405–414.
325. Montagu A. (1968) Natural Superiority of Women, The, Altamira Press, 1999.
326. Mooney S. M. *The Evolution of Sex: A Historical and Philosophical Analysis.* Ph. D. thesis. Boston: Boston Univ. 1992.
327. Money J., Schwartz M., Lewis V. G. Adult Erotosexual Status and Fetal Hormonal Masculinization and Demasculinization. *Psychoneuroendocrinology,* 1984, v. 9, p. 405–414.
328. Moore K. L., Persaud T. V. N. *The Developing Human: Clinically Oriented Embryology.* 6th ed., Philadelphia: W.B. Saunders Co., 1998.
329. Moorley F., Smith I. A comparison between reciprocal crosses of Australorps and White Leghorns. Agr. Gaz. N. S. Wales, 1954, v. 65, N 1, p. 17.
330. Morduhai-Boltovskoi F. D. Polyphemidae of the Pontocaspian basin. *Hydrobiologia,* 1965, v. 25, Fasc. 1–2, p. 212–220.
331. Morison I. M., Paton C. J., Cleverley S. D. The imprinted gene and parent-of-origin effect database. Nucleic Acids Res., 2001, v. 29, p. 275–276.
332. Mosiey I. L., Slan E. A. Human sexual dimorphism: Its cost and benefit. *Advances in child development and behavior* (Reese H. W., ed.). N. Y., 1984, v. 18, p. 147–185.
333. Moya-Laraño J, Halaj J., Wisea D. H. Climbing to reach females: Romeo should be small. *Evolution,* 2002, v. 56, No. 2, p. 420–425.
334. Mršic W. Die spätbefruchtung und deren Einfluss auf Entwicklung und Geschlechtsbildung. Arch. Micr. Anat., 1923, v. 98, p. 129–209.
335. Mršic W. Über die Eireifung bei der Forelle und deren Bedeutung für die übliche Methode der künstlichen Laichgewinnung. *Arch. Hidrobiol.,* 1930, v. 21, p. 649–678.
336. Mueller H. C., Meyer K. Evolution of reversed sexual dimorphism in Falconiforms. *Curr. Ornithol.* 1985, v. 2, p. 65-101.
337. Mulcahy D. L. Optimal sex ratio in Silene alba. *Heredity,* 1967, v. 22, N 3, p. 411–423.
338. Muller H. J. Some genetic aspects of sex. *Amer. Natur.,* 1932, v. 66, p. 118–138.
339. Murphy C. G. Interaction-independent sexual selection and the mechanisms of sexual selection. *Evolution,* 1998, v. 52, p. 8–18.
340. Murray J. Multiple mating and effective population size in Cepaea nemoralis. *Evolution,* 1964, v. 18, p. 283–291.

341. Navarrete-Palacios E., Hudson R., Reyes-Guerrero G., Guevara-Guzman R. (2003) Lower olfactory threshold during the ovulatory phase of the menstrual cycle. *Biol. Psychol.,* 2003, Jul 63, N 3, p. 269–79.
342. Newton J., Marquiss M. Sex ratio among nestlings of the European sparrowhawk. *Am. Nat.,* 1978 (cit.: Maynard Smith J., 1978).
343. Noddings N. Variability: A pernicious hypothesis. *Review of Educational Research,* 1992, v. 62, N 1, p. 85–88.

344. Oliveira-Pinto AV, Santos RM, Coutinho RA, Oliveira LM, Santos GB, et al. (2014) Sexual Dimorphism in the Human Olfactory Bulb: Females Have More Neurons and Glial Cells than Males. PLoS ONE 9(11): e111733. doi:10.1371/journal.pone.0111733
345. Olsen N. J., Kovacs W. J. Gonadal Steroids and Immunity. *Endocrine Reviews,* 1996, v. 17, N 4, p. 369-384.
346. Olsson M., Shine R. Facultative sex allocation in snow skink lizards (*Niveoscincus microlepidotus*). *J. of Evolutionary Biology,* 2001, v. 14, p. 120–128.
347. Orians G. H. On the evolution of mating systems in birds and mammals. *Am. Nat.,* 1969, v. 103, p. 589–603.

348. Paessler K. (2015) Sex Differences in Variability in Vocational Interests: Evidence from Two Large Samples. Eur. J. Pers., 29: 568–578. doi: 10.1002/per.2010.

349. Paglia C. *Sexual Personae: Art and Decadence from Nefertiti to Emily Dickinson.* Yale Univ. Press, 1990, p. 235.

350. Pahnish O. F., Stanley E. B., Bogart R., et al. *Animal Sci.*, 1961, v. 20, p. 454.

351. Parker G. A., Baker, R. R., Smith, V. G. F. The origin and evolution of gamete dimorphism and the male-female phenomenon. *J. Theor. Biol.*, 1972, v. 36, p. 529.

352. Parkes A. S. Studies on the sex-ratio and related phenomena 6. The effect of polygyny. *Ann. Appl. Biol.*, 1925, v. 12, p. 3–13.

353. Parkes A. S. The mammalian sex-ratio. *Biol. Rev.*, 1926, v. 2, p. 1–44.

354. Patterson C. B., Emlen J. M. Variation in nestling sex ratios in the yellow-headed blackbird. *Am. Nat.*, 1980, v. 115, p. 743–747.

355. Perret M. Influence of social factors on sex ratio at birth, maternal investment and young survival in a prosimian primate. *Behav. Ecol. Sociobiol.*, 1990, v. 27, p. 447–454.

356. Pethybridge R. J, Ashford J. R., Fryer J. G. Some features of the distribution of birthweight of human infants. British Journal of Preventive and Social Medicine 1974, v. 28, p. 10–18.

357. Petri M., Howard D., Repke J. *Rheum. Dis. Clin. North Am.* 1994, v. 20, p. 87.

358. Phillips S., Birkenmeier E., Callahan R., Eicher E. Male and Female Mouse DNAs can be Discriminated using Retroviral Probes. *Nature,* 1982, v. 297, N 5863. p. 241–243.

359. Picchioni M. M.; Murray R. M. (2007). Schizophrenia. BMJ. v. 335(7610), p. 91–95. PMID 17626963.

360. Pillard R. C., Weinrich J. D. Evidence of familial nature of male homosexuality. *Archives of General Psychiatry.* 1986, v. 43, p. 808–812.

361. Porter R. H. Olfaction and human kin recognition. Genetica. 1999, v. 104, p. 259–263.

362. Porter R., Beers M. H., Berkow R. *The Merck manual of diagnosis and therapy.* Rahway, NJ: Merck Research Laboratories, 2006, 290 pp.

363. Pratt N. C., Lisk R. D. Effects of social stress during early pregnancy on litter size and sex ratio in the golden hamster (*Mesocricetus auratus*). *J. Reprod. Fertil.* 1989, v. 87, p. 763–769.

364. Pratt N. C., Huck U. W., Lisk R. D. Offspring sex ratio in hamsters is correlated with vaginal pH at certain times of mating. *Behav. Neural. Biol.*, 1987, v. 48, p. 310–316.

365. Pratt N. C., Huck U. W., Lisk R. D. Do pregnant hamsters react to stress by producing fewer males? *Anim. Behav.*, 1989, v. 37, p. 155–157.

366. Puts D. A. Mating context and menstrual phase affect women's preferences for male voice pitch *Evol. Hum. Behav.,* 2005, v. 26, p. 388–397.

367. Quinn G. B., Axelrod J., Brodie B. B. Species and sex differences in metabolism and duration of action of hexobarbital (Evipal). *Fed.Proc.*, 1954, v. 13, p. 395–396.

368. Radulescu AR, Mujica-Parodi LR (2013) Human gender differences in the perception of conspecific alarm chemosensory cues. PLoS One 8(7): e68485.

369. Rasmussen K. *Sci. Agr.*, 1941, v. 21, p. 759.

370. Reich T., Cloninger C. R., Van Eerdewegh P. et al. Secular Trends in the Familial Transmission of Alcoholism. *Alcoholism: Clinical and Experimental Research*, 1988, v. 12, 4, p. 458–464.

371. Rensch B. *Evolution Above the Species Level.* Columbia Univ. Press, New York, 1959.

372. Renner S. S., Ricklefs R. E. *Am. J. Bot.*, 1995, v. 82, p. 596-606.

373. Reyes F. I., Boroditsky R. S., Winter J. S., Faiman C. Studies on human sexual development. II. Fetal and maternal serum gonadotropin and sex steroid concentrations. *J. Clin. Endocrinol. Metab.*, 1974, v. 38 p. 612–617.

374. Rice W. R. Degeneration of a Nonrecombining Chromosome. *Science.* 14 Jan. 1994, v. 263, p. 230–232.

375. Ridley M. *The Cooperative Gene*, The Free Press, New York, pp. 108,111, 2001.

376. Ridley M. *The Red Queen: Sex and the Evolution of Human Nature.* Macmillan Pablishing Co., 1994, 405 p.

377. Riede W. Beitrage zum Geschlechts- und Anpassungs-problem. *Flora,* 1925, v. 18/19, p. 421–452.

378. Riley M., Halliday J. *Birth Defects in Victoria 1999-2000*, Melbourne, 2002.

379. Risch N., Spiker D., Lotspeich L., et al. A genomic screen of autism: Evidence for a multilocus etiology. American Journal of Human Genetics, 1999, v. 65, p. 493–507.

380. Robert K. A, Thompson M. B, Seebacher F. Facultative sex allocation in the viviparous lizard *Eulamprus tympanum*, a species with temperature-dependent sex determination. *Aust. J. Zool.* 2003, v. 51, p. 367–370.

381. Roberts E., Card L. *The influence of broodiness in domestic fowl.* Proc. 5-th Worlds Poultry Congr., 1933, v. 2, p. 353.

382. Robinette W. L., Geshwiler J. S., Low J. B. et al. *J. Wildl. Manage.*, 1957, v. 21, p. 1.

383. Rokitansky K. E. Die defecte der Scheidewande des Herzens. Wien, 1875.

384. Rommelse N. N., Altink M. E., Arias-Vasquez A., Buschgens C. J., Fliers E., et al. (2008). Differential association between MAOA, ADHD and neuropsychological functioning in boys and girls. American Journal of Medical Genetics, Part B: Neuropsychiatric Genetics, 147-B, p. 1524–1530.

385. Rorie R. W. Effect of timing of artificial insemination on sex ratio. *Theriogenology.* 1999, v. 52(8), p. 1273-80.

386. Rothman P. Genius, gender and culture: Women mathematicians of the nineteenth century. *Interdisciplinary Science Rev. L.*, 1988, v. 13, N 1, p. 64–72.

387. Rowe L., & Houle D. (1996) The lek paradox, condition dependence and genetic variance in sexually selected traits. Proceedings of the Royal Society, London – Biological Sciences, 263, 1415–1421.

388. Ruse M. *Homosexuality.* 1988, Oxford: Basil Blackwell.

389. Russell *Breeding statistics.* Maine agric. exp. stat. Ann. Rep., 1891

390. Rychlewski J., Kazimierez Z. Sex ratio in seeds of Rumex acetosa L. as a result of sparse or abundant pollination. *Acta Biol.* Cracov, Ser. Bot., 1975, v. 18, p. 101–114.

391. Saeki I., Kondo K., Himeno K. et al. Evidence of heterosis on sexual maturity and egg production in reciprocal crosses of Japanese Nagoyas and White Leghorns. *Jap. J. Breed.*, 1956, v. 6, N 1, p. 65.

392. Saladin K. S. *Anatomy and physiology.* Third Ed. McGraw Hill, 2004, p. 1024-1027.

393. Sato H. Sexual dimorphism and development of the human cochlea. *Acta Otolaryngologica*, 1991, v. 111, p. 1037–1040.

394. Scheinfield A. *Women and men.* New York, 1944, 405 p.

395. Schirmer W. Über den Einflussgceschlectsgebundner Erhaulagen and die Sanglingsterblichkeit. *Arch. Rass. und Ges. Biol.*, 1929, v. 21, N 18, p. 353–393.

396. Schalk G., Forbes M. R., Male biases in parasitism of mammals: effect of study type, host age and parasyte taxon. *Oicos*, 1997, v. 78 p. 67-74.

397. Schwinger E., Ites J., Korte B. Studies on frequency of Y-chromatin in human sperm. *Hum. Genet.*, 1976, v. 34, N 3, p. 265–270.

398. Searle A. G. Spontaneous frequencies of point mutations in mice. *Humangenetik*, 1972, Bd. 16, N 1/2, p. 33–38.

399. Seiler J. Geschlechtschromosomenuntersuchungen an Psychiden. I. Experimentelle Beeinflussung der geschlechtsbestimmenden Reifeteilung bei Talaeporia tubulosa Retz. *Arch. Zellforsch.*, 1920, v. 15, p. 249–268.

400. Selander R.K. *Sexual selection and dimorphism in birds, sexual selection and the descent of man.* (B.Campbel, ed.). Chicago, Aldine, 1972, p. 180–230.

401. Sex-ratio. *Encyclopaedia Britannica*, v. 20. London, 1960.

402. Shapiro S., Schlesinger E. R., Nesbitt R. E. *Infant, Perinatal, Maternal, and Childhood Mortality in the United States.* Harvard Univ. Press, Cambridge, Mass, 1968.

403. Sheldon B. C., West S. A. Maternal dominance, maternal condition, and offspring sex ratio in ungulate mammals. *Am. Nat.*, 2004, v. 163, p. 40–54.

404. Shettles L. B. Sex preselection. *Infertility,* 1978, v. 1, p. 127–135.

405. Shields S. The variability hypothesis: The history of a biological model of sex differences in intelligence. *SIGNS: J. of Women in Culture & Society*, 1982, v. 7(4), p. 769–797.

406. Shimmin L. S., Chang B. H-J, Li W-H. Male-driven Evolution of DNA Sequences. *Nature,* 22 April 1993, v. 362, p. 745–747.

407. Shine R. Ecological causes for the evolution of sexual dimorphism: A review of the evidence, *Quart. Rev. Biol.*, 1989, v. 64, p. 419-464.

408. Simpson A. J. G., Camargo A. A. Evolution and inevitability of human cancer. *Sem. Cancer Biol.*, 1998, v. 8, p. 439–446.

409. Slatkin M. Ecological causes of sexual dimorphism, *Evolution*, 1984, v. 38, p. 622-630.

410. Smirnov N. N. The taxonomic significance of the trunk limbs of the *Chydoridae. Hydrobiologia*, 1966, v. 27, N 3–4, p. 337–347.

411. Smith C. G. Age incidence of atrophy of olfactory nerves in man. *J. Comp. Neurol.*, 1942, v. 77(3), p. 589–596.

412. Snyder R. G. The sex ratio of offspring of pilots of high performance military aircraft. *Hum. Biol.*, 1961, v. 33, N 1, p. 1–10.

413. Snyder R. L. *The biology of population growth.* Croom Helm, London, 1976.

414. Sorri M., Rantakallio P. Prevalence of hearing loss at the age of 15 in a birth cohort of 12,000 children from northern Finland. *Scandinavian Audiology*, 1985, v. 14, p. 203-207.

415. Spiess E. B., Bowbal D. A. Minority mating advantage of certain eye color mutants of Drosophila melanogaster. IV. Female discrimination among three genotypes. *Behav. Genet.*, 1987, v. 17, p. 291–306.

416. Spitzer A. *Arch. Pathol. Anat.*, 1923, v. 243, p. 81–272.

417. Springer, S. P., Deutsch G. *Left Brain, Right Brain*, N-Y., W. H. Freeman & Co., 1989.

418. Stehlik I., Friedman J., Barrett S. C. Environmental influence on primary sex ratio in a dioecious plant. *Proc. Natl. Acad. Sci. USA.* 2008 Aug 5; v. 105(31) p. 10847-52.

419. Stern C. *Principles of human genetics.* Freeman, W.H. & Co, San Francisco & London, 1960.

420. Stevenson A. C., Bobrov M. *J. Med. Genet.*, 1967, v. 4, p. 190.

421. Stini W. A. Nutritional stress and growth: sex differences in adaptive response. *Amer. Journ. Phys. Anthropol.*, 1969, vol. 31, No 3, p. 417–426.

422. Stinson S. Sex differences in environmental sensitivity during growth and development. *Yearbook Phys. Anthropol.*, 1985, v. 28, p. 123-147.

423. Streicher E., Garbus B. A. The effect of age and sex on the duration of hexobarbital anesthesia in rats. *J.Geront.*, 1955, v. 10, N 4, p. 441–444.

424. Stump J. B. *What's the Difference? How Men and Women Compare*. N.Y., William Morrow & Co., Inc., 1985.

425. Surani M.A.H., Barton S.C., Norris M.L. Development of reconstituted mouse eggs suggest imprinting of the genome during gametogenesis. *Nature*, 1984, v. 308, p. 548–550.

426. Swaab D. *We Are Our Brains: From the Womb to Alzheimer's*. 2015, Allen Lane, 448pp.

427. Symons D. *The Evolution of Human Sexuality*. New York: Oxford University Press, 1979.

428. Szasz T. S. *Psychiatric Justice*. New York, Macmillan, 1965, 283 pp.

429. Székely T, Freckleton R. P., Reynolds J. D. Sexual selection explains Rensch's rule of size dimorphism in shorebirds. *PNAS*, 2004, v. 101, N. 33, p. 12224-12227.

430. Szemere G. Adatok a nemek öröklődésének kerdeseher. *Biol. Közlemenyck* 1958, v. 6, N 2, p. 115–120.

431. Taylor T. J. *Twin Studies of Homosexuality, Part II Experimental Psychology Dissertation (unpublished)*, 1992, Univ. of Cambridge, UK. < http://www.tim-taylor.com/papers/twin_studies/index.html >

432. Ter-Avanesian D. V. Significance of pollen amount for fertilization. *Bull. Torrey Bot.Club.*, 1978, v. 105, N 1, p. 2–8.

433. Terman C. R., Birk A. Compensation by *Drosophila* populations to differential sex mortality. *Bull. Ecol. Soc. Amer.*, 1965, v. 56, p. 130–131.

434. Thomas C. *Report on the Ibo-speaking peoples of Nigeria*, Pt. 1, London, 1913.

435. Thury. Memoire sur la loi de la production des sexes. (Cit. by Parkes A.S., 1926).

436. Tomlinson J. The advantage of hermaphroditism and parthenogenesis. *J.Theoret. Biol.*, 1966, v. 11, p. 54–58.

437. Trevarten C. Cerebral embryology and the split brain. In: *Hemispheric disconnection and Cerebral function.* (Kinsbourne M., Smith W. L., eds.), Springfield, 1974.

438. Trivers R. L. *Social Evolution*. Benjamin Cummings, Menlo Park, 1985.

439. Trivers R. L., Willard D. E. Natural selection of parental ability to vary the sex ratio of offspring. *Science*, 1973, v. 179, N 4068, p. 90–92.

440. Trofimova I. N. Are men evolutionary wired to love the "Easy" buttons? *Nature Proceedings*, 2011, hdl:10101/npre.2011.5562.1, p. 1–16.

441. Trofimova I. N. Understanding misunderstanding: a study of sex differences in meaning attribution. Psychological Research, 2012, DOI: 10.1007/s00426-012-0462-8

442. Tubaro P. L., Bertelli S. *Biol. J. Linn. Soc.,* 2003, v. 80, p. 519–527.

443. Vandenberg S. G., McKusick V. A., McKusick A. B. Twin data in support of the Lyon hypothesis. *Nature*, 1962, v. 194, N 4827, p. 505–506.

444. Van der Broek J. M. Change in male proportion among newborn infants. *Lancet* 1997, v. 349, p. 805.

445. Veciana J. M., Aponte M., Urbano D. University students' attitudes towards entrepreneurship: A two countries comparison. *International Entrepreneurship and Management Journal*, 2005, 1, p. 165–182.

446. Venstrom D. Amoore J. E. Olfactory threshold in relation to age, sex or smoking. *J. Food Sci.*, 1968, v. 33, p. 264–265.

447. Vericad-Corominas J. R. Estudio faunistico y biologico de los mamiferos montaraces del Pirineo. *Publ. Centro Pirenaico Biol. exp. Jaca.*, 1970, v. 4, p. 7–231.

448. Verme L., Ozoga J. Sex ratio of white-tailed deer and the estrus cycle. *J. Wildlife Management*, 1981, v. 45, p. 710–715.

449. Vickers A. D. Delayed fertilization and the prenatal sex ratio of the mouse. *J. Reprod. Fertil.*, 1969, v. 20, p. 63–76.

450. Villee C. A. *Biology*, W. B. Saunders Co. Philadelphia, London and Toronto, 1977.

451. Volkman-Rocco B. The effect of delayed fertilization in some species of the genus *Tisbe (Copepoda, Harpacticoida)*. *Biol. Bull.*, 1972, v. 142, p. 520–529.

452. Vorontsov N. N. *The Evolution of the Sex Chromosomes. Cyto-taxonomy and Vertebrate Evolution* (Chiarelli A. B. ed.), 1973, p. 619–657.

453. Vranckx R., Muylle L., Cole J., Moldenhaser R., Peetermans M. *Vox Sang.* 1986, v. 50, p. 220.

454. Vrijenhoek R. C. Animal clones and diversity. *Bioscience*, 1998, v. 48, p. 617–628.

455. Vuorenkoski V., Lenko H. L., Tjernlund P., e.a. Fundamental voice frequence during normal and abnormal growth, and after androgen treatment. *Arch Dis Child.* 1978, v. 53, N 3, p. 201–209.

456. Waber D. Sex differences in cognition: a function of maturation rate? *Science*, 1976, v. 192, p. 572–573.

457. Wada J. A., Clark R., Hamm A. Cerebral Hemisphere Asymmetry in Humans. *Arch. Neurol.*, 1975, v. 32, p. 239–246.

458. Wagoner D. E., McDonald I. C., Childress D. The present status of genetic control mechanisms in the house fly, *Musca domestica L.* In: *The use of Genetics in Insect Control* (Pal R., Whitten M. J., eds.), Amsterdam, Elsevier, 1974, p. 183–197.

459. Wai J., Cacchio M., Putallaz M., et al. Sex differences in the right tail of cognitive abilities: A 30 year examination. Intelligence (2010), doi:10.1016/j.intell.2010.04.006.

460. Waldron I. Why do women live longer than man? *Soc. Sci. & Med.*, 1976, v. 10, p. 349–362.

461. Walker S. F. Lateralization of function in the vertebrate brain. *British J. of Psychology*, 1980, v. 71, p. 329–367.

462. Wallace A. R. Darwinism: an exposition of the theory of natural selection, with some of its applications, Macmillan., 1889.

463. Warner D. A., Shine R. Reproducing lizards modify sex allocation in response to operational sex ratios. *Biol Lett.*, 2007, v. 3(1), p. 47–50.

464. Warren D. Inheritance of age at sexual maturity in the domestic fowl. *Genetics*, 1934, v. 19, p. 600.

465. Warren D. The crossbreeding of poultry. *Kansas Agric. Exptl. Sta. Techn. Bull.*, 1942, N 52.

466. Weatherhead P. J., Robertson R. J. Offspring quality and the polygyny threshold: "the sexy son hypothesis." American Naturalist, 1979, v. 113, p. 201–208.

467. Webb T. J., Freckleton R. P. Only Half Right: Species with Female-Biased Sexual Size Dimorphism Consistently Break Rensch's Rule. PLoS ONE, 2007, v. 2, N 9, e897. doi:10.1371/journal.pone.0000897

468. Weininger O. Sex and Character: An Investigation of Fundamental Principles. L.-NY, 1906.

469. Wells J. C. K. Natural selection and sex differences in morbidity and mortality in early life. *J. Theor. Biol.*, 2000, v. 202, p. 65-76.

470. Werren J. H., Charnov E.L. Facultative sex-ratios and population dynamics. *Nature,* 1978, v. 272(5651), p. 349–350.

471. West S. A. *Sex Allocation*. Princeton University Press, Princeton, NJ. 2009.

472. White F. N. Variation in the sex ratio of *Mus rattus* associated with an unusual mortality of adult females. *Proc. Roy. Soc. London*, 1914, v. 87, p. 335–344.

473. Wiley R. H. Animal Behaviour Monographs, 1973, v. 6, p. 85–169.

474. Wilkinson G. S., Presgraves D. C., Crymes L. Male eye span in stalk-eyed flies indicates genetic quality by meiotic drive suppresion. *Nature*, 1998, v. 391, N 6664, p. 276-279.

475. Williams G. Evolution, 1957, v. 11, p. 398-411.

476. Williams G. C. *Sex and Evolution*. Princeton, New Jersey, Princ. Univ. Press, 1975.

477. Willson M. F. Sexual selection, sexual dimorphism and plant phylogeny, *Evol. Ecol.*, 1991, v. 5, N 1, p. 69-87.

478. Wilsnack R. W., Vogeltanz N. D., Wilsnack S. C., et al. Gender differences in alcohol consumption and adverse drinking consequences: Cross–cultural patterns. *Addiction,* 2000, v. 95, p. 251–265.

479. Wilson G. D. *Love and Instinct*, London, Temple Smith, 1981.

480. Wilson G. *The Great Sex Divide*, pp. 41-45. P. Owen (London) 1989; Scott-Townsend (Washington D.C.) 1992.

481. Winge O. The Location of Eighteen Genes in *Lebistes Reticulatus*. *J. Genetics,* 1927, v. 18, p. 1–43.

482. Witelson S. F. Sex and the single hemisphere. *Science*, 1976, v. 193, N 4251, p. 425–427.

483. Witschi E. Die Keimdrüsen von *Rana temporaria. Arch. microsc. Anat.* 1914, v. 86, N 1–2, p. 9–113.

484. What Drives Underage Drinking? An Intern. Analysis Commissioned by the Intern. Center for Alcohol Policies, 2004.

485. White F. N. Variation in the sex ratio of Mus rattus associated with an unusual mortality of adult females. *Proc. Roy.Soc.London*, 1914, v. 87, p. 335–344.

486. Whitton J., Sears C. J., Baack E. J. et. al. The dynamic nature of apomixis in the angiosperms. *Intern. J. of Plant Sci.*, 2008, v. 169, p. 169–182.

487. Wolff C. *Love between women*. London, Duckworth & Co., 1971, pp. 230.

488. *World Health Organization (reports). "Congenital malformations",* Geneve, 1966, p. 128.

489. Wrangham R. & Petersen D. (1997) Demonic Males: Apes and the Origins of Human Violence. London: Bloomsbury Publishing Plc.

490. Yokoyama S., Radlwimmer B. F. The molecular genetics and evolution of red and green color vision in vertebrates. *Genetics Society of America*. 2001. v. 158, p. 1697-1710.

491. Young P. T. Sex differences in handwriting. *J. Appl. Psychol.*, 1931, v. 15, N 5, p. 71–78.

492. Zechner, U., Wilda, M., Kehrer-Sawatzki, H., Vogel, W., Fundele, R., & Hameister, H. (2001). A. high density of Xlinked genes for general cognitive ability: A run-away process shaping human evolution? Trends in Genetics, 17, 697–701.

493. Zimmer, C. *Evolution: The Triumph of an Idea*. HarperCollins, New York, pp. 230, 231, 2001.

494. Zorn B., Sucur V., Stare J., Meden-Vrtovec H. Decline in sex ratio at birth after 10-day war in Slovenia. *Hum. Reprod.* 2002, v. 17: p. 3173–3177.

495. Zuckerman M. (1994). Behavioral expressions and biosocial bases of sensation seeking. Cambridge: Cambridge University Press.

REFERENCES IN RUSSIAN

496. Afonkin S. Yu. (2010) *Secrets of human heredity*. Crown print, St.-Petersburg, 77 pp.
497. Aghajanian A. K. Origin and Evolution of the muskrat. In: *Muskrat: Morphology, systematics, ecology*. Nauka, Moscow, 1993, p. 7-19.
498. Aghajanian A. K. Origin and Evolution. In: *Water vole: The image of the species*. M., Nauka, 2001, p. 22-54.
499. Agapova G. A. Biodiversity of northern salmon *Oncorhynchus gorbuscha (Walbaum)* // Readings in memory of V. J. Levanidov. 2011, Vol. 5, Vladivostok Dal'nauka, p. 11–16.
500. Adrianovsky A. F. *Materials for the study of diverticular disease of digestive tract*. Cand. diss. M., 1969.
501. Akimushkin I. I. *World of Animals*. M., Molodaja gvardija (Young Guard), 1973.
502. Akopian A., Theory of sexual dihronomorfizm - methodological basis of gender approach. VII Russian Scientific Forum "Men's Health and Longevity". Moscow, February, 2009.
503. Aksyanova G. A. Manifestation of sexual dimorphism in the anthropological image of the people of Northern Eurasia. *Bulletin of Archaeology, Anthropology and Ethnography*, 2011, № 2 (15), p. 173-141.
504. Alexandrov B. V. Radiographic study of variation and the nature of inheritance of the vertebrae. Genetics, 1966, v. 2, № 7, p. 52-60.
505. Alexakhin I. V., Tkachenko A. V. Two-channel management principle. Systems Research. Yearbook. Institute of Science and Technology History. USSR Academy of Sciences, Moscow, Nauka, 1977, p. 17.
506. Alekseeva T. N. The problem of biological adaptation of man. *Priroda (Nature)*, 1975, № 6, p. 38.
507. Altukhov Yu. P., Varnavskaya N. V. Adaptive genetic structure and its connection with intrapopulation differentiation by sex, age and growth rate of Pacific salmon *Oncorhynchus nerka* (Walb). *Genetics*, 1983, v. 19, N. 5, p. 796–807.
508. Anisimov V. N., Soloviev M. V. *Evolution of concepts in gerontology*. St. Petersburg, 1999.
509. Anishchenko T. G. Sexual aspects of stress and adaptation. *Uspehi Sovrem. Biologii*, 1991, v. 3, p. 460–475.
510. Antonian S. A. Some factors influencing the change in sex ratio. *Biological Journal of Armenia*, 1974, v. 27, № 10, p. 83-87.
511. Arkatov V. V., Andreev V. S., Ratkin A. V. Genetic control of the formation of flower color in sweet peas. *Genetics*, 1976, v. 12, № 8, p. 30-37.
512. Arkhipov I. *Vokrug sveta (Around the World)*, 1978, № 10, p. 60-62.
513. Aslanian M. M. Features of inheritance and embryonic development in pigs crossing. Nauchn. Proceedings. Higher. school, 1962, № 4, p. 179-184.
514. Astaurov B. L. *Problems of General Biology and Genetics*, M., Science, 1979, 230 pp.
515. "Asia and Africa Today", 1970, № 5, p. 53.

516. Bagrunov V. P. *Sex differences in the species and the individual variability of the human psyche*. Authoref. Dis. Cand. Science, L., 1981.
517. Ballyuzek F. V., Skvirsky V. Ya, Skorobogatov G. A. Researchers say (to the point of sensations): on drinking water and plumbing, on medical silver and alcohol, about inactivity and diet,... St. Petersburg, 2009.
518. Bednij M. S. *A boy or a girl? (Medico-demographic analysis)*. Moscow, Mysl, 1987.
519. Belkin A. I. Third gender. Destiny of Nature's stepsons. M., Olympus, 2000, 432 pp.
520. Bianchi V. L. *Asymmetry of the brain of animals*. L., 1985, 250 pp.
521. Blinkov S. M., Glezer I. I. *Human brain in figures and tables*. L., 1964, 180 pp.
522. Boiko E. I. *Reaction time of man*. M., Medicine, 1964, 300 pp.
523. Bolshakov V. N., Kubantsev B. S. *Gender structure of populations of mammals and its dynamics*. M., Nauka, 1964, 233 pp.
524. Borodin P. M., Gorlov I. P. Effect of stress on genetic variation. In: *Microevolution*, M., 1984, p. 100–101.
525. Borodin P. M., Schuler L., Belyaev D. K. Problems in genetics of stress. Part I. Genetic analysis of the behavior of mice in a stressful situation. *Genetics*, 1976, v. 12, № 12, p. 62-72.
526. Bragina N. N., Dobrokhotova T. A. (1988) The functional asymmetry of the human. 2nd ed., M.: Medicine, 240 p., p. 55.
527. Brook S. P. *World population: ethnic and demographic reference*. Moscow, Nauka, 1981, 880 pp.
528. Bunak V. V., Nesturh M. F., Roginsky Ya. Ya. *Anthropology*, M., 1941, 300 pp.
529. Burakovsky V. I., Kolesnikov S. A. *Particular surgery diseases of the heart and blood vessels*. M., Medicine, 1967, 190 pp.

530. Chaylahyan M. H., Khryanin V. N. *Sex of plants and its hormonal regulation*. Moscow, Nauka, 1982, 173 pp.

531. Chetverikov S. S. Main factor of evolution of insects. *News of the Moscow Entomol. Soc.*, 1915, v. 1, p. 14-24.

532. Con I. O. *Sociology of personality*. M .: Politizdat, 1967, 383 p.

533. Davidovsky I. V. *Gerontology*. M., Medicine, 1966, 310 pp.

534. Denisova Z. V. *Mechanisms of emotional behavior of the child*. L., Science 1978, 143 pp.

535. Dobrynin A. J. Reciprocal crosses of Moscow hens and Leghorns. Proc. Of The Institute of Genetics, AN USSR, 1958, № 24, p. 307-320.

536. Dubinin N. P., Glembotskij Ya. L. *Genetics of populations and selection*. Moscow, Nauka, 1967, 487 pp.

537. Dzhagaryan A. D. *Atlas of Cardiac Surgery*. Yerevan, the Armenian State. publ., 1961, 311 pp.

538. Dzhaparidze D. I. *Sex in plants*. Tbilisi, Izd-vo AN GSSR, 1963, Part 1, 307 p.

539. Dzhaparidze D. I. *Sex in plants*. Tbilisi, Metsniereba, 1965, Part 2, 302 p.

540. Ebers G. *Uarda*. M. Nauka (Science), 1965, p. 76.

541. Einstein A. Physics and reality. M., 1965, p. 109, 350 pp.

542. Epidemiology of cancer in the USSR and the USA. 1979. M. Medicine. 384 pp.

543. Eremeeva V. D., Hrizman T. P. *Boys and girls—two different worlds. Neuropsychologists to teachers, educators, parents, school psychologists*. Moscow, Linka-Press, 1998, 184 p.

544. Feinberg L. From ape to man. *Science and Life*, 1982, № 5, p. 72-85.

545. Frankfurt U. I., Frank A. *Josiah Willard Gibbs.*, 1964, p. 76, 280 pp.

546. Gallbladder. In: *Bolshaja Meditsinskaja Encyclopedia*, M., Sovetskaja Encyclopedia, 1978, v. 8, p. 184-204.

547. Gavrilov L. A., Gavrilova N. S. The Biology of longevity. Moscow, Nauka, 1991, 280 pp.

548. Gavrilova N. S., Semenova V. G., Gavrilov L. A. Data files on life expectancy in humans. In: *Problems of biology of aging*. Moscow, Nauka, 1983, p. 71-76.

549. Gilinskiy Y. I. *Deviantology: sociology of crime, drug addiction, prostitution, suicide and other "deviations"*. 2nd ed., 2007, St. Petersburg: Yurtsd. Center Press.

550. Ginsburg V. *Elements of anthropology for health professionals*. L., Medgiz, 1963, 216 pp.

551. Golubovsky M. D. Organization of the genotype and shape of genetic variation of eukaryotes. In: Methodological problems in medicine and biology. Novosibirsk, Nauka (Science), 1985, p. 135-152.

552. Grant B. *The evolution of organisms*. Wiley, New York. 1980, 407 pp.

553. Grishko I. The problem of sex in hemp. Tr. Institute of cannabis. Kiev, 1935, v. 8, p. 197-241.

554. Gubler E. V., Genkin A. A. *Application of nonparametric statistics in biomedical research*. L., Science, 1973.

555. Gurevich G. A. Report on the Management Section of the Council on Cybernetics, 1966.

556. Isaev D. N., Kagan, V. K. Sex education of children. M., 1998.

557. Ivanov V. *Odd and even. Brain asymmetry and sign systems.* , 1978, 190 pp.

558. Ivanova P. G. Influence of age of sexual cells of animals on the quality of offspring. Ushenie zapiski LSU, ser. biol., 1953, № 165, issue 33, p. 3.

559. Julien Sh.-A. History of North Africa. M., 1961, v. 2, p. 272.

560. Kamalian V. S. Effect of altitude on the sex ratio of offspring. *Izvestija AS ArmSSR, Biolog. and agricultural Sciences*, 1958, v. II, № 4,.p. 97-101.

561. Kamalian B. S. Influence of age of the parents on sex ratio of offspring. *Zh. gen. Biol.*, 1962, v. 23, № 6, p. 455-459.

562. Kardo-Sysoeva E. On ginadromorfizm in *Salix cinereae L*. Tr. Leningr. Soc. of Naturalists, 1924, v. 54, № 3, p. 41-44.

563. Kirpichnikov V. S. Autosomal genes in *Lebistes Reticulatus* and the problem of genetic sex determination. Biol. Journ., 1935, v. IV, № 2, p. 343-354.

564. Kirsanov Z. I., Rogozin A. P. Detection of sex and age of the writer by the handwriting of the manuscript. In: Legal cybernetics, M., 1973, p. 161-176.

565. Kolesov D. V., Selverova I. B. *Physiological and pedagogical acpekts of puberty*. M., Pedagogy, 1978, 280 pp.

566. Kon I. S. *Sociology of personality*. M., Politizdat, 1967, 383 pp.

567. Kon I. S. Adam, Eve and age - the tempter. Literaturnaja Gazeta, 1979, № 1.

568. Kon I. S. *Child and society*. M., Pedagogika, 1988, 264 pp.

569. Kon I. S. The normalization of homosexuality. Sexology and sexual pathology, 2003, № 2, p. 2-12.

570. Kochetova T. V. The evolutionary approach in social-psychological research: gender differences as the leading determinants of status preferences. *Social Psychology and Society*, 2010, № 1, p. 78–90.

571. Korytin S. A. *Behavior and sense of smell of carnivorous mammals*. Ed. 2, 2007, 224 p.

572. Krechetovich L. M. *Questions on the evolution of vegetation*. M., MOIP, 1351 p.

573. Krimsky L. D. Pathological anatomy of congenital heart disease and complications after surgical treatment. M., Medicine, 1963.

574. Kuznetsov A. V., Sigaeva V. A., Kuznetsova I. e. a. Rabbit sperm is capable of binding foreign DNA. Probl. Of reproduction. 1996, № 1, p. 7-10.

575. Kuptsov A. I. Unisex female sunflower. Soc. Crop Science, 1935, № 14, p. 149-150.

576. Kurbanov R. Sex ratio of the silkworm. Cand. diss. Tashkent, 1973, p. 147.

577. Kurbatov A. D. Effect of the development on the formation of gender and other characteristics of farm animals. Avtoref. Dis. Doct., Pushkin A-D, 1965.

578. Kukharenko V. I., On the primary sex ratio in humans. Genetics, 1970, v. 6, № 5, p. 142-149.

579. Kukharenko V. I. Investigation of prenatal sex ratio in man by short-term tissue culture. *Genetics*, 1971, v. 7, № 8, p. 166-169.

580. Kushakevich S. History of development of the genital glands in *Rana Esculenta*. St. Petersburg, 1910.

581. Lazutkin A. P. Postgosudarstvennaya paradigm of equity control. Krasnoyarsk, SibGTU 2011, 242 p. ISBN 978-5-8173-0485-5.

582. Lek D. The number of animals and its regulation in nature. M., Publ. Foreign Lit., 1957, 400 pp.

583. Levin V. L. On the different damage of tissues of males and females. Experiments with white mice. "Dokl. AN USSR, 1949, v. 66, № 4, p. 749-751.

584. Levin V. L. Some features of the action of strychnine on the isolated tissues and whole organisms of male and female rodents. "Dokl. AN USSR, 1951a, v. 78, № 1, p. 165-168.

585. Levin V. L. On the damage of different tissues of males and females. Experiments with white rats. "Dokl. AN USSR, 1951b, v. 78, № 4, p. 817-819.

586. Lisitsyn Y. P., Kopit N. Ya. *Alcoholism (socio-hygienic aspects)*. M., Medicine, 1978, 232 pp.

587. Loginov A. A. *Man and Woman. The beginning of family life* (Kochetov A. I. ed.), Minsk, 1989, p. 231-257.

588. Luchnikova E. M., Petrova V. V. Compensatory increase in the proportion of males in child generation influenced by their deficiency in the parent generation. (Experiments on simulated populations of Drosophila). *Vestn. Lenigradskogo universiteta*, 1972, v. 1, № 3, p. 143-150.

589. Maksimovskiy L. F. Capabilities to target the formation of the mammals offspring sex ratio. *Agricultural biol.*, 1988, № 1, p. 10–19.

590. Malinowski A. Unfinished ideas of some Soviet geneticists. Priroda (Nature), 1970, № 2, p. 79-83.

591. Malinowski A. On the theoretical biology. Preprint. Methodological issues of biophysics. I All-Union Biophysical Congress. Pushchino, 1982.

592. Mamzina E. A. *Influence of the nature of the pairing of animals on the sex ratio of offspring*. Diss. Cand. Biol. Science, L., 1955.

593. Manashev G. G. *Laws of formation of multi-rooted human teeth*. Authoref. Diss. Doct. Med. Sciences, Krasnoyarsk, 2005, 244 p.

594. Manuilova E. F. Cladocerans (*Cladocera*). Determinants of the fauna of the USSR issued by Zool. Inst. AN USSR, 88, M.-Leningrad, 1965, p. 1-327.

595. Markaryan P. N. Semyaproduktsiya, fertility and quality of offspring depending on the different treatment of sheep. Avtoref. Dis. Cand. Yerevan, 1965.

596. Markel A. L. Some features of the selection process in the conditions of stress stimulation. In: *Microevolution*, 1984, p. 115-116.

597. Michurin I. V. Results of sixty years of work. Moscow, Izdatel'stvo AN USSR, 1950, 550 pp.

598. Miller T. On the sex ratio among the produced calves. *Archives of Veterinary Science*, 1909, Vol. 9.

599. Milovanov V. K. *Biology of reproduction and artificial insemination of animals*. M. Sel'khozgiz, 1962, 696 pp.

600. Minina E. G. *Offset of sex in plants by the influence of environmental factors*. M., AN SSSR, 1952, 199 pp.

601. Minina E. G. Significance of shift of sex [ratio] in plants for breeding (on the links of heterosis and polyploidy with sexualization). *Zhurn. gen. biology*, 1965, v. 26, № 4, p. 416-427.

602. Mordukhai-Boltovskoy F. D. Biology and systematics of polifemid in Ponto-Caspian bassein. Vopr. Hydrobiology. 1st All-Union Congress. of Hydrobiological Society. M., Nauka, 1965, p. 301.

603. Naugolnykh V. I. On the resistence of the leaves of dioecious plants to poisonous substances. *Dokl. AN USSR*, 1947, v. 57, № 4, p. 403-406.

604. Neck. In: *Bolshaja Meditsinskaja Encyclopedia*, M., Sovetskaja Encyclopedia, 1986, v. 27, p. 406.

605. Nikityuk B. A. In: *Evolution of the development rates of individual animals*. M., Nauka, 1977, p. 83-94.

606. Novoselsky S. A. *Problems of demographic and health statistics*. M., Gosstatizdat, 1958, p. 1-197.

607. Panteleyev P. A. On the Role of the Hypotheses in Zoological Investigations. *Vestnik zoologii*, 2003, v. 37, № 2, p. 3–8.

608. Panteleyev P. A. Rodentology. M., T-vo naushih izdanij KMK. 2010, 221 p., p. 74.

609. Pavlovskii O. M. Generalized Photoportrait: What Does It Tell? *Nauka i Zhizn'*, 1980, N. 1, p.84–90 (in Russian).

610. Pavlovsky O. M. Biological age in humans. Environmental aspects. In: *Progress in Science and Technology*, Ser. Anthropology, v. I, Moscow, VINITI, 1985, p. 5-53.

611. Pokhilenko A. P. (2012) Evaluation of morphological variability of Megaphyllum Sjaelandicum (Diplopoda, Julida) populations. Ecology and noosferology, v. 23, № 1–2.

612. Pulse. In: *Bolshaja Meditsinskaja Encyclopedia*, M., Sovetskaja Encyclopedia, 1962, v. 27, p. 515.

613. Rajewski P. M., Sherman A. L. Importance of gender in the epidemiology of malignant tumors (systemic-evolutionary approach). In: *Mathematical treatment of medical-biological information*. M., Nauka, 1976, p. 170-181.

614. Ratkin A. V. Andreev B. C., Arkatov V. Genetic control of the formation of flower color in sweet peas. Genetics, 1977, v. 13, № 9, p. 1534-1542.

615. Ratkin A. V., Zaprometov M. N., Andreev B. C., et al. Studies of the antotsianid and flavonoid biosynthesis in the flowers of sweet peas. Zh. gen. biology, 1980, v. 41, № 5, p. 685-699.

616. Razumov V. V.One more time about the philosophy of medicine. *Fundamental research*, 2011, № 11, p. 433-439. p.438.

617. Roginskij Ja. Ja., Levin M. G. *Antropologija (Anthropology)*. Moskva: Vyssaja Skola, 1963.

618. Rott N. N. Sex determination in amphibians. Bulletin Moscow Soc. Explorers Of Nature. Dep. Biology, 1963, v. 68, p. 118-134.

619. Rozanova M. A. Issues of sex in higher plants. In: *Theoretical basis of plant breeding*. Leningrad, State Publishing House, Agricultural state farms and kolkhoz lit., 1935, v. I, p. 145-162.

620. Sadowsky E. Sexual dimorphism and individual development characteristics of the coordination abilities of elite martial arts athletes. Theory and Practice of Physical Culture, 1999, № 8.

621. Salganik R. I. The role of genetic induction in normal and pathological processes. Vestn. AMS USSR, 1968, № 8, p. 3-10.

622. Schuler L. D., Borodin P. M., Belyaev D. K. Problems of genetics of stress. Genetics, 1976, v. 12, № 12, p. 72-82.

623. Sherman A. L. Some problems of medical information processing and computer diagnosis of congenital heart diseases. Cand. Dis. M., 1970.

624. Shmalgausen I. I. Principles of comparative anatomy of vertebrates. M., 1947.

625. Shmalgausen I. I. Fundamentals of the evolutionary process in the light of cybernetics. In: Problems of Cybernetics, v. 4, Fizmatgiz, 1954, p. 121-149.

626. Shmalgausen . I. I. *Issues of Darwinism*. L., Science, 1969, 493 pp.

627. Shubin, I. G., Shubin N. G. Sexual dimorphism and its peculiarities in mustelids. Zhurn. gen. Biology, 1975, v. 36, № 2, p. 283-290.

628. Sidorov B. N., Sokolov P. N. *Purely feminine forms of castor oil plant*. M., Izd-vo AN USSR, 1945, p. 283-290.

629. Simonov P. V., Rusalova M. N., Preobrazhenskaya L. A., Vanetsian G. L. Novelty factor and asymmetry of brain activity. "Zh. Higher. nerve. activities. 1995, v. 45, N. 1, p. 13-17.

630. Skorobogatov G. A. Mathematical modeling of global socio-economic processes. Reports of Methodology FIAN Seminar. Moscow, FIAN, 2005, v. 16, p. 1-62.

631. Smirnov N. N. Structure of limbs and its significance for ecology and systematics of the family *Chydoridae*. "Issues. Hydrobiology, 1st Congress of the All-Union. Hydrobiological Society. M., Nauka, 1965a, p. 385.

632. Smirnov N. N. Lifecycle of some *Chydoridae*. Zool. Journ., 1965b, v. 44, № 9, p. 1409-1411.

633. Solovenchuk L. L., Bondarenko L. V. Genetic monitoring of human populations on normal phenes. Human Genetics and Pathology: Materials of the Ist final Conference. Research Institute of Medical Genetics. Tomsk, 1989.

634. Sukachev V. N. Dendrology with the basics of forest geobotany. L. Goslestehizdat, 1938, 295 pp.

635. Svetlov P. G. On various endurance to starvation and other harmful factors of males and females of *Drosophila Melanogaster*. *Dokl. AN USSR*, 1943a. v. 41, № 8, p. 354-357.

636. Svetlov P. G. Ontogonez of sex differential of sensitivity in *Drosophila Melanogaster. Dokl. AN USSR*, 1943b, v. 41, № 9, p. 410-412.

637. Svetlov P. G. Sensitivity of the intestinal epithelium in males and females of Drosophila Melanogaster to the damaging effect of lactic acid. *Dokl. AN USSR*, 1945, v. 48, № 5, p. 377-379.

638. Svetlov, P. G., Ivanov K. V. Sex differences in endurance to damaging effects of Cyclopoida. *Dokl. AN USSR*, 1949, v. 68, № 6, p.1143-1146.

639. Svetlov P. G., Svetlova M. G Sex differences in resistance to damaging agents in dioecious plants. *Dokl. AN USSR*, 1950a, v. 70, № 4, p. 741-744.

640. Svetlov P. G., Svetlova M. G. Origins of damage sex differences in the ontogeny of dioecious flowering plants. *Dokl. AN USSR*, 1950b, v. 70, № 5, p. 925-928.

641. Svetlov P. G., Czekanowskaja O. V. On sex differences in sensitivity to the harmful factors of imaginal discs of larvae of Drosophila melanogaster. *Dokl. AN USSR*, 1945, v. 46, № 7, p. 321-325.

642. Takhtajan A. L. Issues of evolutionary morphology of plants. L., Leningrad State University, 1964, 214 pp.

643. Ter-Avanesyan D. V. The role of flower pollen quantity in fertilization of plants. Tr. of Sci. Botany, Genetics and Breeding, 1949, v. 28, p. 119-133.

644. Ter-Avanesyan D. V. *Pollination and hereditary variation*. M., Soviet Science, 1957, 250 pp.

645. Tishkina E. I. Binary conjugated differentiations of culture: elite and mass. Proc. of Ural State. Univ., 2010, № 2 (76), p. 23–30.

646. Tkachenko B. I. (ed.) *Physiological basis of human health*. St. Petersburg, Arkhangelsk, 2001.

647. U Han. *Biography of Chu Juanchan*. M., Nauka, 1980, p. 227.

648. Ulomsky C. N. On the fauna and ecology of the lower crustaceans in rice fields of Stalinabad. In: Works on Hydrobiology. Proc. Of the Inst. of zool. and parazitol. AN Tadzh. SSR, 1959, v. 112, p.25-65.

649. Urlanis B. Ts. *Sociology in the USSR*. Moscow, Mysl, 1965, v. 2, p. 10-73.

650. Vavilov N. I. The law of homologous series in hereditary variability. Dokl. on III All-Russian breeding congress in Saratov, June 4, 1920 Saratov: Gubpoligrafotdel, 1920, p. 16.

651. Weininger O. *Sex and Character. The fundamental research*. 1902.

652. Velikanova M. S. *Paleoanthropology of the Prut-Dniester interfluve*. Moscow, Science, 1975, 282 p.

653. Vendrovsky B. Current status of the problem of the evolution of sex. *Uspekhi Sovrem. Biologii*, 1933, v. 2, № 3, p. 12-28.

654. Vinogradova T. V., Semenov V. V. Comparative study of cognitive processes in both men and women: the role of biological and social factors. Questions of psychology, 1992, p. 63-71.

655. Vishnevsky A. A., Galankin N. K. Congenital malformations of the heart and large vessels. M., Medicine, 1962, 280 pp.

656. Vorontsov N. N., Lyapunova E. A., Ivanitskaya E. Y. Variability of sex chromosomes of mammals. Genetics, 1978, v. 15, p. 1432-1446.

657. Vysotsky D. L. (2004) *Elements of biological concepts: the theory of building in applications and examples*. Nauka, 570 pp, with.p. 234-242.

658. Willie K. Dete V. *Biology*. 1975, M., Mir.

659. Yusupov R. M. About Sexual Dimorphism and Significance of Female Samples of Skulls in Anthropology. In: References on History and Culture of Bashkiria, Ufa, 1986, p. 51–56.

660. Zavadovsky M. Sex of animals and its transformations. M. - L., 1923.

661. Zhedenov V. N. *Lungs and heart of animals and humans*. M.: Meditsina, 1954, 350 pp.

662. Zhukov D. A. *Biology of conduct. Humoral mechanisms*. St. Petersburg, Resch (Speech), 2007, 466 pp.

663. Zhukovsky P. M. *Cultivated plants and their relatives*. L.: Kolos, 1964, 596 pp.

664. Zhukovsky P. M. *Botanica*. Vysshaya Schkola (High School), 1967, 667 pp.

665. Zhuchenko A. A., Korol A. B. *Recombination in evolution and breeding*. M., Nauka (Science), 1985, p. 317-318.

666. Zhuravlyova M. I., Soboleva T. S. About some gender questions in investigation of psychology of creativity. MCE, 2005, vol. 1, p. 165–171.

TEXTBOOKS

1. Vasilchenko G. S. *General sexopathology.* M., Medicine. 1977, 488 pp.
2. Vasilchenko G. S. (Ed.) *General sexopathology.* A guide for physicians. M., Medicine, 2005, 512 pp. ISBN 5-225-04820-X
3. Sokolova E. I. *The two sexes: why?* All-Russian Institute of Advanced Qualification of Engineers Teachers and Vocational Training Specialists. St. Petersburg, 1992, 41 p.
4. Borinskaya S. A. The Evolutionary theory of sex. *"Biology in School"*, 1995, № 6, p. 12-18.
5. Tkachenko A. A., Vvedensky G. E., Dvorianshikov N. V. *Forensic sexology. Guide for physicians.* M., Medicine, 2001.
6. Tyugashev E. A. *The Economics of Family and Household.* Textbook. Novosibirsk, SibUPK, 2002, p. 33-38.
7. Ilyin E. P. *Differential psychophysiology of men and women.* 2003.
8. Tyugashev E. A., Popkova T. V. *Sem'evedenie.* Textbook. Novosibirsk, SibUPK, 2003, 180 p., p. 38.
9. Tyugashev E. A. *Sem'evedenie.* Textbook. Novosibirsk, SibUPK, 2006, 194 p.
10. Nartova-Bochaver S. K. *Differential Psychology: Study Guide.* M., Flinta, Moscow Psychological and Social Institute, 2003, p. 158-179.
11. Zhukov D. A. *Biology of conduct. Humoral mechanisms.* St. Petersburg, Resch (Speech), 2007, 466 pp.
12. Dikevich L. L. *Gender Psychology.* Teaching aid for students majoring in "Psychology", Smolensk Humanitarian University, Smolensk, 2008, 18 p., p. 2.
13. Podkolzin M. M. *Human Ecology.* Methodical complex. Specialty 020800.62 - "Environment and Natural Resources", Volgograd State University, Volgograd, 2009, 46 p., Seminar number 4.
14. Paliy A. *Differential psychology.* Introductory Textbook. K. Academy, 2010, 432 c.

INFORMATION ON THE INTERNET

< http://www.geodakian.com > Theory's official web site.

Subject Index

Index Latin

npliance